Airport Operations

About the Authors

Norman J. Ashford was Professor of Transport Planning at the Loughborough University of Technology, England, from 1972 to 1997. He holds bachelor's, master's, and doctoral degrees in civil engineering. Dr. Ashford worked as a civil engineer in Canada and taught at the Georgia Institute of Technology and Florida State University. He served as the Director of the Transportation Institute for the State of Florida. Dr. Ashford runs an aviation consulting company and has been active in the areas of airport planning, design, operations, and privatization for more than 100 airports in more than 40 countries.

H. P. Martin Stanton (deceased) was an airport operations expert of international renown who worked for the International Civil Airports Association in Paris, the Frankfort Airport Authority, and the Ministry of Civil Aviation Britain, among other organizations. He was a qualified pilot and an air traffic controller.

Clifton A. Moore (deceased) was President of Llanoconsult Inc., an airport consulting service. His background featured almost 40 years of wide-ranging experience, including as CEO of the Southern California Regional Airport Authority. Mr. Moore played a central role in the development of the Los Angeles International Airport terminal and the modernization of LAX in the early 1980s. He was World President of the International Civil Airports Association for 8 years.

Pierre Coutu, A.A.E., Ed.D., is Founder and President of Aviation Strategies International (ASI), an international network-based consulting firm in aviation business strategy and executive-level training/coaching. Dr. Coutu is the Programme Executive for the Global ACI-ICAO Airport Management Professional Accreditation Programme (AMPAP), an executive development training initiative for which ASI is the designated administrator, responsible for its design, development, and international deployment since 2007. He also co-chairs the World Aviation Governance Forum (WAGF), a joint undertaking of the United Nations' UNITAR-CIFAL and ASI. With a career in the aviation business than spans almost 40 years, Dr. Coutu's experience includes working with the Canadian Transportation Ministry in various capacities in airport management and development, and serving as Executive Vice President and COO of the International Aviation Management Training Institute (IAMTI) from 1987 to 1998, during which time the Institute graduated close to 5,000 aviation executives from 150 different countries. Dr. Coutu also teaches airport management within aviation MBA programs at various universities in North America, Europe, and Southeast Asia.

John R. Beasley graduated from the University of Oxford with a First Class Honours degree in Natural Sciences (Physics). He is a Chartered Physicist, a Member of the Institute of Directors, and a Member of the Association of German Engineers (Verein Deutscher Ingenieure). Mr. Beasley has worked for a number of technical consultancy practices within the defense sector, specializing in airborne systems. In 2001 he founded Analytical Decisions Ltd., a company providing operational analysis, procurement support, and project management services to the civil aviation sector, advising major airports and airlines. Since 2010 Mr. Beasley has worked for BAA plc, focusing on future baggage-handling systems and processes.

Airport Operations

Norman J. Ashford

H. P. Martin Stanton

Clifton A. Moore

Pierre Coutu

John R. Beasley

Third Edition

New York Chicago San Francisco
Lisbon London Madrid Mexico City
Milan New Delhi San Juan
Seoul Singapore Sydney Toronto

1 2 3 4 5 6 7 8 9 0   DOC/DOC   1 8 7 6 5 4 3 2

ISBN 978-0-07-177584-7
MHID 0-07-177584-6

Sponsoring Editor	**Copy Editor**
Larry S. Hager	James K. Madru
Editing Supervisor	**Proofreader**
Stephen M. Smith	Carol Shields
Production Supervisor	**Indexer**
Pamela A. Pelton	Robert Swanson
Acquisitions Coordinator	**Art Director, Cover**
Bridget L. Thoreson	Jeff Weeks
Project Manager	**Composition**
Sandhya Gola,	Cenveo Publisher Services
Cenveo Publisher Services	

Printed and bound by RR Donnelley.

McGraw-Hill books are available at special quantity discounts to use as premiums and sales promotions, or for use in corporate training programs. To contact a representative, please e-mail us at bulksales@mcgraw-hill.com.

This book is printed on acid-free paper.

Contents

Preface xvii
Acknowledgments xix

1 **The Airport as an Operational System** 1
  1.1 The Airport as a System 1
  1.2 National Airport Systems 4
  1.3 The Function of the Airport 8
  1.4 Centralized and Decentralized Passenger
      Terminal Systems 10
  1.5 The Complexity of the Airport Operation ... 15
  1.6 Management and Operational Structures ... 17
  References 29

2 **Airport Peaks and Airline Scheduling** 31
  2.1 The Problem 31
  2.2 Methods of Describing Peaking 33
      The Standard Busy Rate 33
      Busy-Hour Rate 36
      Typical Peak-Hour Passengers 36
      Busiest Timetable Hour 37
      Peak Profile Hour 37
      Other Methods 38
      Airport Differences 38
      Nature of Peaks 39
  2.3 Implications of Variations in Volumes 42
  2.4 Factors and Constraints on Airline
      Scheduling Policies 44
      Utilization and Load Factors 45
      Reliability 45
      Long-Haul Scheduling Windows 45
      Long-Haul Crewing Constraints 48
      Short-Haul Convenience 48
      General Crewing Availability 48
      Aircraft Availability 48
      Marketability 49
      Summer-Winter Variations 50
      Landing-Fee Pricing Policies 50
  2.5 Scheduling Within the Airline 52
  2.6 Fleet Utilization 56
  2.7 IATA Policy on Scheduling 58

2.8   The Airport Viewpoint on Scheduling   59
2.9   Hubs   60
References   61

3   Airport Noise Control **63**
3.1   Introduction   63
3.2   Aircraft Noise   64
Single-Event Measures   65
Cumulative-Event Measures   65
3.3   Community Response to Aircraft Noise   68
3.4   Noise-Control Strategies   69
Quieter Aircraft   70
Noise-Preferential Runways   70
Operational Noise-Abatement
Procedures   71
Runway Operations   73
Insulation and Land Purchase   73
3.5   Noise Certification   74
3.6   Noise-Monitoring Procedures   76
3.7   Night Curfews   82
3.8   Noise Compatibility and Land Use   83
References   87
Further Reading   87

**4   Airport Influences on Aircraft
Performance Characteristics** **89**
4.1   Introduction   89
4.2   Aircraft   90
4.3   Departure Performance   100
4.4   Approach and Landing Performance   105
4.5   Safety Considerations   107
4.6   Automatic Landing   110
4.7   Operations in Inclement Weather   114
4.8   Specific Implications of the Airbus A380
(New Large Aircraft)   116
References   117

5   Operational Readiness **119**
5.1   Introduction   119
5.2   Aerodrome Certification   119
5.3   Operating Constraints   122
Visibility   122
Crosswind Effects   124
Bird-Strike Control   128
5.4   Operational Areas   132
Pavement Surface Conditions   132

5.5   Airfield Inspections   138
5.6   Maintaining Readiness   140
      Maintenance Management   140
      Preventive Maintenance   142
      Electrical Maintenance   142
      Operational Readiness: Aircraft Rescue
         and Firefighting   146
      Safety Aspects   147
      Airfield Construction   148
      Conclusion   150
References   150

6   **Ground Handling** **153**
    6.1   Introduction   153
    6.2   Passenger Handling   153
    6.3   Ramp Handling   158
    6.4   Aircraft Ramp Servicing   162
          Fault Servicing   162
          Fueling   162
          Wheels and Tires   163
          Ground Power Supply   163
          Deicing and Washing   164
          Cooling/Heating   164
          Other Servicing   166
          Onboard Servicing   166
          Catering   166
    6.5   Ramp Layout   166
    6.6   Departure Control   169
    6.7   Division of Ground Handling
          Responsibilities   171
    6.8   Control of Ground Handling Efficiency   174
    6.9   General   175
    References   179

7   **Baggage Handling** **181**
    7.1   Introduction   181
    7.2   Context, History, and Trends   181
    7.3   Baggage-Handling Processes   183
          Overview   183
          Bag Drop   184
          Hold Baggage Screening   188
          Bag Storage   189
          Flight Build and Aircraft Loading   189
          Arrivals Reclaim   191
          Transfer Input   193
          Interterminal Transfers   193

7.4   Equipment, Systems, and Technologies   194
       Baggage-Handling-System
           Configurations   194
       Check-in and Bag Drop   194
       Sorting   196
       Hold-Baggage Screening   197
       Bag Storage   198
       Flight Build   200
       Reclaim   203
7.5   Process and System Design Drivers   204
       Appearance Profiles   204
       Bags per Passenger   204
       Transfer Ratios   205
       Processing Times   206
7.6   Organization   206
       Staffing   206
7.7   Management and Performance Metrics   207
       Overall   208
       Baggage System   209
       Arrivals Delivery Performance   209
   References   211

8   **Passenger Terminal Operations**   213
8.1   Functions of the Passenger Terminal   213
8.2   Terminal Functions   219
8.3   Philosophies of Terminal Management   221
8.4   Direct Passenger Services   222
8.5   Airline-Related Passenger Services   228
8.6   Airline-Related Operational Functions   230
       Flight Dispatch   230
       Flight Planning   231
       Aircraft Weight and Balance   232
       Takeoff   232
       In Flight   232
       Landing   234
       Balance/Trim   235
       Loading   235
       Flight-Crew Briefing   235
       Flight Watch (Flight Control)   239
8.7   Governmental Requirements   239
8.8   Non-Passenger-Related Airport
       Authority Functions   240
8.9   Processing Very Important Persons   241
8.10 Passenger Information Systems   241

8.11   Space Components and Adjacencies 246
8.12   Aids to Circulation 249
8.13   Hubbing Considerations 254
References 255

9   Airport Security **257**
9.1   Introduction 257
9.2   International Civil Aviation Organization
       Framework of International
       Regulations 258
9.3   Annex 17 Standards 259
9.4   The Structure of Planning for Security 260
9.5   Airport Security Program 261
9.6   U.S. Federal Involvement in
       Aviation Security 262
9.7   Airport Security Program: U.S. Structure ... 262
9.8   Airport Security Planning Outside
       the United States 264
9.9   Passenger and Carry-On Baggage
       Search and Screening 265
       Centralized and Decentralized
          Screening 266
       Security Screening Checkpoint 269
9.10   Baggage Search and Screening 275
9.11   Freight and Cargo Search and Screening ... 277
9.12   Access Control Within and Throughout
       Airport Buildings 277
9.13   Vehicle Access and Vehicular Identification   279
9.14   Perimeter Control for Operational Areas ... 279
       Fencing 279
       Access Gates 280
9.15   Aircraft Isolated Parking Position and
       Parking Area 281
9.16   Example of a Security Program for a
       Typical Airport 281
       Security Program for (Official Name of
          Airport Goes Here) 282
9.17   Conclusion 287
References 287

10   Cargo Operations **289**
10.1   The Cargo Market 289
       Gross Domestic Product 289
       Cost 290

Technological Improvements 290
Miniaturization 290
Just-in-Time Logistics 291
Rising Consumer Wealth 291
Globalization of Trade and Asian
    Development 291
Loosening of Regulation 291
Cargo Types 292
Patterns of Flow 293
10.2 Expediting the Movement 295
10.3 Flow Through the Terminal 297
10.4 Unit Load Devices (IATA 1992, 2010) 301
10.5 Handling Within the Terminal 304
Low Mechanization/High
    Manpower 304
Open Mechanized 305
Fixed Mechanized 305
10.6 Cargo Apron Operation 307
10.7 Facilitation (ICAO 2005) 314
10.8 Examples of Modern Cargo Terminal
      Design and Operation 317
10.9 Cargo Operations by the Integrated
      Carriers 320
References 323

11   Aerodrome Technical Services 325
11.1 The Scope of Technical Services 325
11.2 Safety Management System 326
11.3 Air Traffic Control 329
Fundamental Changes 329
Function of ATC 331
International ATC Collaboration 333
Flight Rules 336
General Flight Rules 336
Visual Flight Rules 337
Instrument Flight Rules 337
Classes of Airspace 338
Separation Minima 339
Operational Structure 343
Operational Characteristics
    and Procedures 345
11.4 Telecommunications 350
Fixed Services 351
Mobile Services 352
Radio Navigation Services 355

Satellite Navigation 358
Broadcast Services 360
11.5 Meteorology 361
Function 361
World Area Forecast System 362
Meteorologic Observations
and Reports 362
Aircraft Observations and Reports 365
Terminal Airport Forecasts 366
Significant Weather Forecasts
and Charts 366
Upper-Air Grid-Point Data
Forecasts 368
SIGMETs/AIRMETs 369
Weather Information Support for
General Aviation 370
Climatologic Information 371
Services for Operators and
Flight Crew Members 371
Information for ATC, SAR, and AIS 372
Use of Communication 372
Trends in Meteorologic Services 372
11.6 Aeronautical Information 373
Scope 373
Urgent Operational Information 374
Availability of Information 375
11.7 Summary 375
References 376

12 **Airport Aircraft Emergencies** **377**
12.1 General 377
12.2 Probability of an Aircraft Accident 377
12.3 Types of Emergencies 379
12.4 Level of Protection Required 380
12.5 Water Supply and Emergency
Access Roads 383
12.6 Communication and Alarm Requirements ... 384
12.7 Rescue and Firefighting Vehicles 385
12.8 Personnel Requirements 388
12.9 The Airport Emergency Plan 389
Command 394
Communications 398
Coordination 398
12.10 Aircraft Firefighting and Rescue
Procedures 398

12.11   Foaming of Runways 400
12.12   Removal of Disabled Aircraft 402
12.13   Summary 404
References 410

13   Airport Access **411**
13.1   Access as Part of the Airport System 411
13.2   Access Users and Modal Choice 414
13.3   Access Interaction with Passenger
         Terminal Operation 418
         Length of Access Time 419
         Reliability of Access Trip 419
         Check-in Procedures 419
         Consequences of Missing a Flight 422
13.4   Access Modes 424
         Automobile 424
         Taxi 427
         Limousine 429
         Rail 430
         Bus 434
         Dedicated Rail Systems 435
13.5   In-Town and Other Off-Airport Terminals ... 437
13.6   Factors Affecting Access-Mode Choice 438
13.7   General Conclusions 439
References 440

14   Operational Administration and Performance **441**
14.1   Strategic Context 441
14.2   Tactical Approach to Administration
         of Airport Operations 447
14.3   Organizational Considerations 449
14.4   Managing Operational Performance 452
         Planning for Performance 452
         Operations Program Execution 455
         Operations Program Control 457
         Internal Assessment 458
         External Assessment 459
         Airport Economic Regulatory
           Oversight 462
         Industry Benchmarking 463
14.5   Key Success Factors for High-Performance
         Airport Operations 464

15   Airport Safety Management Systems **465**
15.1   Safety Management System Framework ... 465
         Regulatory Framework 466

ICAO Standards and Recommended
    Practices 467
ICAO Annex 14 Aerodromes (Volume 1:
    Aerodrome Design and Operations) 467
ICAO's Stance on the Implementation
    of a Member State's Safety
    Program 468
15.2 Safety Management Systems and
    Aerodromes 469
    Introduction of SMSs to
        Aerodromes 469
    Assessment of the Current Safety Level
        (Where Are We At Now?) 472
    Acceptable Level of Safety 473
15.3 SMS Manual 475
    Overview 475
    The Key Elements of an SMS Manual 477
    Policy, Organization, Strategy,
        and Planning 478
    Risk Management 481
    Safety Assurance 484
15.4 Implementation 485
    Issues 485
    Guidance and Resources 490
15.5 Key Success Factors in Airport
    SMS Implementation 493
    Integration 493
    Communication Technology 494
References 495

16 **Airport Operations Control Centers** **497**
16.1 The Concept of Airport Operations
    Control Centers 497
    Introduction 497
    Origins to the Present 499
    Management Philosophy 502
    Strategic Significance 502
    Regulatory Requirement for AOCCs ... 504
16.2 Airport Operations Control System 505
    AOCS Dynamics 505
    AOCS Users 508
16.3 The Airport Operations Coordination
    Function 508
    Purpose 508
    Applications 511

16.4 Airport Performance-Monitoring
        Function 513
            Purpose 513
            Application 515
16.5 Design and Equipment Considerations 516
            Physical Layout 516
            AOCC Systems and Equipment 518
            Ergonomics 519
16.6 Organizational and Human
        Resources Considerations 520
            AOCC Management Structure and
                Reporting Relationships 520
            Staffing and Key Competencies 522
16.7 Leading AOCCs 524
            Auckland Airport: A Focus on
                Customer Service 525
            Beijing Capital International Airport:
                Tightly Aligned on Best Practices 525
            Dublin Airport: Real-Time Automated
                Level-of-Service Measurement 528
            Fort Lauderdale–Hollywood
                International Airport: Self-Audit
                and Improvement Plan 528
            Kuala Lumpur International Airport:
                Monitoring a Network of Airports ... 530
            Los Angeles International Airport: Most
                Recent and Comprehensive 530
            Munich Airport: Direct Impact on
                Minimum Connecting Times 533
            Zagreb Airport: Proving the Concept
                for Small and Medium Airports 533
16.8 Best Practices in Airport Operations
        Control Center Implementation
        (Key Success Factors) 534

17 The Airport Operations Manual 537
17.1 The Function of the Airport
        Operations Manual 537
17.2 A Format for the Airport
        Operations Manual 538
17.3 Distribution of the Manual 544
17.4 U.S. Example: Federal Aviation
        Administration Recommendations
        on the Airport Certification Manual
        (FAA 2004) 545

Suggested Airport Certification
Manual: FAA Format 546
References 552

18 **Sustainable Development and Environmental
Capacity of Airports** **553**
18.1 Introduction 553
The Sustainable-Development
Challenge 555
18.2 The Issues 555
Noise Impacts 555
Local Air Quality 557
Airport Carbon Management 558
Energy 562
Water Use 563
The Management of Solid Wastes 565
Surface and Groundwater Pollution ... 567
Adapting to a Changing Climate 569
Biodiversity 572
18.3 Environmental Management Systems 572
18.4 Conclusion 575
References 576

Index **579**

Preface

As the world enters the second decade of the twenty-first century, the point of deregulation of the air transport industry is now 35 years in the past. Since this landmark step, the world of civil aviation has changed irrevocably. Privatization and liberalization of air transport have occurred on a worldwide scale. There is now widespread privatization of many of the larger and medium-sized airports; in parallel with this, there has been a significant withdrawal of national governments from the operation and even the ownership of airports. Across the world, governments frequently have introduced requirements of competition at airports in the areas of passenger, freight, and aircraft handling. In most regions, deregulation also has resulted in the development of low-cost airlines with their special requirements in both facilities and procedures. Over the last 20 years, there also has been very rapid growth in passenger and freight traffic at many Asian airports associated with the sustained economic growth of a number of the large Asian economies.

Other substantial changes since the 1980s include the introduction of airline alliances. These have greatly influenced the way in which the carriers now want to use the airports, requiring alliance positioning for maximum commercial benefit. Airline equipment also has changed with the introduction of long-range very high-capacity aircraft. With the support and at the urging of the International Air Transport Association (IATA), electronic facilitation has been brought about by the spread of the Internet with its effects on online booking, ticketing, check-in, flight tracking, and passenger handling with respect to delays and cancellations. E-documentation has been introduced in the carriage of freight to reduce paper documentation.

Another of the substantial changes since deregulation is the intense increase in security requirements following the Lockerbie and September 2001 atrocities and subsequent terrorist attacks on aircraft and airports. Airports, some of which once had perfunctory security checks, are now continuously closely monitored by both national governments and international regulators to ensure that the security measures in place discourage terrorist activity and conform to international requirements.

In the area of environmental impact, the introduction of Federal Aviation Administration (FAA) Stage 4 (ICAO Chapter 4) aircraft and the banning of old FAA Stage 2 and 3 (ICAO Chapter 2 and 3) aircraft have brought about a general alleviation in the noise impact around airports. Concern at the beginning of the twenty-first century is more about carbon footprint, global warming and rising sea levels, water and air pollution, and sustainable development.

In order to reflect the evolution of regulatory guidance and best practices, two new chapters (Chapters 15 and 16) have been added dealing with safety management systems and airport operations control centers.

Overall, this new edition, a significant updating of the earlier editions, seeks to describe the status of civil air transport at airports from the viewpoint of the situation found at the time of publication.

Norman J. Ashford
Pierre Coutu
John R. Beasley

Acknowledgments

In the preparation of the text, we have been greatly indebted to the following individuals who have given freely of their time to assist in the gathering and verification of data. Without their help, it would not have been possible to complete this book.

Peter Adams, Aviation Strategies International, Australia

Paulo Barradas, CONSULSADO, Portugal

Paul Behnke, Aviation Strategies International, USA

Monica Tai Chew, Aviation Strategies International, Canada

Gabriela Cunha, AEROSERVICE, Brazil

Allan Dollie, Aeroasset Systems, Ltd., United Kingdom

Hilary Doyle, American Airlines, United Kingdom

Frank Elder, Feather Aviation, United Kingdom

Marcello Ferreira, AEROSERVICE, Brazil

Paul Luijten, Schiphol Amsterdam Airport, Netherlands

Maria K. R. Luk, Hong Kong International Airport, China

Stan Maiden, formerly BAA plc, United Kingdom

Carol McQueen, American Airlines, United Kingdom

Stanislav Pavlin, University of Zagreb, Croatia

Inna Ratieva, ResultsR, Netherlands

Antonio Gomes Ribeiro, Architectos Associados, Portugal

Norman Richard, Edmonton Airports, Canada

Mario Luiz F. De M. Santos, AEROSERVICE, Brazil

Ian Stockman, Cranfield University, United Kingdom

Vojin Tosic, University of Belgrade, Serbia

Peter Trautmann, Bavarian Airports, Germany

Jaap de Wit, University of Amsterdam, Netherlands

Stefan Wunder, FRAPORT, Germany

Wasim Zaidi, Aviation Strategies International, Canada

The international organizations Airports Council International, Montreal, International Air Transport Association, Montreal, and International Civil Aviation Organization, Montreal, all have provided great support in the rewriting of this book, as have the national organizations such as the Federal Aviation Administration (United States) and the Civil Aviation Authority (United Kingdom). We also would like to acknowledge the generous assistance of other colleagues from BAA plc who have contributed to the production of this work.

The following airports and other organizations kindly participated in responding to surveys that were used in compiling various data tables or in providing other types of information referred to throughout the book:

Abha Regional Airport, Saudi Arabia

Adelaide International Airport, Australia

Aéroport de Bamako Senou, Mali

Aéroports de Paris, France

AT&T, USA

Austin-Bergstrom International Airport, USA

Banjul International Airport

Beijing Capital International Airport, China

Belgrade Nikola Tesla International Airport, Serbia

Brazzaville Maya International Airport, Congo

Brussels Zaventem National Airport, Belgium

Cape Town International Airport, South Africa

Capital Airports Holding Company, China

Cheddi Jagan International Airport, Guyana

Christchurch International Airport, New Zealand

Comox Valley Airport, Canada

Conakry Airport, Guinea

Dakar Leopold Sedar Senghor International Airport, Senegal

Damman King Fahd International Airport, Saudi Arabia

Detroit Metropolitan Wayne County Airport, USA

Dubai International Airport, UAE

Dublin International Airport, Ireland

Edmonton International Airport, Canada

Faro Airport, Portugal

Flughafen Graz Betriebs GmbH, Austria

Fort Lauderdale–Hollywood International Airport, USA

Fraport AG, Germany

Gan International Airport, Maldives

Greenville-Spartanburg International Airport, USA

Harry Mwaanga Nkumbula International Airport, Zambia

Hartsfield-Jackson Atlanta International Airport, USA

Ibrahim Nasir International Airport, Maldives

Jomo Kenyatta International Airport, Kenya

Kabul International Airport, Afghanistan

Kaunas International Airport, Lithuania

Khartoum International Airport, Sudan

Kinshasa N'djili International Airport, DR Congo

Kuala Lumpur International Airport, Malaysia

L. F. Wade International Airport, Bermuda

Langkawi International Airport, Malaysia

Liege Airport, Belgium

Lisbon International Airport, Portugal

Los Angeles World Airports, USA

Lucknow Choudhary Charan Singh Airport, India

Lynden Pindling International Airport, Bahamas

Madrid Barajas International Airport, Spain

Malaysia Airports Holdings Berhad, Malaysia

Marshall Islands International Airport, Marshall Islands

Metropolitan Washington Airports Authority, USA

Mostar International Airport, Croatia

Munich Airport, Germany

Murtala Muhammed International Airport, Nigeria

Nashville International Airport, USA

New Plymouth Airport, New Zealand

Newark Liberty International Airport, USA

Niamey Diori Hamani International Airport, Niger

Norman Manley International Airport, Jamaica

North Las Vegas Airport, USA

Paris Charles de Gaulle International Airport, France

Paris Orly International Airport, France

Piarco International Airport, Trinidad

Porto Airport, Portugal

Pula Airport, Croatia

Queen Alia International Airport, Jordan

Rajiv Gandhi International Airport, India

San Antonio Airport, USA

Sanandaj Airport, Iran

Sarajevo International Airport, Bosnia Herzegovina

SITA, Switzerland

Shanghai International Airport Ltd., China

Sunshine Coast Airport, Australia

Taoyuan International Airport, Taiwan

Tocumen International Airport, Panama

Vancouver International Airport, Canada

Windsted Corporation, USA

Yerevan Zvartnots International Airport, Armenia

Zagreb Airport, Croatia

We are also indebted to many other organizations and individuals associated with the aviation industry, too numerous to name.

Airport Operations

CHAPTER 1

The Airport as an Operational System

1.1 The Airport as a System

The airport forms an essential part of the air transport system because it is the physical site at which a modal transfer is made from the air mode to the land modes or vice versa. Therefore, it is the point of interaction of the three major components of the air transport system:

- The airport, including its commercial and operational concessionaires, tenants, and partners, plus, for these discussion purposes, the airways control system
- The airline
- The user

The planning and operation of airports must, if they are to be successful, take into account the interactions among these three major components or system actors. For the system to operate well, each of the actors must reach some form of equilibrium with the other two. Failure to do so will result in suboptimal conditions, exemplified by a number of undesirable phenomena that are indicators of inadequate operation. Each phenomenon can, in a state of unrestrained competition, lead to an eventual decline in the scale of operation at the airport facility or at least a loss of traffic total share as traffic is attracted elsewhere. In the absence of a competitive option, total demand levels will be depressed below levels achievable in the optimal state. Suboptimality can become manifest in a number of ways:

- Deficit operations at the airport
- Deficit operations by the airline(s) at the airport
- Unsatisfactory working conditions for airline and airport employees

- Inadequate passenger accommodation (low levels of service [LOS])
- Insufficient flight supply
- Unsafe operations
- High operational costs to users
- Inadequate support facilities for airlines
- High delay levels for airlines and passengers
- Inadequate access facilities
- Sluggish passenger demand

Figure 1.1 displays a simple hierarchical system diagram of the primary interactions among the airport, the airline, and the user.

The figure attempts to show how these interactions produce the prime parameters of operational scale, passenger demand, airport

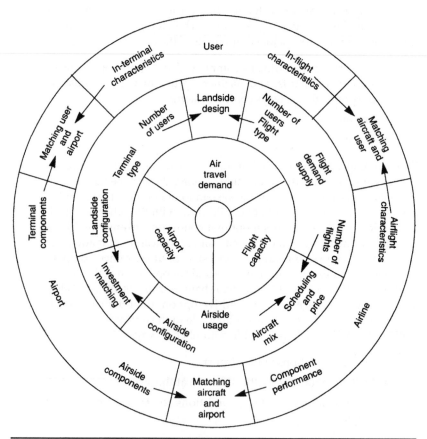

Figure 1.1 A hierarchical system of airport relationships.

capacity, and flight capacity. Although the simplified diagram helps with conceptualization of the main factors of airport operation, large airports are in fact very complex organizational structures. This is not surprising because a large airport might well be one of the largest generators of employment in a metropolitan region. The very largest hub airports, such as Chicago O'Hare, Los Angeles, and London Heathrow may well have a total site employment in the region of 100,000 workers (TRB 2008). To place this in some context of urban scale, the number employed at a large airport can more than equal the number of workers in a city of over half a million population. Large systems such as this are necessarily more complex than the simple trichotomy expressed in Figure 1.1. A more complete listing of involvement for a large airport is contained in Table 1.1. This table includes an important fourth actor, the nonuser, who can have an important impact on airport operation and who is greatly affected by large-scale operations.

Principal Actor	Associated Organizations
Airport operator	Local authorities and municipalities
	Central government
	Concessionaires
	Suppliers
	Utilities
	Police
	Fire service
	Ambulance and medical services
	Air traffic control
	Meteorology
Airline	Fuel supplies
	Engineering
	Catering/duty-free
	Sanitary services
	Other airlines and operators
Users	Visitors
	Meeters and senders
Nonusers	Airport neighbor organizations
	Local community groups
	Local chambers of commerce
	Environmental activist groups
	Antinoise groups
	Neighborhood residents

TABLE **1.1**   Organizations Affected by the Operation of a Large Airport

1.2   National Airport Systems

Modern airports, with their long runways and taxiways, extensive apron and terminal areas, and expensive ground handling and flight navigation equipment, constitute substantial infrastructure investments. All over the world, airports have been seen as facilities requiring public investment, and as such, they are frequently part of a national airport system designed and financed to produce maximum benefit from public funding. Each country, with its own particular geography, economic structure, and political philosophy, has developed a national airport system peculiar to its own needs. This overall national system is important to the individual airport in that the national framework determines the nature of current and future traffic handled at the facility in terms of such parameters as volume, international/domestic split, number of airlines served, and growth rates. Since 1987, outside the United States, a very large number of the larger airports have been privatized and, to a large degree, have become deregulated in matters relating to competition and services offered. Those which were rigidly controlled within a central government structure have necessarily become more open to adaptation to a deregulated aviation industry. Two different national systems will be discussed here briefly: those of the United States and those in the United Kingdom.

The United States is a very large industrialized nation with just over 19,800 airports (including heliports, STOLports, seaplane bases, and joint-use civil-military airports), of which approximately 14,600 are closed to the public or of limited public use (Figure 1.2). Of the

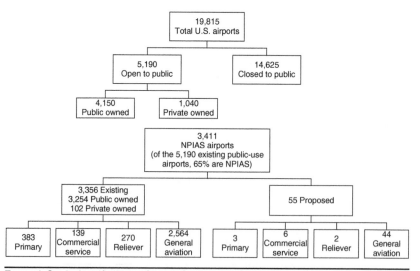

Figure 1.2   A classification of the U.S. airport system. (*Source: FAA, NPIAS.*)

Airport	Hub Type Classification	Percentage of Annual Passenger Boardings	Common Name
Commercial Service Publicly owned airports that have at least 2,500 passenger boardings each calendar year and receive scheduled passenger service	**Primary** Have more than 10,000 passenger boardings each year	**Large** 1% or more	Large hub
		Medium At least 0.25% but less than 1%	Medium hub
		Small At least 0.05% but less than 0.25%	Small hub
		Nonhub More than 10,000 but less than 0.05%*	Nonhub primary
	Nonprimary	**Nonhub** At least 2,500 and no more than 10,000	Nonprimary Commercial service
Nonprimary (except commercial service)			Reliever
			General aviation
Other than passenger classification			Cargo service

*Nonhub airports—Locations having less than 0.05 percent of the total U.S. passengers, including any nonprimary commercial airport, are statutorily defined as nonhub airports. For some classification purposes, primary locations are separated within this type, although more than 100 nonhub airports are currently classified as nonprimary commercial service airports.

Source: FAA, NPIAS 2009.

TABLE **1.2**   Definitions of U.S. Airport Categories

remainder, over 550 facilities provide primary or other commercial services for passenger transport aircraft (Table 1.2). Of the more than 2,800 reliever and general-aviation airports not served by air carriers, some have more flight operations than a number of the airports served by major or commuter airlines; for example, Phoenix Deer Valley has over 480,000 operations per year, and Los Angeles Van Nuys has over 400,000 operations per year. Publicly owned airports in the National Plan of Integrated Airport Systems (NPIAS) are eligible for federal aid in the construction of most facilities required at the airport, other than those related to commercial activities. Ownership of the large and medium-sized airports is almost entirely in the hands of local communities. The two large airports in the Washington, DC, area were formerly operated directly by the federal government, but since June 1987, they have been operated by the Metropolitan Washington Airports Authority. In the U.S. system, privately owned commercial transport airports are few in number and are barely significant to the national system. In 2013, there was still some pressure for some of the largest facilities to be privatized, but there is little political support for such action. The scale of monies paid to each publicly owned airport from the Airport and Airways Trust Fund is related to the functional role of the facility. The system consists of geographically widely separated hub, regional, and municipal airports that have an opportunity to expand as traffic increases. As a consequence, the functional classification system of the U.S. National Airport System plan is a relatively free structure related to passenger throughput and aviation activity. Airports are free to move their classification, which is a de facto demand classification.

The United Kingdom presents an entirely different conceived system that has developed for a relatively small country that has, in addition to a few major airports, a large number of regional facilities spread throughout the country. By the early 1970s, the smaller airports, which had been turned back to civilian use after the war, were owned by local municipal governments. Economic appraisal of the performance of the British airport system indicated that most of the airports were loss-making facilities that were unlikely to be able to recoup past investments made in the hope of attracting traffic (Doganis and Thompson 1973; Doganis and Pearson 1977). To encourage the development of a national airport system in which a few major hubs would cater to international traffic and the rest of the airports would assume a subsidiary role, the British government developed an airports policy to guide central-government investment in the airport system (HMSO 1978). The policy recognized four distinct categories of airports and indicated that government approval in the form of finance and planning permission to develop facilities would be based on this categorization. These categories are *gateway international airports*, which supported a wide range of international and intercontinental services; *regional airports*, which provided short-haul international and domestic services; *local airports*, which provided third-level services (e.g., scheduled passenger services with aircraft having fewer than 25 seats); and *general-aviation airports* (Table 1.3). This four-tier system worked for a number of years, coming under increasing pressure as more and more passengers wished to fly direct into European and North African destinations in two-engine aircraft without having to transit through the hub airports. Also, the development of aircraft technology in the early 1980s brought North American destinations within the range of two-engine extended-range (ER) aircraft such as those available from Airbus and Boeing. Following deregulation of the aviation industry in the United States and abolition of the Civil Aeronautics Board, in 1987 Britain went a step further by privatizing all publicly owned airports having annual revenue in excess of £1 million. This meant that the British Airports Authority (BAA), which had more than three-quarters of all British airport

Gateway International Airports

Airports supplying a wide range and frequency of international services, including intercontinental services and a full range of domestic services.

Regional Airports

Airports catering to the main air traffic demand of individual regions. They are concerned with the provision of a network of short-haul international services (mainly to Scandinavia and other parts of Europe) and a range of charter services and domestic services, including the links with gateway airports.

Local Airports

Airports providing third-level services (e.g., scheduled passenger services operated by aircraft with fewer than 25 seats), catering privately for local needs, concentrating on general aviation with some feeder services and some charter flights.

General Aviation Airports

Airports concerned primarily with the provision of general aviation facilities.

Source: CAA.

TABLE 1.3 The British National Airport System (as of 1978)

traffic and whose seven airports included Heathrow and Gatwick, was floated on the London stock market in that year as BAA plc. All other airports at that time became private companies, but with all shares entirely owned by local governmental authorities. One by one the airports were sold to the private sector, and since the general privatization, most other airports have been transferred into private hands. By 2013, only a few, notably including Manchester, remain in the public sector. Newcastle airport was owned 51 percent by seven local authorities and 49 percent by the Copenhagen Airport. BAA plc, still carrying approximately 70 percent of all passengers uplifted in Britain, was bought by the Spanish company Grupo Ferrovial in July 2006, changed its name to BAA Airports Ltd., and ceased to be traded on the London Stock Exchange, where it had been part of the FTSE Index. By 2013, airports that were nominally regional facilities, such as Birmingham, Bristol, and Newcastle, were offering scheduled flights to destinations as widespread as Pakistan, the United States, the Caribbean, Dubai, and Mexico. The neat classifications of 1978 had been allowed to blur into a laissez-faire system where airport capacity and ability to generate traffic were the prime determinants of function. Also by this time, the general growth in air passenger traffic had greatly reduced the number of loss-making British regional airports. The pragmatic approach to the national plan for airports was recognized in a white paper (Department of Transport 2003) that examined the potential for individual growth and development of the major British airports. It is apparent that the policy for regulated development of air services by declaring and promoting gateways had an enduring effect on the manner in which capacity supplied to the British airport system and the demand that it created. While probably not as strong as the policy of privatization, which promoted the path to airport profitability, the structure of the British Airports Authority is still very much influenced by the policy set out in 1978.

1.3   The Function of the Airport

An airport is either an intermediate or terminal point for an aircraft on the air portion of a trip. In simple functional terms, the facility must be designed to enable an aircraft to land and take off. In between these two operations, the aircraft may, if required, unload and load payload and crew and be serviced. It is customary to divide the airport's operation between the airside and landside functions shown in the simplified system diagram in Figure 1.3. More detailed system diagrams are provided for passenger and cargo processing in Figures 8.1 and 10.7. The overall system diagram indicates that after approach and landing, an aircraft uses the runway, taxiway, and apron prior to docking at a parking position, where its payload is processed through the terminal to the access/egress system. Departing passengers make their way through the landside operation to the departure gates.

The airport passenger and freight terminals are themselves facilities that have three distinct functions (Ashford, Mumayiz, and Wright 2011):

Change of mode. To provide a physical linkage between the air vehicle and the surface vehicle designed to accommodate the operating characteristics of the vehicles on landside and airside, respectively.

Processing. To provide the necessary facilities for ticketing, documentation, and control of passengers and freight.

Change of movement type. To convert continual shipments of freight by trucks and of departing passengers by car, bus, taxi, and train to aircraft-sized batches that generally depart according to a preplanned schedule and to reverse this process for arriving aircraft.

Many small airports that provide little more than a simple passenger terminal for low-volume passenger operations provide very little more than a passenger terminal facility. The operation of the airport is not significantly more complex than that of a railroad station or an interurban bus station. Medium- or large-scale airports are very much more complex and require an organization that can cope with such complexity. Airports of a significant size must have an organization that can either supply or administer the following facilities:

- Handling of passengers
- Servicing, maintaining, and engineering of aircraft

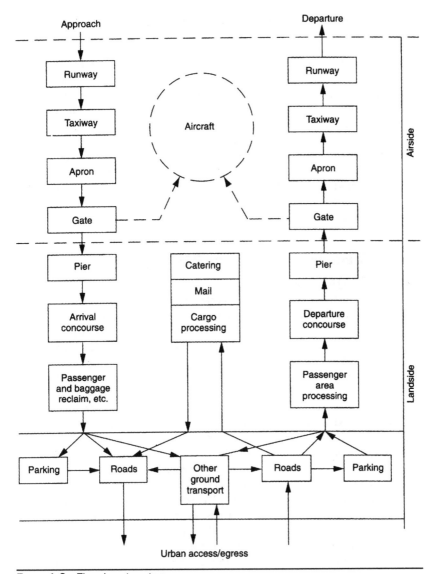

FIGURE 1.3   The airport system.

- Airline operations, including aircrew, cabin attendants, ground crew, and terminal and office staff
- Businesses that provide services to passengers and are necessary for the economic stability of the airport (e.g., concessionaires, leasing companies, etc.)
- Aviation support facilities (e.g., air traffic control, meteorology, etc.)

- Government functions (e.g., agricultural inspection, customs, immigration, health)

Large international hubs are complex entities that have all the problems of any large organization with many employees. In some cases, the airport itself is a large employer. In other cases, the authority acts more as a broker of services, resulting in a low level of direct authority employment Regardless of the chosen modus operandi, overall staffing levels at airports will be high, and there will be complex interactions among the various employing organizations. Inefficient operations and operational failures in such large systems, whether through incompetence, disorganization, or industrial action, involve huge expenses in terms of additional wages, lost passenger time, and the costs of delayed freight.

1.4   Centralized and Decentralized Passenger Terminal Systems

The way in which an airport terminal system is operated and the administrative structure of the operating company may be affected by the physical design of the airport itself. It is convenient to classify airports into two broad and very different operational types: *centralized* and *decentralized*. Most older terminals were designed using the centralized concept, where processing was carried out in the main terminal building, and access to the aircraft gates was attained by piers and satellites or by apron transport. Many airports still operate quite satisfactorily using centralized facilities (e.g., Tampa and Amsterdam Schiphol). Other airports started life as centralized facilities but became decentralized when additional terminals were added to cope with increased traffic (e.g., London Heathrow, Paris Orly, and Madrid Barajas). Some airports were designed ab initio as decentralized facilities operating with a number of unit terminals, each with a complete set of facilities (e.g., Dallas–Fort Worth, Paris Charles de Gaulle, Kansas City, and New York JFK). A hybrid form of centralization/decentralization occurs with extensive remote pier developments (e.g., Atlanta and Hong Kong) and remote satellites (e.g., Pittsburgh and Kuala Lumpur). Figure 1.4 shows examples of centralized and decentralized layouts, respectively.

Until the early 1960s, passenger traffic through even the world's largest airports was so small that centralized operation was commonplace. There can be economies of scale on the use of fixed equipment such as baggage systems, check-in desks, and mobile apron equipment. Similar economies are found with the airport company, airline, and airport tenants' staff requirements. It also has been found that fewer security personnel are required in centralized designs. Where an airport company has chosen a centralized operation, there is a tendency for the administration to be very closely involved in the day-to-day operations of the terminal area.

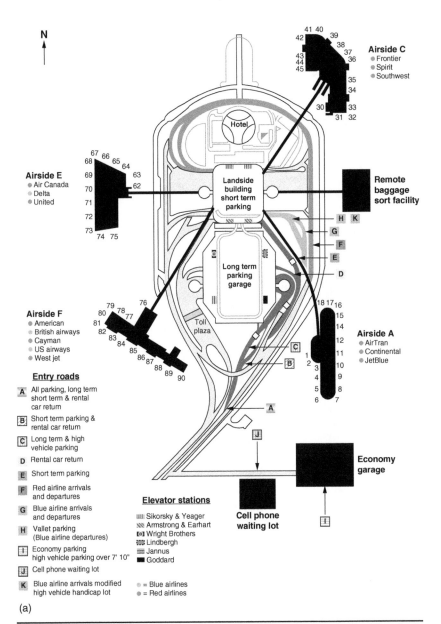

FIGURE 1.4   (a) Schematic of Tampa International Airport—a centralized terminal layout. (*Courtesy: Hillsborough County Airport Authority.*) (b) Schematic of Shanghai International Airport—a decentralized terminal layout. (*Adapted from Shanghai Pudong International Airport.*)

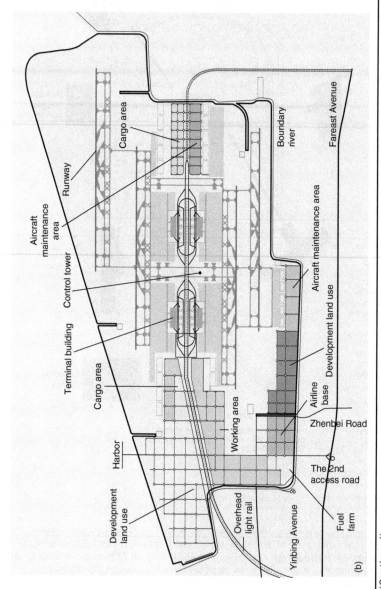

Figure 1.4 (*Continued*)

12

As traffic grew at the major hubs, the physical size of some centralized facilities also grew as terminal additions were constructed. Some facilities became extraordinarily large in scale. For example, before the redevelopment of Chicago O'Hare, the distance between extreme gates in the single terminal was just under 1 mile (1.6 km). Parking facilities also grew in scale. Passengers could face very long walks, either when interlining or as originating or terminating passengers.

In order to overcome the problem of unsatisfactory walking distances, decentralized designs were developed to keep walking distances down to the region of 984 feet (300 m), as recommended by the International Air Transport Association (IATA), with a maximum distance between curbside and the further check-in counter set at 328 feet (100 m). Decentralization was carried as far as the gate arrival concept of Dallas–Fort Worth (DFW) and Kansas City, where the total walking distances from car to aircraft originally were set at 328 feet (100 m). The advantages of decentralization are significant. Terminals are kept on a human scale, passenger volumes never become uncomfortably high, and walking distances are kept short. Parking lots are kept small and walking distances reasonable. Such areas are easier to supervise and thus safer from a crime viewpoint, and curbside drop-offs are simple to design. Operationally, however, decentralization can lead to higher airport staff requirements because some functions of administration and security must be carried out separately in each terminal. Because the scale of decentralized facilities is very large, each unit requires a full range of passenger and staff facilities. It is possible, therefore, to have poor economy of scale in terms of the fixed facilities, such as baggage rooms, baggage claims, and check-in areas. Similarly apron handling equipment must be duplicated.

For a large airport, the scale of separation between units can be very large. For example, if the 14-unit terminals at DFW had been constructed according to the original master plan, the distance between the two extreme unit terminals would have been 3 miles (5 km). Completely decentralized designs mean that interlining passengers must have some form of transit system to permit reasonable interterminal movement. At DFW, Frankfurt, and Singapore Changi, this is achieved by an automatic transit vehicle; at older terminals, such as London Heathrow and Paris Charles de Gaulle (CDG), this is achieved by a simple bus service. Neither method is particularly convenient if the number of terminals is large, such as at CDG. At decentralized airports, the administrative offices are frequently well separated from the day-to-day operations of the terminals.

An often ignored problem arising from decentralization is the loss of daily capacity when a given terminal area is broken up into a number of subareas that operate independently. Capacity is determined by peak-hour operations, and demand peaks are more easily smoothed for one large terminal than for four smaller terminals. Prior to the opening of Heathrow Terminal 5 in 2008, London Heathrow operated four terminals that were functionally different. With some exceptions, the traffic assignment to the four terminals was as follows:

Terminal 1: Domestic and short-haul European routes (British Airways)
Terminal 2: Short-haul European routes (foreign carriers)
Terminal 3: Long-haul routes (foreign carriers)
Terminal 4: Long-haul routes (British Airways) and some short-haul European routes (foreign carriers)

An internal BAA study found that if all flights could have been assigned to one giant terminal, the required number of aircraft terminal gates would have been lowered significantly (Ashford, Stanton, and Moore 1997). Because of the different peaking characteristics observed in the four different terminals, when the airport is subdivided into four largely independent units, the demands on apron space and apron equipment were greater than those of a single terminal unit.

With the development of airline alliances, which began in the 1980s, at the large airports, the assignment of airlines among multiple-unit

Traffic Type	Passengers
Domestic	19.6 million
European (Chengen)	14.7 million
International	12.3 million
Total	**46.6 million**
Alliance	**Passengers**
One World	31.2 million
Star	5.1 million
Wing	4.0 million
Sky Team	3.8 million
Other	2.5 million
Total	**46.6 million**
One World Alliance	12 airlines
Star Alliance	27 airlines
Sky Team	13 airlines
Annual transfer	21 million
	45 percent
Transfer within same alliance	19 million
	41 percent

TABLE 1.4   Breakdown of Traffic to Be Assigned to Existing and New Terminals at Madrid Barajas Airport, 2000

terminals became even more complicated. At Madrid Barajas, for example, prior to the commissioning of the new terminal in 2006, the airport management studied how best to use the huge new facility in conjunction with the existing three terminals on the site. Much consideration had to be given to the fact that the major-based airline, Iberia, was part of the One World airline alliance, which had 12 members, including American Airlines, British Airways, and Qantas. Since the annual peak transfer rate at Madrid was 45 percent, with 41 percent of transfers being within the same alliance, the decision was made to locate One World in the new terminal and the other major alliances, Star Alliance and Sky Team, in the existing three terminals. The old and new terminal sites were separated by more than 2 miles (3 km). For interlining transferring passengers, a landside bus service was introduced. This allocation of airlines to the various terminals was calculated to minimize the overall inconvenience to Madrid's transfer passengers and was the principal rationale behind the final decision on terminal allocation. The complexity of the traffic mix to be dealt with is shown in Table 1.4.

1.5   The Complexity of the Airport Operation

Until the deregulation and privatization of the air transport industry in the late 1970s and 1980s, it had been seen in many countries almost as a public-service industry that required support from the public purse. Subsidies to aviation were provided in a number of different ways by different countries. Early post–World War II airports had very little commercial activity, and the services provided by the airport were very basic. Airports such as Shannon in the Irish Republic, and Amsterdam Schiphol in the Netherlands, were among the first to develop income from commercial activities. By the 1970s, the commercial revenues had become very important in terms of total income and at many of the largest European airports provided virtually all the profit, aeronautical income barely covering covered by aeronautical expenditures.

The larger airports became complex businesses with functions that extended well beyond the airfield or "traffic" side of operations. As airports increase in size, in terms of passenger throughput, the nonaviation revenues become increasingly important (Ashford and Moore 1999). It is also clear that in most countries, airports maintain economic viability by developing a broadly based revenue capability. In general, the organizational structure of the airport company changes to reflect the increasing importance of commercial revenues with increasing passenger throughput. As the relative and absolute sizes of the nontraffic element of the airport's revenue increase, much more attention must be paid to developing commercial expertise, and some of the largest airports have developed considerable in-house expertise in maximizing commercial revenues.

Among the nonaeronautical activities found at airports are (ICAO 2006; Ashford and Moore 1999):

- Aviation fuel suppliers
- Food and beverage sales (i.e., restaurants, bars, cafeterias, vending machines, etc.)
- Duty-paid shopping
- Banks/foreign exchange
- Airline catering services
- Taxi services
- Car rentals
- Car parking
- Advertising
- Airport/city transport services (i.e., buses, limousines, etc.)
- Duty-free shopping (e.g., alcohol, tobacco, perfume, watches, optical and electronic equipment)
- Petrol/automobile service stations
- Hairdressing/barber shop
- Internet services
- Casinos/gaming machines
- Cinema
- Vending machines for other than food
- Hotels/motels
- Freight consolidators/forwarders/agents
- Art exhibitions
- Music concerts
- Souvenir shops

The degree to which airports go to develop nonaeronautical activities is likely to depend on the destination of the revenues generated from such activities. At most airports, these go directly to the airport and add to the airport's profitability. As such, the airport company has a great incentive to generate as much of this type of income as possible, and the energy of the company is directed in this direction. There are a number of situations that can act as a disincentive to the airport company:

- Where income from nonaeronautical sources goes directly to the national treasury
- Where the government gives the duty-free franchise to the government-owned airline

- Where the U.S. airport is operated on a residual cost basis, and income from nonaeronautical sources is used to reduce landing fees for the airlines and does not accrue to the airport

Under these circumstances, there is little incentive for the airport management to attempt to increase nonaeronautical income, and in the absence of outside efforts, this side of the business is likely to stagnate.

1.6   Management and Operational Structures

Prior to deregulation of the airlines, which started in 1979 in the United States, the widespread model for operation of a transport airport was to be run as a department of either the central government or the local government. After deregulation, most central governments divested themselves of the operation of airports wherever possible. By 2013, many central governments still were involved in the running of airports, but it had become less common in the economies of the developed world. Countries such as the Netherlands, Spain, and Germany still had mostly government-owned facilities.

There is no single form of administrative structure that is ideal for every airport. Airports differ in their type and scale of throughput, vary in their interrelationships with other governmental and quasi-governmental bodies, and also must fit within differing matrices of allied and associated organizations at central, regional, and local governmental levels. Organizational structures also must be recognized as evolutionary in nature, depending on the previously existing structure and on the pressures for change, some of which arise from the personalities and abilities of individuals with directorial responsibilities in the existing organization. In any event, organizational structure will undergo radical reform should an airport become privatized. After privatization, there are many models of airport administration, the most common of which are summarized as follows with examples:

Government-owned airports
- Within a local government department: *Sacramento Airport, United States*
- Autonomous airport authority: *INFRAERO, Brazil; Dublin Airport Authority, Ireland*
- Within a multimodal transport authority: *Port Authority of New York and New Jersey, United States*
- Within a civil aviation department: *Abu Dhabi*
- Privatized company with shares owned by the local authority: *Manchester Airport, United Kingdom*

Privatized airports
- Single private airport: *Knock Airport, Ireland; Punta Cana Airport, Dominican Republic*
- Partially government-owned airport: *Newcastle Airport, United Kingdom (51 percent local authorities, 49 percent Copenhagen Airport)*

- Part holding of a multiairport operator: *Cardiff Airport, Luton Airport, United Kingdom*[1]
- Subsidiary company to a conglomerate: *London Heathrow, part of BAA Airports Ltd. (wholly owned by Ferrovial, Spain)*

Concessions

- Government-owned but leased on concession: *Lima, Peru (concession to FRAPORT and two minor partners)*
- Public/private consortium using build-own-operate-transfer (BOOT): *Athens Spata Airport (55 percent Greek government, 45 percent consortium led by Hochtief)*

The structure of an airport's organization depends on the role the airport company assumes in the operation of the facility. This may vary from a largely brokering function with minimal operational engagement in many on-airport activities (the U.S. model) to direct line involvement in many of the airport's functions (the European model). It also should be borne in mind that in common with other commercial and governmental organizations, the management structure may be divided into staff and line functions. The way these functions are accommodated also varies among airports. Staff departments are those which provide direct managerial support to the airport director or general manager. Often relatively small in size in terms of personnel, they are engaged in decision making that has an impact on the whole organization. Line departments, on the other hand, are the portions of the organization engaged in the day-to-day operation of the facility. In comparison with the staff departments, they usually require heavy staffing. The manner in which staff and line departments report to the airport director differs greatly among airports. Figure 1.5 shows three different formalized structures that cover the range of what might occur at any particular airport.

Option A is the structure where staff and line departments all report directly to the airport director. This is the normal situation at a small airport where staff functions are not excessive, and the airport director, as a matter of course, is closely involved with day-to-day operations. Option B is likely to occur at larger airports. The increasingly busy line departments report through a deputy director, while the staff departments are in a close supportive role to the director. At larger airports, option C is likely to occur, with line and staff departments reporting through two separate deputy directors. Examples with minor variants are described later in this chapter.

Figures 1.6 and 1.7 show the organizational structures of two autonomous Western European airports. Both structures reflect the

---

[1]The ownership and operation of Cardiff, Luton, and Belfast International airports are complicated. In 2013, these were owned by Abertas Infrastructuras, Spain (90 percent), and AENA, the Spanish Airports Authority (10 percent). The Spanish consortium had earlier bought the British airport operator TBI, which had acquired these three airports. By 2013, Abertas also owned three airports in Bolivia plus Skavsta Airport, Stockholm, Sweden, and Orlando Sanford, United States. Additionally, the company managed three other U.S. airports: Atlanta, GA, Macon, GA, and Burbank, CA.

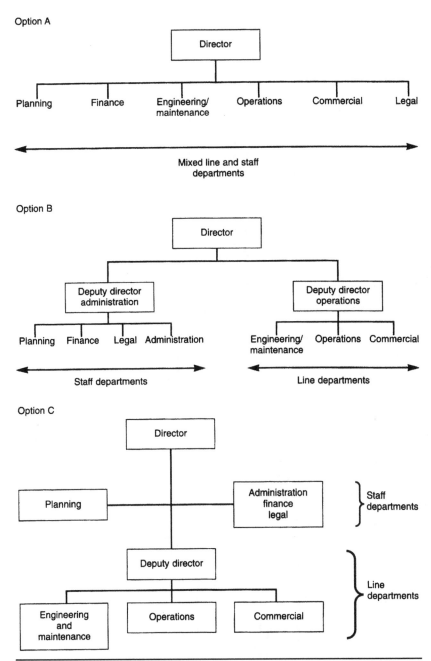

Figure 1.5 Schematic positions of line and staff departments in the structure of airport administrations.

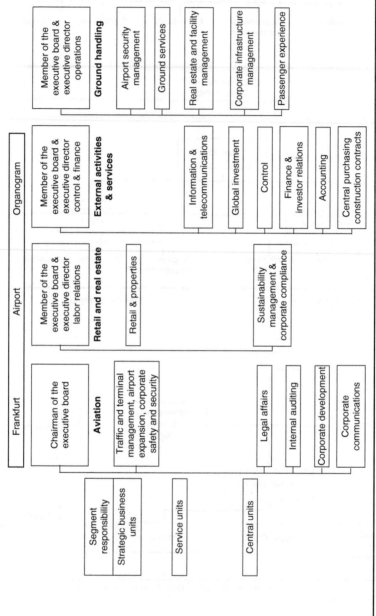

Figure 1.6 Administrative and staff structure, Frankfurt Airport (FRAPORT), 2011. (*Courtesy: FRAPORT.*)

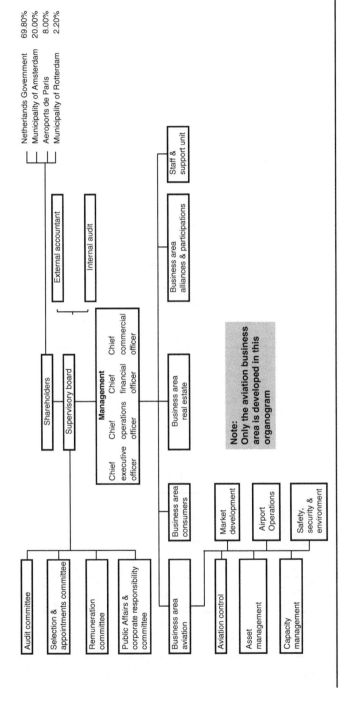

FIGURE 1.7 Administrative and staff structure, Schiphol Amsterdam Airport, 2011. (*Courtesy: Schiphol Amsterdam Airport.*)

21

fact that the organization is involved with the operation of a single airport on the European model and that some of the ground handling is carried out by airport authority employees. Until the 1990s, almost all the ground handling at these airports was carried out by airport authority staff only.

A very different form of functional arrangement exists in U.S. airports, where the airport authority requires that all the operational aspects of passenger and freight handling are carried out by the airlines and handling companies. The organization of the Los Angeles World Airports organization shown in Figure 1.8, and those of Sacramento and San Francisco airports are shown in Figures 1.9 and 1.10.

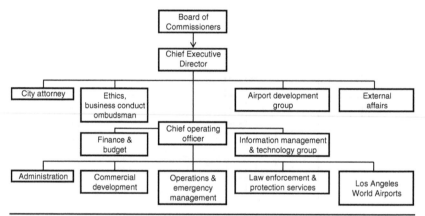

FIGURE 1.8   Organizational structure of Los Angeles World Airports (LAWA), 2011. (*Courtesy: Los Angeles World Airports.*)

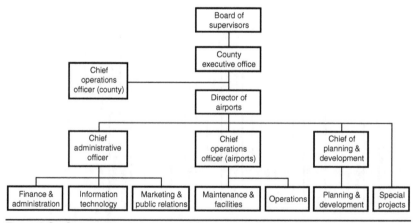

FIGURE 1.9   Organizational structure of Sacramento Airport, 2011. (*Courtesy: Sacramento Airport.*)

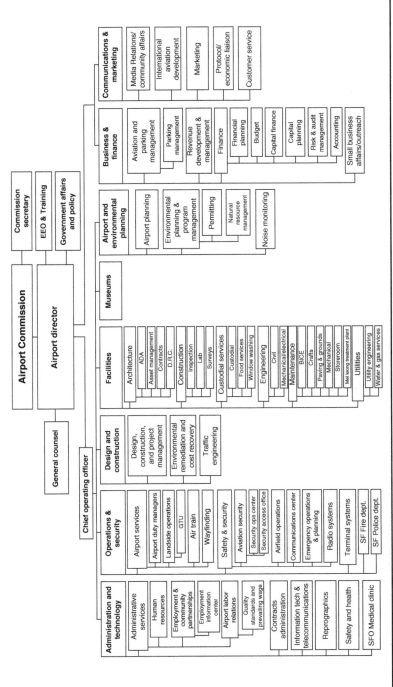

Figure 1.10 Organizational structure of San Francisco Airport, 2011. (*Courtesy: San Francisco International Airport.*)

In many countries, governmental or quasi-governmental authorities are charged with the operation of a number of airports (e.g., the Port Authority of New York and New Jersey [PANYNJ], Aeroports de Paris [AdP] in France, and Aeroportos e Navagaceo Aerea [ANA] in Portugal). The organizational structure of such multiairport authorities is usually designed for achieving system-wide objectives. Therefore, policies are directed by an overall executive, to whom the usual staff functions give support. Individual airports become operational elements in the overall structure. Wiley developed a typical pro forma organization structure for a three-airport authority within a multimodal authority (Figure 1.11). His model was based on his administrative experience in the PANYNJ (Wiley 1981).

Practical examples of structures of this nature are shown in Figures 1.12 and 1.13, which show the actual organizational structures of the PANYNJ and the equivalent of the Civil Aviation Authority

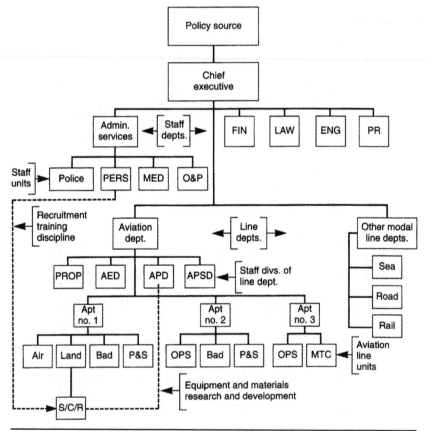

Figure 1.11   Pro forma organogram for a three-airport multimodal planning and operating authority. (*Wiley 1981.*)

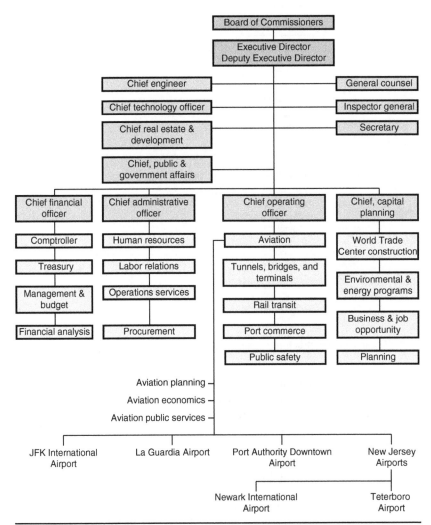

FIGURE 1.12 Organizational structure of the Port Authority of New York and New Jersey, 2011.

(ANA) of Portugal. In each case, the structure of the organization permits the development of system-wide policies affecting a number of airports; this is a requirement that is clearly not necessary in the case of an autonomously operated single airport. The PANYNJ structure is especially interesting because of the multimodal interests of the authority. Aviation constitutes only one department within the complex structure, even though this department operates the three very large airports of the New York metropolitan area.

With the privatization of many of the larger airports since 1987, a number of private companies now own airports on a multinational

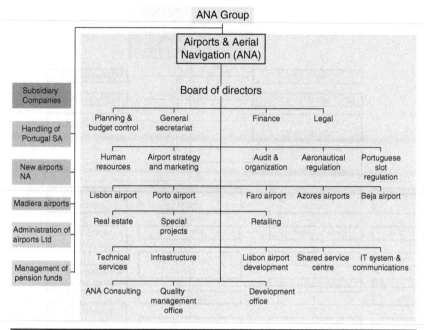

FIGURE 1.13   Organizational structure of the Portuguese civil aviation system, 2011.

basis. The organograms of these organizations become very complex, as can be seen from the example shown in Figure 1.14. This indicates the structure of an organization involved in the ownership, management, or operation of some 30 airports in North and South America and Europe in 2011.

The variety of structures shown in this chapter, when taken in conjunction with the great differences in roles undertaken by the various airport companies, means that it is not possible to determine or even impute any strong relationship between passenger throughput and the size of the airport company staff. Where the airport company "brokers out" most activities, the airport staff requirements will be low. As more activities are undertaken by the airport itself, the staff requirements naturally increase. Figure 1.15 shows the annual passenger throughput for a number of airports in the 1990s plotted against the airport staff at that time. As expected, there was a large variation about any single "fitted" line, indicating that there is only a weak correlation between the two variables. However, if the data points are split into two categories, North American airports and other airports, a reasonably strong correlation becomes apparent on a log-log basis. Each point on the graph represents a different operational situation with differing responsibilities. However, the graph does dramatically demonstrate the increased labor requirements of airport companies that retain a portion of handling activities in lieu of

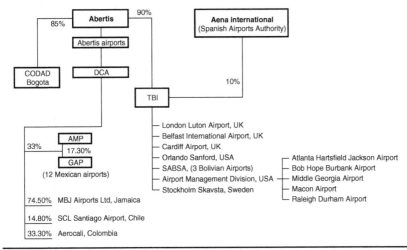

Figure 1.14 The structure of a private company having multinational airport interests, 2011.

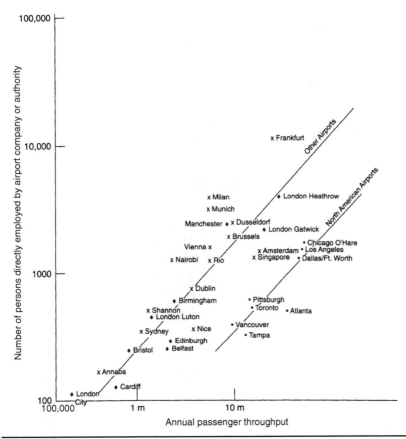

Figure 1.15 Annual passenger throughput in relation to airport company staff.

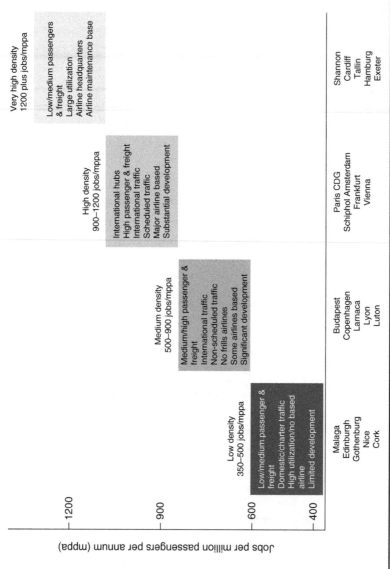

Figure 1.16 Onsite employment types at European airports, 2003. (*York Aviation 2004.*)

delegating these to others. In the European Union in the early 1990s, legislation required larger airports to provide competition for the handling of passenger and freight operations. As a result, the very large handling operations at European airports were broken up, and much handling was devolved from the airport companies to other companies specializing in airport handling.

A study carried out for the Airports Council International examined a range of European airports in 2003 (York Aviation 2004). It examined the level of employment at a number of airports and grouped the range of airports examined into four categories:

- Low density
- Medium density
- High density
- Very high density

Figure 1.16 shows the four classifications along with the types of activities likely to occur within each class and the range of employment observed within each category. The low-, medium-, and high-density airports show increases in employment related to growth in passenger and freight traffic. Very high-density employment is not related to high workload units (WLUs)[2] of traffic but rather to specialized large employment centers such as airline maintenance bases and airline headquarters.

References

Ashford, N., and C. E. Moore. 1999. *Airport Finance*, 2nd ed. Bournemouth, UK: Loughborough Airport Consultancy.

Ashford, N., S. A. Mumayiz, and P. H. Wright. 2011. *Airport Engineering: Planning, Design, and Development of 21st Century Airports*. Hoboken, NJ: Wiley.

Ashford, N., H. P. M. Stanton, and C. E. Moore. 1997. *Airport Operations*, 2nd ed. New York: McGraw-Hill.

Department of Transport. 2003. *The Future of Air Transport*. London: Her Majesty's Stationer's Office.

Doganis, R., and R. Pearson. 1977. *The Financial and Economic Characteristics of the UK Airport Industry*. London: Polytechnic of Central London.

Doganis, R., and G. Thompson. 1973. *Economics of British Airports*. London: Polytechnic of Central London.

HMSO. 1978. *Airports Policy*. Command Paper 7084. London: Her Majesty's Stationery Office.

ICAO. 2006. *Airport Economics Manual*, 2nd ed. Document 9562. Geneva: International Civil Aviation Organization.

TRB. 2008. *ACRP Synthesis 7, Airport Economic Impact Methods and Model: A Synthesis of Airport Practice*. Washington, DC: Transportation Research Board.

Wiley, John R. 1981. *Airport Administration*. Westport, CT: Eno Foundation.

York Aviation. 2004. *The Social and Economic Impacts of Airports in Europe*. Geneva: Airports Council International.

---

[2]Workload unit (WLU): 1 WLU equals one departing or arriving passenger or 100 kg of freight.

CHAPTER 2

Airport Peaks and Airline Scheduling

2.1 The Problem

Airport operators usually speak enthusiastically about the business of the airport in terms of the throughput of passengers and cargo, as represented by the annual number of passengers processed or the annual turnover of tons of air freight. This is entirely understandable because annual income is determined to a large degree by these parameters; moreover, the numbers are impressively large. However, it is important, when considering annual figures, to bear in mind that whereas annual flows are the prime determinant of revenue, it is the peak flows that determine to a large degree the physical and operational *costs* involved in running a facility. Staffing and physical facilities are naturally keyed much more closely to hourly and daily requirements than to annual throughput.

In common with other transportation facilities, airports display very large variations in demand levels with time. These variations can be described in terms of

- Annual variation over time
- Monthly peaks within a particular year
- Daily peaks within a particular month or week
- Hourly peaks within a particular day

The first of these is extremely important from the viewpoint of the planning and provision of facilities. Air transportation still is considered the fastest growing mode, and there is little indication that this situation is likely to change. Consequently, operators of airport facilities often are faced with growing volumes that approach or exceed capacity. During the period 1970–2010, the average worldwide rate of air passenger growth was close to 7 percent. Even during the difficult period 2000–2010, which included the oil price increases during the period 2005–2008 and the subsequent recession, the average world air passenger growth rate also was approximately 5 percent. Air transport passenger travel is expected to continue to grow at 5.1 percent between the years 2010 and 2030 and air cargo to grow at 5.6 percent during the same period (Boeing 2011). Although the operator must be closely involved in the long-term planning of the airport, it is not the function of this text to deal with planning aspects that are covered elsewhere (Ashford, Mumayiz, and Wright 2011; Horonjeff et al. 2010). Emphasis here will be on the short-term considerations of day-to-day operation. Therefore, the discussion will concentrate on monthly, daily, and hourly variations of flows. In the operational context, this is natural because many of the marginal costs associated with the day-to-day provision of staffing and rapidly amortized equipment are not really related to long-term variations in traffic but rather to variations within a 12-month span.

At most, if not all, air transport airports, the major consideration must be passenger flow. At many of the larger airports, cargo operations are becoming increasingly important in part because cargo transport continues to outstrip passenger traffic in terms of growth rate. In the planning and operation of air cargo facilities, however, it must be noted that the peaks for air cargo operations do not coincide with those for air passenger transport. The two submodes usually can be physically separated to the necessary degree, even though proximity of the freight and passenger aprons is desirable because much freight is carried in the bellies of passenger transport aircraft. The particular problems of cargo operations will be discussed in Chapter 10.

When considering the characteristics of the peaking of passenger flows, it is always important to bear in mind that the "passenger" is not a homogeneous entity. Passenger traffic is built up from the individual journey demands of many passengers. These passengers are traveling under different conditions, they have different needs, and consequently, they place different overall demands on the system. Not surprisingly, this is reflected in different peaking characteristics, depending, for example, on whether the passenger is domestic or international, scheduled or charter, leisure or business, full fare or special fare.

Complicating the whole matter of peaking is the fact that unlike the situation in most other modes of transportation, where the passenger is dealing with only one operator, in air transport, there is the complex interrelationship of the passenger, the airport, and often several airlines. In the matter of peaking, the aims of the airline and the airport operator do not necessarily coincide. The airport operator would like to spread demand more evenly over the operating day in order to decrease the need for the supply of facilities governed by the peak. The airline, on the other hand, is looking to maximize fleet utilization and to improve load factors by offering services in the most attractive time slots. There is therefore a potential conflict between the airline satisfying its customer, the passenger, and the airport attempting to influence the demands of its principal customer, the airline.

2.2 Methods of Describing Peaking

Even the busiest airport operates over a wide range of traffic flows. Many of the world's largest air transport terminals are virtually deserted for many hours of the year; these same facilities only a few hours later may be operating at flows that strain or surpass capacities. Few facilities are designed to cope with the very highest flow volume that occurs in the design year of operation. Most are designed such that for a few hours of the year there will be an acceptable level of capacity overload. Different airport and aviation authorities approach this problem in different ways. Figure 2.1 shows one of the characteristics of traffic peaking for a typical airport, that is, the curve of passenger traffic volumes in ranked order of magnitude. It can be seen that for a few hours per year there are very high peak volumes of traffic. Operational practice tends to accept that for a few hours of each year facilities must be operated at some level of overload (i.e., volumes that exceed physical and operational capacity) with resulting delays and inconvenience. To do otherwise and to attempt to provide capacity for all volumes would result in uneconomical and wasteful operation.

The Standard Busy Rate

The *standard busy rate* (SBR) measure or a variation of it is a design standard that has been used in the United Kingdom and elsewhere in Europe, most notably by the former British Airports Authority (BAA).

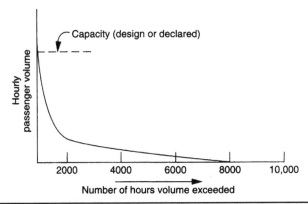

FIGURE 2.1 Typical distribution of hourly passenger traffic volumes at an air transport airport throughout the year.

It is frequently defined as the thirtieth highest hour of passenger flow, or that rate of flow that is surpassed by only 29 hours of operation at higher flows. The concept of the thirtieth highest hour is one that is well rooted in civil engineering practice in that this form of design criterion has been used for many years to determine design volumes of highways. Design for the SBR ensures that facilities will not operate at or beyond capacity for more than 30 hours per year in the design year, which is felt to be a reasonable number of hours of overload. The method does not, however, take explicit note of the relationship of the SBR to the actual observed annual peak volume. In practice, this relationship is likely to be on the order of

$$\text{Absolute peak-hour volume} = 1.2 \times \text{SBR} \qquad (2.1)$$

but there is no guarantee that this will be so.

Table 2.1 shows that in terms of aircraft movement, the ratio of the SBR to the absolute peak increases with increasing annual

Annual Movements	Peak Day		Peak Hour		SBR	
	To Average Day Ratio	Number	To Peak Day Ratio	Number	To Peak Hour Ratio	Number
10,000	2.666	73	0.1125	8	0.688	6
20,000	2.255	124	0.1051	13	0.732	10
30,000	2.045	168	0.1011	17	0.759	13
40,000	1.907	209	0.0983	21	0.779	16
50,000	1.807	248	0.0961	24	0.794	19
60,000	1.729	284	0.0944	27	0.807	22
70,000	1.666	320	0.0930	30	0.819	24
80,000	1.613	354	0.0918	32	0.828	27
90,000	1.568	387	0.0908	35	0.837	29
100,000	1.529	419	0.0898	38	0.845	32
110,000	1.494	450	0.0890	40	0.852	34
120,000	1.463	481	0.0883	42	0.859	36
130,000	1.435	511	0.0876	45	0.865	39
140,000	1.409	541	0.0869	47	0.871	41
150,000	1.386	570	0.0863	49	0.876	43
160,000	1.365	598	0.0858	51	0.881	45
170,000	1.345	626	0.0853	53	0.886	47
180,000	1.326	654	0.0848	55	0.891	49

TABLE 2.1   Relationship Between Annual, Peak-Hour, SBR and Peak Day Aircraft

Annual Movements	Peak Day		Peak Hour		SBR	
	To Average Day Ratio	Number	To Peak Day Ratio	Number	To Peak Hour Ratio	Number
190,000	1.309	681	0.0844	57	0.895	51
200,000	1.293	708	0.0840	59	0.899	53
210,000	1.278	735	0.0836	61	0.903	55
220,000	1.264	762	0.0832	63	0.907	57
230,000	1.250	788	0.0828	65	0.910	59
240,000	1.237	814	0.0825	67	0.914	61
250,000	1.225	839	0.0821	69	0.917	63
260,000	1.214	864	0.0818	71	0.920	65
270,000	1.203	890	0.0815	73	0.924	67
280,000	1.192	914	0.0812	74	0.927	69
290,000	1.182	939	0.0810	76	0.929	71
300,000	1.172	964	0.0807	78	0.932	72
310,000	1.163	988	0.0804	79	0.935	74
320,000	1.154	1012	0.0802	81	0.938	76
330,000	1.146	1036	0.0799	83	0.940	78
340,000	1.137	1060	0.0797	84	0.943	80
350,000	1.130	1083	0.0795	86	0.945	81
360,000	1.122	1106	0.0793	88	0.948	83
370,000	1.114	1130	0.0791	89	0.950	85
380,000	1.107	1153	0.0788	91	0.952	87
390,000	1.100	1176	0.0786	92	0.954	88
400,000	1.094	1199	0.0785	94	0.957	90
410,000	1.087	1221	0.0783	96	0.959	92
420,000	1.081	1244	0.0781	97	0.961	93
430,000	1.075	1266	0.0779	99	0.963	95
440,000	1.069	1288	0.0777	100	0.965	97
450,000	1.063	1311	0.0776	102	0.967	98
460,000	1.057	1333	0.0774	103	0.969	100
470,000	1.052	1354	0.0772	105	0.971	102
480,000	1.047	1376	0.0771	106	0.972	103
490,000	1.041	1398	0.0769	108	0.974	105
500,000	1.036	1420	0.0768	109	0.976	106

Source: UK Civil Aviation Authority.

TABLE 2.1   *(Continued)*

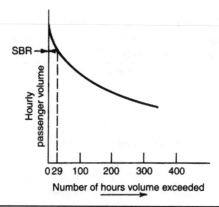

FIGURE **2.2**   Location of the standard busy rate.

volume. This reflects the fact that as the traffic of an airport develops, extreme peaks of flows tend to disappear.

The table indicates that use of the SBR method in low-volume airports could result in high (peak/SBR) ratios that, in turn, could lead to severe overcrowding for a few hours per year. The location of the standard busy hour is shown in Figure 2.2.

Busy-Hour Rate

A modification of the SBR that also has been used for some time is the *busy-hour rate* (BHR), or the 5 percent busy hour. This is the hourly rate above which 5 percent of the traffic at the airport is handled. This measure was introduced to overcome some of the problems involved with using the SBR, where the implied level of congestion at the peak was not the same from airport to airport. The BHR is easily computed by ranking the operational volumes in order of magnitude and computing the cumulative sum of volumes that amount to 5 percent of the annual volume. The next ranked volume is the BHR. This is shown graphically in Figure 2.3.

Typical Peak-Hour Passengers

The Federal Aviation Administration (FAA) uses a peak measure called the *typical peak hour passengers* (TPHP) that is defined as the peak hour of the average peak day of the peak month. In absolute terms, this approximates very closely the SBR. To compute the TPHP from annual flows, the FAA recommends the relationships shown in Table 2.2. Stated in this form, it is apparent that the peak is more pronounced with respect to annual flows at small airports. As airports grow larger, the peaks flatten, and the troughs between peaks become less pronounced.

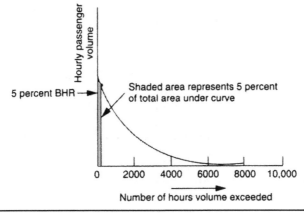

FIGURE **2.3**   The 5 percent busy-hour rate.

Total Annual Passengers	TPHP as a Percentage of Annual Flows
30 million and over	0.035
20,000,000–29,999,999	0.040
10,000,000–19,999,999	0.045
1,000,000–9,999,999	0.050
500,000–999,999	0.080
100,000–499,999	0.130
Under 100,000	0.200

Source: FAA.

TABLE **2.2**   FAA Recommended Relationships for TPHP Computations for Annual Figures

Busiest Timetable Hour

This simple method is applicable to small airports with limited databases. Using average load factors and existing or projected timetables, the *busiest timetable hour* (BTH) can be calculated. The method is subject to errors in forecasting, the rescheduling and reequipping vagaries of the airlines and variations in average load factors.

Peak Profile Hour

Sometimes called the *average daily peak*, the *peak-profile-hour* (PPH) method is fairly straightforward to understand. First, the peak month is selected. Then, for each hour, the average hourly volume is computed across the month using the actual length of the month (i.e., 28, 30, or 31 days as applicable). This gives an average hourly volume for an "average peak day." The peak profile hour is the largest hourly value in the average peak day. Experience has shown that for many airports, the PPH is also close to the SBR.

Other Methods

Although many outside the United States use some form of the SBR method to define the peak, there is little uniformity in method. In West Germany, for example, most airport authorities have used the thirtieth highest hour. Prior to introduction of the BHR, the BAA used the thirtieth highest hour or the PPH, whereas most other British airports used the thirtieth highest hour. In France, Aeroports de Paris based its design on a 3 percent overload standard. (In Paris, studies have shown that the thirtieth busy hour tends to occur on the fifteenth busiest day.) Dutch airports use the sixth busiest hour, which is approximated by the average of the 20 highest hours.

Airport Differences

The shape of the volume curve shown in Figure 2.1 differs among airports. The nature of these differences can be seen by examining the form of the curves for three airports with widely differing functions, as shown in Figure 2.4.

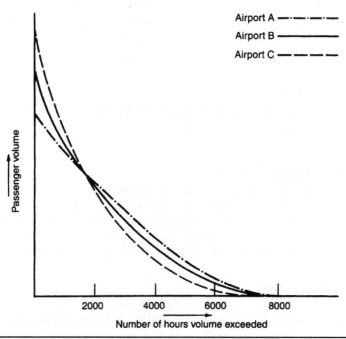

FIGURE 2.4   Variation of passenger volume distribution curves for airports with different traffic characteristics.

Airport A   A high-volume airport with a large amount of short-haul domestic traffic (typical U.S. or European hub).

Airport B   A medium-volume airport with balanced international/domestic traffic and balanced short-haul/long-haul operations (typical Northern European metropolitan airport).

Airport C   A medium-volume airport with a high proportion of international traffic concentrated in a vacation season (typical Mediterranean airport serving resort areas).

Airport C will carry a higher proportion of its traffic during peak periods and, therefore, there is a leftward skew to the graph in comparison with Airport B. A typical U.S. or European hub, on the other hand, with larger amounts of domestic short-haul traffic carries more even volumes of passengers across the period 0700 to 1900 hours, decreasing the leftward skew of the graph.

Nature of Peaks

Airport traffic displays peaking characteristics by the month of the year, by the day of the week, and by the hour of the day. The form and time of the peaks very much depend on the nature of the airport traffic and the nature of the hinterland served.

The following factors are among the most important affecting peaking characteristics:

1. *Domestic/international ratio.* Domestic flights will tend to operate in a manner that reflects the working-day pattern because of the large proportion of business travelers using domestic flights.

2. *Charter and low-cost carrier (LCC)/scheduled ratio.* Charter flights are timetabled for maximum aircraft usage and are not necessarily operated at the peak periods found most commercially competitive by scheduled airlines. Low-cost carriers also strive for maximum aircraft usage and tend to schedule flights in hours not commercially attractive to full-fare passengers.

3. *Long-haul/short-haul.* Short-haul flights are frequently scheduled to maximize the usefulness of the day either after or prior to the flight. Therefore, they peak in early morning (0700 to 0900) and late afternoon (1630 to 1830). Long-haul flights are scheduled mainly for a convenient arrival time, allowing for reasonable rest periods for travelers and crew and to avoid night curfews.

4. *Geographic location.* Schedules are set to allow passengers to arrive at a time when transportation and hotels are operating and can be used conveniently. For example, the six- to eight-hour eastward transatlantic crossing is most conveniently scheduled for early-morning arrivals at the European airports, avoiding curfews. Allowing for the time differences between North America and Europe, this means an evening departure from the eastern seaboard.

5. *Nature of catchment area.* The nature of the region served has a strong influence on the nature of traffic peaking throughout the year. Areas serving heterogeneous industrial-commercial metropolitan areas such as Chicago, Los Angeles, London, and Paris show steady flows throughout the year, with surges at the Christmas, Easter, and summer holiday periods reflecting increased leisure travel. Airports in the vicinity of highly seasonal vacation areas, such as the Mediterranean and the Caribbean, display very significant peaks in the vacation months.

Figure 2.5 shows the monthly variations in traffic at several airports serving widely different geographic areas in both the northern and southern hemispheres. Daily variations in the peak week are shown in Figure 2.6. The analysis is carried further by Figure 2.7, which shows hourly passenger movements for two congested airports. London Heathrow's runways are operating

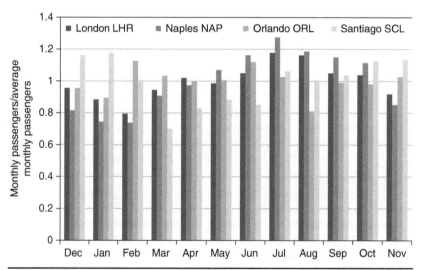

Figure 2.5 Monthly variations in passenger traffic at selected airports. (*Source: Reporting airports.*)

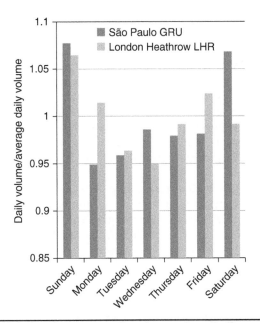

FIGURE 2.6 Variations in passenger flows in a peak week. (*Source: BAA and INFRAERO.*)

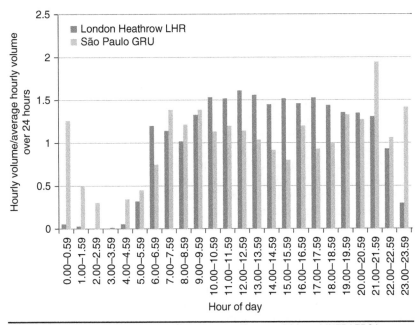

FIGURE 2.7 Variations in hourly traffic volumes. (*Source: BAA and INFRAERO.*)

most of the day at capacity, whereas the capacity limitation at São Paulo Guarulhos is in the terminals. The extreme peaks in São Paulo's flows are associated with severe crowding and a lowering of level of service (LOS).

Despite the difference between peaks caused by the many factors that affect peaking, there is in some aspects, in fact, great overall similarity between airports. It is therefore possible to deduce general relationships between peak and annual flows at airports largely because no airport is entirely unifunctional, just as no town is entirely industrial, governmental, educational, or leisure-structured in its makeup.

Figure 2.8 shows the relationship between peak flows as represented by SBR and annual flows for a number of rather diversely selected airports. Also shown on this graph are the FAA peak/annual recommended ratios, as embodied in the TPHP concept. The great similarity of the two approaches becomes apparent when they are presented graphically.

2.3 Implications of Variations in Volumes

It can be easily demonstrated that the demand for peak-hour schedules affects the amount of infrastructure that must be supplied by the airport. Whereas the need to implement service in an off-peak period will not necessarily involve the airport in significant marginal costs, at a crowded airport, the decision to take another service in the peak hour might well add significant marginal costs. There are, however, economies of scale that result from peak-hour operations.

Figure 2.9 shows the relationship between passenger flows and air-carrier aircraft operations. It can be seen that whereas passenger volumes vary significantly between peak and nonpeak hours, the same scale of variation is not observed in aircraft movement volumes. This reflects the fact that during off-peak periods, aircraft operate at lower load factors than during the attractive peak-hour slots. The implications in terms of costs and revenues need to be considered. Services such as ramp handling, emergency services, air traffic control, runway and taxiway handling, and even some terminal services (e.g., announcements and baggage check) are based on the aircraft unit rather than on the number of passengers it carries. In off-peak hours, these services are provided at a less economic rate per passenger than during peak periods owing to low load factors during off-peak periods. Therefore, the airport is faced with a dilemma. Although peak operations would appear to involve high marginal costs in terms of infrastructure, operation at close to peak volumes is highly economic once this infrastructure is provided. There is even a temptation for

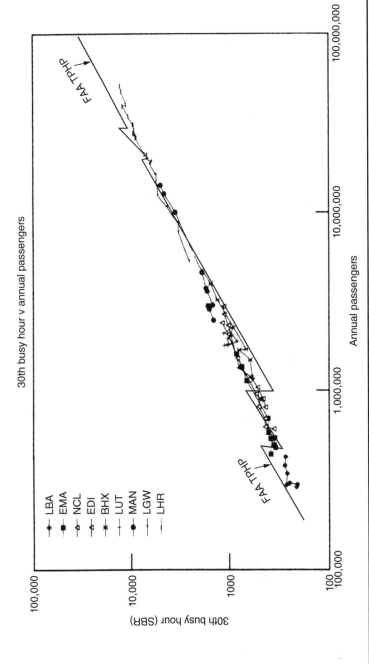

Figure 2.8 Relationships among standard busy rate (SBR), typical peak-hour passenger volume, and annual passenger volume.

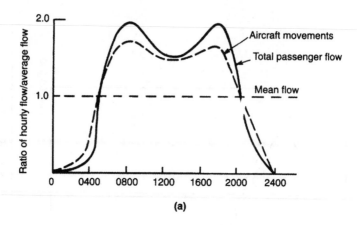

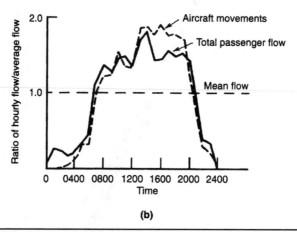

Figure 2.9 (a) Idealized relationship between air carrier movements and passenger flow. (b) Observed relationship between air traffic movements and passenger flow at Chicago O'Hare Airport. (*Source: FAA.*)

the airport to operate at flow levels above the design rate. This inevitably leads to reduced LOS in terms of processing delays and overcrowding.

2.4 Factors and Constraints on Airline Scheduling Policies

The development of a schedule, especially at a major hub with capacity problems, is a complex problem for the airline. The process involves considerable skill and a clear understanding of company policies and operating procedures. Among the factors to be considered, the following are most important.

Utilization and Load Factors

Aircraft are expensive items of equipment that can earn revenues only when being flown. Clearly, all other factors being equal, high utilization factors are desirable. However, utilization alone cannot be used as the criterion for schedule development; it must be accompanied by high load factors. Without the second element, aircraft would be scheduled to fly at less than breakeven passenger payloads, which typically are close to 70 percent on long-haul operation of a modern wide-bodied aircraft.

Reliability

No airline would attempt to schedule using the sole criterion of maximizing utilization of aircraft. Utilization can be maximized, however, subject to the double constraints of load factors and punctuality. As attempted utilization increases, the reliability of the service will suffer in terms of punctuality. Schedule adherence is a function of two random variables: equipment serviceability and late arrivals or departures of aircraft owing to en-route factors.

Computer models are used to predict the effect of schedules on punctuality, and the result is compared with target levels of punctuality set in advance for each season.

Long-Haul Scheduling Windows

A schedule must take into account the departure and arrival times at the various airports at origin, en route, and at destination. In 2012, Qantas offered a service between London and Sydney that called at Frankfurt, Singapore, and Melbourne. Leaving London at 1830, the flight first called at Frankfurt 2115/2350, local time, avoiding the landing ban at Frankfurt from 0100/0400. The next stop on the following day was Singapore, 1800/1945, on the evening of the next day, followed by a call at Melbourne, 0500/0645, the morning of the day after that. The final leg of the flight landed at Sydney at 0810, well after the end of the night curfew, which ran from 2300 to 0600. If the same service were to be attempted with a schedule to land at Sydney at least an hour and 20 minutes before the beginning of the curfew, it would have to leave London at 0800 two days before. This is a poor time to begin a long flight because of problems accessing London Airport at such an early hour. Departure times must be set recognizing that many passengers must travel from city centers to the airport and must arrive at the airport some reasonable time before the scheduled time of departure. The landing time at Sydney also gives too small a margin for error.

Figure 2.10 provides examples of scheduling windows for flights to and from London. Eastbound transatlantic flights from New York JFK to London Heathrow take approximately seven hours, and there

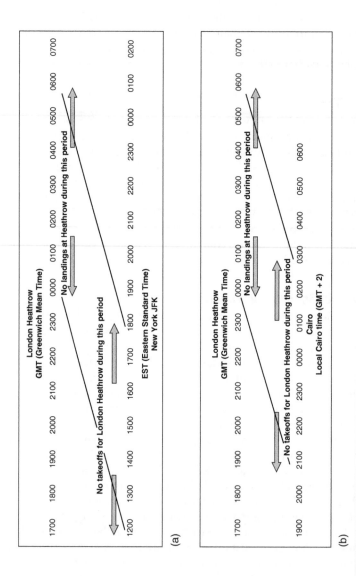

Figure 2.10 Scheduling windows for eastbound and westbound flights into London Heathrow Airport.

is a time difference of five hours between the two cities. The Heathrow night jet ban, which has few exceptions, commences at 0000 hours and ends at 0600 hours. Eastbound flights are therefore scheduled to take off before 1200 hours or after 1800 hours. Zurich has a no-exception night jet ban between 2300 hours and 0600 hours. Eastbound flights from New York JFK must leave either before 0900 hours or after 1600 hours.

Westbound flights into London Heathrow also must be scheduled to arrive outside the hours of curfew. Flights from Cairo, two hours ahead of London in time, must leave before 2050 hours to arrive before 000 hours or after 0250 hours to arrive after 0600 hours. For flights between Cairo and Zurich, the departures are restricted to hours outside the hours between 1950 and 0250 hours.

In 2008, 53 African states complained that the night bans in Europe discriminated against their services to Europe, which were precluded by night bans on landing. It was claimed that the night bans such as those in Zurich severely restricted their services by making early-morning connections in Europe possible only with very unsatisfactory takeoff times at the African departure airport (ICAO 2008; MPD 2005).

Airport (Runway) Slots

Runway takeoff and landing slots also must be considered. In many airports, especially in Europe, North America, and Asia, existing runways are running near to capacity during peak periods of the day. This capacity is limited owing to the necessary safety margins required in the separation of arriving and departing aircraft. Many airports near their slot capacity are coordinated. This means that a regulatory authority such as the FAA or the Civil Aviation Authority (CAA) has to determine and allocate a number of slots as being available to arriving or departing flights. Actual coordination is carried out semiannually at International Air Transport Association (IATA) slot conferences. A carrier often will have the right only to its historical slots, provided that these are being used. Consequently, at a coordinated airport, any carrier will be uncertain whether it will be possible to move from its historic slots or gain more slots. This situation poses problems to schedulers, who must make assumptions on the likely slots available to them.

Terminal Constraints

Another constraint faced by schedulers is that of airport passenger apron and terminal capacity. Many airports are operating only slightly below the capacities of these facilities often as built 20 or more years earlier. In the case of terminals, authorities often limit the number of passengers that can pass through a terminal during a half-hourly period, stating that this flow is the "declared capacity." This obviously sets a limit on the number of arrivals, departures, or combinations of aircraft apron movements that can be scheduled in capacity-constrained periods, presenting schedulers with yet another hurdle.

Long-Haul Crewing Constraints

On long-haul flights, crews may not be used continuously. Typically, a maximum tour of duty could be 14 hours, which includes 1½ hours of pre- or postflight time; there is also a required minimum rest period (usually 12 hours). Therefore, crews are changed at *slip* airports, and timing must be arranged so that fresh crews are available at these airports to relieve incoming flights that will be continuing their journeys.

Short-Haul Convenience

Because short-haul flights frequently carry large numbers of business travelers, departure and arrival times are critical to marketing the flights. Short-haul flights that cannot provide a one-day return journey suitably scheduled around the business working day are difficult to market.

General Crewing Availability

In addition to the special problems associated with layovers of long-haul flight and cabin crews at slip airports, all schedules must be built around the availability of maintenance, ground, air, and cabin crews. There is clearly a very strong interrelationship between the numbers of various crew personnel required and the operations to be scheduled, especially in terms of mixed short- and long-haul flights.

Aircraft Availability

Airlines must schedule the use of their aircraft in a manner that reflects the needs of routine maintenance checks. The individual manufacturers provide advice on aircraft maintenance programs, but each operator needs approval of its continuous inspection program from its appropriate airworthiness regulatory authority, for example, the FAA, Transport Canada, or the European Aviation Safety Agency (EASA). Most aircraft maintenance organizations use an approach based on Boeing's Maintenance Steering Group 3 (MSG-3) recommendations, which require four different kinds of checks (Kinnison 2004):

> *A Check.* This is a light check carried out every 500 to 800 hours, usually overnight at an aircraft stand.
>
> *B Check.* This is also a light check usually carried out overnight at the aircraft stand, usually every three to five months.

C Check. This is a heavy-maintenance check carried out in a hangar at approximately 15 to 21 months.

D Check. Also known as a *heavy maintenance visit* (HMV), this check is carried out every four to six years and requires several months in a hangar. On a large aircraft, as many as 100 technicians may be involved.

Irregular unavailability also can occur when nonroutine maintenance, such as cabin upgrades, is required or when there is a change of livery or change of ownership. Depending on the fleet type of the aircraft, its age, and the purpose for which it is being used, availability of particular aircraft type will differ. Other factors affecting availability could include

Geographic location. Operation in the temperate zones in northern Europe, northern United States, or Canada or in hot and dusty desert conditions requires different routine maintenance regimes.

Number of operational cycles or operational hours. Short-haul operations will average perhaps one landing every 2 hours; many long-haul operations have only one landing every 12 to 15 hours. However, aircraft cannot necessarily always be considered long- or short-haul vehicles just by type; for example, charter companies operate B757s from Europe to the Middle East, and British Airlines (BA) operates A318s from Britain to North America. Some companies such as BA operate the same aircraft on both short- and long-haul routes: The BA B767 fleet has two maintenance schedules, one for aircraft operating within Europe and another for those which operate long-haul flights.

Style of operation. There is an increasing trend to use complex maintenance scheduling, wherein some of the A checks and some of the B checks are carried out simultaneously, and all the B checks are completed within the scheduled framework of checks A-1 through A-10. Similarly, the C check can be segmented in such a way that part of the check can be carried out within the time frame allotted to the A and B checks. Such practice shortens the time that aircraft have to be withdrawn from the active fleet for maintenance purposes. Therefore, it is ... generally not possible now to state hard and fast guidelines for the actual timing of maintenance checks. The roles of the scheduling and aircraft maintenance departments in an airline are to develop jointly a schedule that fits the needs of operations provision and maintenance requirements.

Marketability

The scheduled times of departure or arrival must be marketable by the airline. Connections are especially important at major transfer points, such as Atlanta, London, and Singapore. Whenever possible, passengers avoid long layovers at an airport. Other factors that the airline considers are that departure and arrival times at major generating hubs must be at times when public transport is operating and may have to coincide with hotel check-in and check-out times and room availability. It is also important to have continuity of flight times across the days of the week if the flight operates several times a week.

Summer-Winter Variations

Where there is a large amount of seasonal traffic, usually vacation-related, there can be substantial differences in scheduling policies between summer and winter operations. The large variation in demand that can occur at airports serving such areas as Florida and the Bahamas, the Caribbean, and the Mediterranean resort areas is substantial, and this will affect the schedules of airports with which their services link. Seasonal variations also are large at airports such as Munich that serve ski resorts.

Landing-Fee Pricing Policies

At some airports, an attempt has been made to vary landing and aircraft-related fees in order either to use a pricing policy to spread peaking or to recoup extra finance for operations carried out in the uneconomic night hours. An example of the former policy is that which was used by the former BAA, which at a stroke adopted punitive peak-hour tariffs at London Heathrow to encourage airlines to transfer operations from Heathrow to Gatwick airports and to move operations from the peak period. Under this policy, a typical turn-around of a long-range B747 at Heathrow during the peak period was 2.8 times the cost for an operation outside the peak tariff times and 183 percent of the cost that would have been involved had the operation taken place at the less popular London Gatwick airport at the same peak time. The effect of this peak tariff was not large, as can be seen in Table 2.3, which shows the observed operational impact of this particular differential tariff.

In general, there is little evidence to indicate that airlines do reschedule significantly to avoid such tariffs. Airline operators claim that there are far too many other constraints precluding massive rescheduling outside peak-demand periods and that, therefore, such tariffs are almost entirely ineffective in achieving their proclaimed purpose.

The truth would appear to lie somewhere between these two positions. Where there is no differential peak pricing, airlines have no particular incentive, other than congestion-induced delay costs,

	London Heathrow Airport: Peak Passengers as a Percentage of Total Passengers			
	July	August	September	October
1976 (prepeak tariffs)	30.7	30.8	30.4	30.5
1977 (postpeak tariffs)	29.7	26.3	24.5	24.3

Source: BAA.

TABLE 2.3   Effect of Peak Tariffs on Traffic

to move operations from the congested peak period. On the other hand, the commercial viability of a flight and its ability to conform to bans and curfews might necessitate operations in peak hours. High differentials for peak operations might appear at first to be a reasonable step for the operator to take to spread congestion. However, any such action should be evaluated in light of the impact on the based carriers whose operations inevitably represent a very large proportion of the airport's total movements. The short-term economic gain to the airport could put a long-term economic strain on the finances and competitiveness of the based carriers. Withdrawal of services, movement of the airline base, or even collapse of the carrier will have a serious financial impact on the airport.

The second type of tariff that was instituted to support uneconomic operations during slack night hours is exemplified by a surcharge on handling fees formerly levied at Rome for arrivals and departures between 1900 and 0700. This amounted to a 30 percent surcharge if operations occurred within the period. A tariff of this nature has the bizarre effect that a transiting aircraft arriving and departing in a period that is partly within the surcharge period can in fact halve the surcharge by remaining on the stands for five more minutes, consequently using more airport resources.

There is in fact a very wide variation in the manner in which airports structure landing fees. Table 2.4 shows that for the major airports, landing fees are often computed from some combination of

- Aircraft weight
- Apron parking requirements
- Passenger load
- Noise level created
- Emissions charge
- Security requirement
- Peak surcharge

Airport	Country	Landing Charge		Parking Charge		Passenger Charge	
		Fixed	Variable	Fixed	Variable	Arrival	Departure
Dusseldorf	Germany		X		X		X
Faro	Portugal		X		X		X
Miami	USA		X		X		X
Montego Bay	Jamaica		X		X		X
Orlando	USA		X		X		X
Bridgetown	Barbados		X		X	X	X
Athens	Greece		X		X		X
Manchester	UK		X		X		X
London Gatwick	UK		X		X		X
Madrid	Spain		X		X		X
Amsterdam	Netherlands		X		X		X
Brussels	Belgium		X		X		X
Frankfurt	Germany		X		X		X

Source: IATA.

TABLE 2.4   Aeronautical Fee Structures at Selected International Airports, 2010

There is a very large variation among airports when it comes to the cost of a turnaround. Tables 2.5 illustrates, for a selected number of airports, the large range of charges involved in the turnaround of a Boeing 737 under identical base assumptions. These data are further illustrated by Figure 2.11, which graphically shows the variation among the same airports (Stockman 2010). It is clear that most of the variation is caused by the introduction of passenger charges and government tax.

2.5   Scheduling Within the Airline

As a fundamental element in the supply side of air transport, the question of scheduling involves a large number of persons and sections within the structure of the airline itself. A typical functional interaction chart is shown in Figure 2.12. The commercial economist takes advice from market research and interacts with the various route divisions, which control the operations of the various groupings of the airline routes. In some airlines, there are no route divisions. The commercial economist and route divisions are both part of the commercial department in this case. In advising schedules planning,

Noise Scheme	Security Charge	Runway Charge	Peak Surcharge	Emissions Charge	Lighting Surcharge	Baggage Charge	Other Terminal Charges
X	On ticket			X			X
	On ticket					X	X
	X						X
	X						
	X						X
	On ticket						
	On ticket					X	X
X	X		X			X	X
X	Included in pass. charge		X	X			X
X	On ticket						
X	On ticket						X
X	On ticket						X
X	On ticket						X

which is concerned with the overall planning of the airline's schedule, the commercial economist will take note of a number of factors that affect the decision as to whether or not to attempt to incorporate a service within a schedule. These factors could include some of the following:

- Historical nature of the route
- Currently available route capacity
- Aircraft type
- Fare structure (i.e., standby, peak, night, etc.)
- Social need for route and subsidies
- Political considerations (in the case of the flag airline of a country)
- Competition
- Requirements for special events

Once the decision has been made that a service should be incorporated into the schedule, the schedules planning section of the airline, which is frequently divided into long- and short-haul functions, will

a. Turnaround Charges Classified by Area Incurred

	USD KUL	USD MEX	USD DXB	USD MAD	USD WAW	USD AKL	USD SVO	USD DEL	USD LHR	USD JNB	USD ORD	USD NRT
Landing	$80	$217	$414	$824	$1,361	$1,126	$1,390	$508	$1,076	$2,579	$963	$3,991
Parking	$0	$0	$0	$0	$0	$0	$0	$0	$101	$0	$0	$203
Infrastructure/services	$27	$276	$106	$489	$238	$0	$0	$172	$86	$0	$0	$1,267
Passenger	$2,369	$2,474	$2,693	$2,323	$2,350	$3,011	$3,072	$3,867	$6,222	$5,102	$6,879	$4,174
TAX	$0	$0	$0	$0	$0	$0	$0	$0	$11,117	$0	$4,336	$0

b. Turnaround Charges Classified Description of Method of Charging

	USD KUL	USD MEX	USD DXB	USD MAD	USD WAW	USD AKL	USD SVO	USD DEL	USD LHR	USD JNB	USD ORD	USD NRT
Landing	$80	$217	$295	$790	$976	$784	$830	$374	$758	$2,579	$963	$1,676
Noise Emissions									$75			
Parking									$101			$203
Infrastructure/services		$175	$106	$331	$111			$172	$86			$1,100
Airbridge	$27	$102	$106	$158	$127							$167
TNAV			$120	$34	$384	$342	$561	$134	$242			$2,315
Passenger	$2,120	$2,306	$2,448	$1,933	$2,350	$2,210	$2,234	$3,867	$6,222	$4,637	$6,879	$3,319
Security	$249	$167	$245	$389		$801	$838			$465		$855
TAX									$11,117		$4,336	
Total without tax	$2,475	$2,967	$3,213	$3,635	$3,949	$4,137	$4,463	$4,547	$7,486	$7,681	$7,842	$9,635
Total with tax	$2,475	$2,967	$3,213	$3,635	$3,949	$4,137	$4,463	$4,547	$18,603	$7,681	$12,178	$9,635

Courtesy: Ian Stockman.

Table 2.5 International Airport Charges for Selected Airports, 2010 (U.S. Dollars)

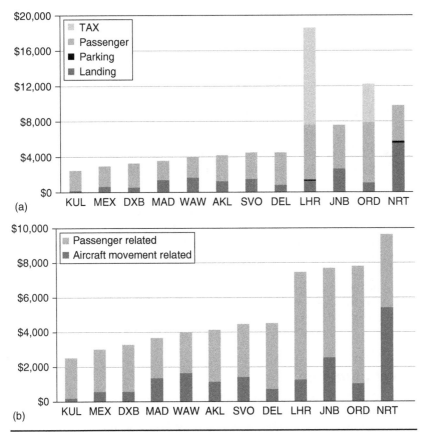

(a)

(b)

FIGURE 2.11 (a) Turnaround charges by type. (b) Turnaround charges, passenger-related/aircraft-related. (*Courtesy: Ian Stockman.*)

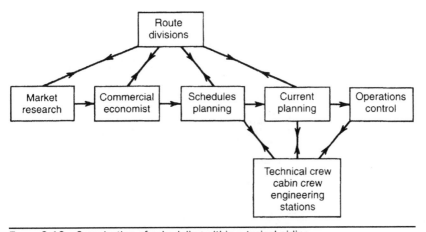

FIGURE 2.12 Organization of scheduling within a typical airline.

examine the scheduling implications of planning the service. At this stage, the factors that affect the planning include

- Length of haul—long or short
- Availability of aircraft allowing extras for aircraft in maintenance and on standby
- Acceptability of service schedule to the airport
- Availability of technical, flight, cabin, and engineering crews
- Clearances with countries concerned where there are no bilateral agreements to overfly or use airports for technical stops

When schedules planning is satisfied that all overall planning considerations have been resolved satisfactorily, the service is passed to current planning, which is charged with the implementation of the particular service schedule. This is done in conjunction with input from the technical, cabin, engineering, and stations staffs. The final implementation of the service is carried out under operations control, which deals with daily operations and the need to provide service in terms of difficulties from crew sickness, fog, ice, delays, aircraft readiness status, and so on.

2.6   Fleet Utilization

Figure 2.13 presents two fleet utilization diagrams for an airline using one smaller fleet of relatively new aircraft and another larger fleet of older aircraft. Several points are noteworthy. First, the aircraft in service are heavily used between the hours of 0700 and 2200; there is little use of these aircraft outside these hours because they constitute part of the short- and medium-haul fleet of this carrier. Second, where

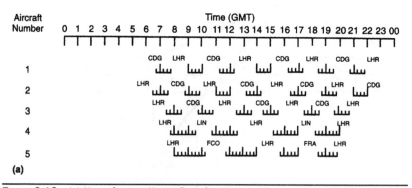

Figure 2.13   (a) Use of a small new fleet for short-haul operations. (b) Use of a large older fleet for short- and medium-haul operations.

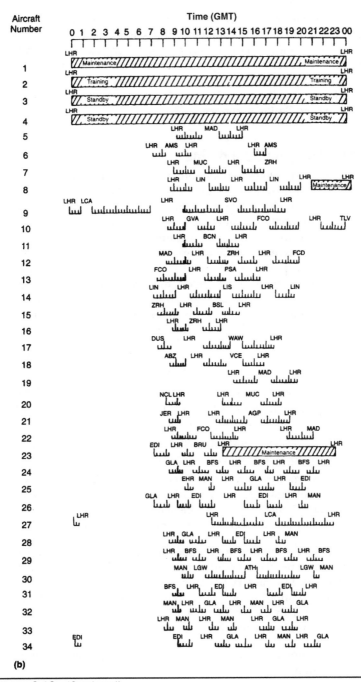

FIGURE 2.13   (*Continued*)

there is a small young fleet, there are no standby aircraft, and maintenance is carried out in the nighttime hours. For older and larger fleets, aircraft are withdrawn from service for maintenance, and several aircraft are held back as standby units for use should an aircraft have to be withdrawn for maintenance or repair. Low-cost carriers tend to have fewer standby aircraft, accepting that equipment failure will result in cancellation of services.

2.7   IATA Policy on Scheduling

The scheduled airlines industry organization, IATA, has developed a general policy on scheduling that is set out in its *Worldwide Scheduling Guidelines* (IATA 2010). At some airports, there are official limitations, and the coordination is likely to be carried out by general governmental authorities. Much more common is the situation where the airlines themselves establish an agreed schedule through the mechanism of the airport coordinator. It is recommended that the airport coordinator be the national carrier, the largest carrier or an agreed co-ordinating agent. Coordination compiles a recognized set of priorities that normally produce an agreed schedule with minimum serious disagreement. These priorities include

1. *Historical precedence.* Carriers have "grandfather rights" to slots in the next equivalent season.

2. *Effective period of movement.* Where two or more airlines are competing for the same movement slot, the airline intending to operate for the largest period of time has priority.

3. *Emergencies.* Short-term emergencies are treated as delays; only long-term emergencies involve rescheduling.

4. *Changes in equipment, routing, and so* on. Applications for a schedule of new equipment at different speeds or adjustments to stage flight times to make them more realistic have priority over totally new demands for the same slot (Ashford et al. 2011).

Aviation is a mixture of a number of segments that can be broadly classified into regular scheduled, programmed charter, irregular general aviation, and military operations. It is the function of the airport to arrange to afford appropriate access to any limited facility after consultation with the representative of the categories. IATA's policy states that the aims of coordination are

• To resolve problems without recourse to governmental intervention

• To ensure that all operators have an equitable opportunity to satisfy their scheduling requirements within existing constraints

- To seek an agreed-on schedule that minimizes the economic penalties to the operators concerned
- To minimize inconvenience to the traveling public and the trading community
- To arrange for regular appraisal of declared applied limits

Schedules are set on a worldwide basis at the semiannual IATA scheduling conferences for the summer and winter seasons. More than 100 IATA and non-IATA airlines meet at these huge conferences where, by a process of reiterated presentation of proposed schedules, the airport coordinators eventually are able to set an agreed-on schedule for the airports they represent.

2.8   The Airport Viewpoint on Scheduling

Most large airports with peak-capacity problems have strong and declared policies that affect the manner in which scheduling is carried out. The viewpoint of the airport operator is that which represents not only its own needs but also the interests of air travelers, the airlines as an industry group, and in some cases even the nontraveling public. These interests are protected by obtaining a schedule that provides for the safe and orderly movement of traffic to meet the needs of passengers within the economic and environmental constraints of the airport. The viewpoints of the various interested parties differ substantially. The airport operator seeks an economic and efficient operation within the constraints of the facilities available. Air travelers are looking for travel in reasonably uncongested conditions with a minimum of delays and a high frequency of service at desirable times unmarred by unreliability. As an industry group, the airlines are also seeking efficiency of operation and high frequency and reliability of service. However, each airline quite naturally will desire to optimize its own position and will seek to gain its own best competitive situation. In the case of the airlines, the aims of the individual company are not necessarily the same as the interests of the industry group. At some airports, nontravelers become involved where limits have been set for environmental reasons on the number of air transport services that can be scheduled, such as at London Heathrow, where there was a limit of 275,000 air transport movements per year. This restraint was lifted in the mid-1980s, and by 2010, the number of annual movements had increased to more than 466,000 with very little increase in nighttime movements. The increased capacity was achieved almost entirely by *peak spreading*, or infilling the nonpeak troughs. Similar aircraft movement constraints exist at Ronald Reagan Washington National Airport. It is the custom of many airports to declare their operational capacities at six-monthly or annual intervals. This operational capacity is observed by the scheduling committee, which consists of representatives of the scheduled airlines serving the airport. Normally, the airport operator is not represented directly on this committee. As already stated, the airport's interests are represented by the major carrier at the airport. At Los Angeles, this is United Airlines; at Frankfurt, it is Lufthansa; and at London Heathrow, British Airways. Therefore, the airport that has capacity limitations is often in an arm's-length relationship with the individual airlines seeking additional services.

2.9   Hubs

There is some ambiguity in the term *hub* when used in the context of air transport. Prior to deregulation of the airlines, the FAA used the term to designate large airports serving as the major generator of services, both international and domestic, within the United States. With the advent of deregulation, airlines were able to control LOS in terms of routes and frequencies. This enabled the establishment of what the airlines designated as *hubs*, which provided services both to other major airports also designated as hubs and to smaller airports providing spoke services. The airline hubbing system was associated with much greater frequency of services between hubs and from the spoke airports, supposedly accompanied by higher load factors on the aircraft. Direct services between smaller nonhub airports generally were abandoned. Some airports operate as hubs for one airline only (e.g., Newark, NJ, for Continental and Rome Fiumicino, Alitalia). Others, such as New York JFK, London Heathrow, and Changi Singapore are hubs for two or more airlines. Hubbing presents airlines with the opportunity to use better their aircraft and passengers with many more flight combinations, although these combinations almost always require transfer at the hub. Flights from hubs are usually nonstop, and those to other hubs are usually in larger, more comfortable aircraft than formerly. Flights to spoke airports are often on smaller aircraft with capacities of fewer than 50 persons. The effectiveness of a hub airport depends on

- Its geographic location
- The availability of flights to multiple destinations
- The capacity of the airport system to handle aircraft movements and passenger volumes
- The ability of the terminal layout to accommodate passenger transfers

Hub airports have very different patterns from airports supporting long-haul flights on predominant sectors. When operating as an airport with a two-airline hub operation, Dallas–Fort Worth reported 12 peaks throughout the day during which aircraft are on the ground providing for transfers. Therefore, aircraft ... arrive and depart in 12 arrival and departure waves, which in FAA terminology are described as *banks*. Hub terminals in the United States typically have high terminal usage, with peaks occurring at roughly two-hour intervals between 0700 and 2200 hours. Similar successions of banks have been analyzed at the U.S. Airways hubs at Philadelphia, where 11 blanks were recorded, and at Pittsburgh, PA and Charlotte, NC (Gumireddy and Ince 2004).

References

Ashford, N. J, S. A. Mumayiz, and P. H. Wright. 2011. *Airport Engineering: Planning, Design, and Development of 21st Century Airports*, 4th ed. Hoboken, NJ: Wiley.

Boeing. 2011. *Current Market Outlook, 2010–2030.* Seattle: Boeing Airplane Company.

Gumireddy, L., and I. Ince. 2004. *Optimal Hub Sequencing at U.S. Airways.* Washington: Airline Group of the International Federation of Operational Research Societies (AGIFORS).

Horonjeff, R., F. X. McKelvey, W. J. Sproule, and S. B. Young. 2010. *Planning and Design of Airports*, 5th ed. New York: McGraw-Hill.

IATA. 2011. *Worldwide Scheduling Guidelines.* Geneva: International Air Transport Association.

ICAO. 2008. *Airport Constraints: Slot Allocation and Night Curfew.* Presentation of 53 African States, Conference on the Economics of Airports and Air Navigation Services. Montreal: International Civil Aviation Organization.

Kinnison, H. A. 2004. *Aviation Maintenance Management.* New York: McGraw-Hill.

MPD Group Ltd. 2005. *Assessing the Economic Cost of Night Flight Restrictions.* Final Report TREN/F3/10-2003, MPD Group for the European Commission, Directorate of Air Transport. Brussels: European Union.

Stockman, I. 2010. Unpublished research paper, Cranfield University, Cranfield, UK.

CHAPTER 3
Airport Noise Control

3.1 Introduction

Airport noise is a worldwide problem. It inhibits the development of new airports and can seriously constrain the efficient and economic operation of existing facilities. In 1968, the assembly of the International Civil Aviation Organization (ICAO) recognized the seriousness of the problem and, knowing that the introduction of new, noisier aircraft types could aggravate the situation, instructed its council to call an international conference on the subject of aircraft noise in the vicinity of airports. This took place in 1969, providing the source document for Annex 16 to the Convention on Civil Aviation in 1971. Since then, Annex 16 has been revised through several editions (ICAO 2008a). The annex contains the essential international guidelines for noise control at airports in the form of standards and recommended practices (SARPS). Sovereign governments, such as the United States, have their own regulations (FAA 2012). In some cases, these are more stringent than those of ICAO. However, all such SARPS are designed to combat perhaps the most significant airport problem—noise. *Noise*, which can be defined as unwanted sound, is a necessary by-product of the operation of transportation vehicles.

Aviation is not the only form of transport that generates noise. Automobiles generate it from such sources as the engine, the tires, and the gearbox. Railroad trains generate noise aerodynamically and from rail and wheel contact, suspension, and the traction motors. Aircraft produce noise from their engines and from the aerodynamic flow of air over the fuselage and wings. Airports of themselves generate little noise. It is the noise generated by aircraft in and around airports that causes problems. The scale of noise generation by air transport can be seen in Figure 3.1. Whereas the air mode is not the only generator of transport noise, it can be seen as the source of the loudest and most disturbing noise. ICAO publishes reliable information on the effect of aviation on the environment, particularly noise impact (ICAO 2010a). Because noise at airports is a significant and troublesome problem for nearly all airport authorities and operators,

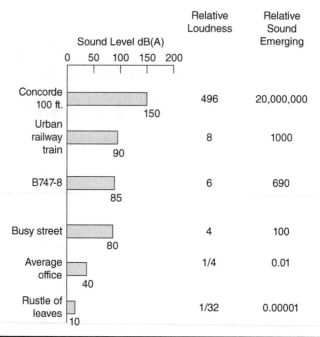

Figure 3.1 A scale of noise and sound.

they find that they must have some knowledge of the technical terms used by noise experts. A few of these are introduced in this chapter.

Loudness is the subjective magnitude of sound, and it is normally considered to double with an increase in sound intensity of 10 dB. The ear, however, is not equally sensitive to all ranges. The audible spectrum of sound is 20 to 20,000 Hz. Maximum sensitivity to sound is perceived around the middle of this range. It has been recognized for some time that the ear is particularly sensitive to frequencies within the A-range. Sound is therefore normally measured using A-weighted decibels dB(A), which reflect the sensitive ranges.

3.2 Aircraft Noise

Aircraft noise can be described by measuring the level of noise in terms of sound intensity. Where a noise-level measure is required, the simple dB(A) method is not entirely satisfactory. Following the introduction of jet aircraft, research carried out at JFK Airport, New York, indicated that the ear summed noise in a much more complicated way than the A-scale weighting of dB(A). As a result, another noise level measure was devised, the *perceived noise level* (PNL), a D-weighted summation that is sufficiently complex to warrant computer calculation.

Single-Event Measures

Noise intensity by itself is not a complete measure of noise. Intensity requires the factor of duration, which has been found to have a strong influence on the subjective response to noise. The principal measures of single-event noise used are *effective perceived noise level* (EPNL or L_{EPN}) and *sound exposure level* (SEL, sometimes abbreviated to LSE or LAE).

Annex 16 to the Convention on Civil Aviation of ICAO recommends use of EPNL, which modifies the PNL figure by factors that account for duration and the maximum pure tone at each time increment. This measure of the single event therefore incorporates measures of sound level, frequency distribution, and duration. Federal Aviation Administration (FAA) practice, on the other hand, uses a measure based on the sound exposure levels, weighted on the A-scale, over the time during which the sound is detectable. The accumulation procedure takes note of the logarithmic nature of sound addition. Both the EPNL and the SEL are used as bases for developing environmental measures of noise exposure.

Cumulative-Event Measures

In the case of noise nuisance generated in the process of airport operation, it is not simply the magnitude of the worst single noise event that gives a measure of environmental impact. Over the operational day of the airport, many noise "events" occur. Therefore, single-event indices are not useful methods of measuring aircraft noise disturbance, which is related to annoyance and interference with relaxation, speech, work, and sleep. Quantifying such interference requires noise measurements in terms of instantaneous levels, frequency, duration, time of day, and number of repetitions. Many surveys have been carried out to correlate community response to all these factors.

Day/Night Average Sound Levels (United States)

The form of cumulative noise event measure used in the United States is the *day/night average sound level* (DNL or L_{DN}), which is computed from

$$L_{DN}(i, j) = \text{SEL} + 10 \log_{10} (N_D + 10N_N) - 49.4 \qquad (3.1)$$

where N_D = number of operations 0700–2200 hours
N_N = number of operations 2200–0700 hours
SEL = average sound exposure level
i = aircraft class
j = operation mode

Partial L_{DN} values are computed for each significant type of noise intrusion using Ashford et al. (2011). They are then summed on an energy basis to obtain the total L_{DN} owing to all aircraft operations:

$$L_{DN} = 10 \, \log_{10} \sum_i \sum_j (10) \frac{L_{DN}(i, j)}{10} \tag{3.2}$$

Noise and Number Index

Another cumulative event measure that is widely quoted in airport noise literature is the *noise and number index* (NNI). This is a rather simple measure that was used widely in the United Kingdom and had limited use elsewhere. Equation (3.3) is the relevant formula for computation:

$$NNI = \bar{L}_{PN} + 15 \, \log N - 80 \tag{3.3}$$

Surprisingly perhaps, the definition of terms within the formula is not completely standardized among users. It is common practice, however, to define N as the number of occurrences of aircraft noise exceeding 80 PNdB, the peak level caused by a Boeing 707 at full power at approximately 13,000 feet (4,000 m) height. $\bar{L}_{PN}$ is the logarithmic average of peak levels. NNI has been replaced in the United Kingdom by L_{den} (see below).

Equivalent Continuous Sound Level L_{EQ}

Usually specified for a relatively long measurement period, the *equivalent continuous sound level* L_{EQ} is defined as the level of equivalent steady sound that, over the measurement period, contains the same weighted sound energy as the observed varying sound. It is stated in mathematical form as

$$L_{EQ} = 10 \, \log_{10} \left\{ (1/T) \int_0^T 10^{L(t)/10} \, dt \right\} \tag{3.4}$$

where $L(t)$ is the instantaneous sound level at time t, and T is the measurement period. In practice, this is the same as

$$L_{EQ} = 10 \, \log_{10} \left\{ (1/T) \sum 10^{SEL\,1/10} \right\} \tag{3.5}$$

which is the summation of the individual aircraft sounds over the measurement period T.

There are several versions of the A-weighted equivalent continuous sound level metrics in use in Europe (L_{EQ} or LA_{EQ}). It is accepted by the World Health Organization (WHO) as the preferred metric for noise disturbance.

Because of shortcomings of the NNI, which was not entirely applicable to all airports, L_{EQ} has replaced the NNI in the United

Kingdom and is shown here only for reference purposes. It has been found at Heathrow that the NNI and L_{EQ} equivalences were

35 NNI   57 L_{EQ}
45 NNI   63 L_{EQ}
55 NNI   69 L_{EQ}

Noise-Exposure Forecast (United States)

Prior to the development of the L_{DN} index, the measure of cumulative noise exposure in the United States was the *noise-exposure forecast* (NEF), which still occurs in much FAA literature. It is computed from

$$NEF = \bar{L}_{EPN} + 10 \log N - K \qquad (3.6)$$

where $\bar{L}_{EPN}$ = average effective perceived noise level that is computed from individual L_{EPN} values. This is the EPNL defined previously.
$K = 88$ for daytime periods (0700–2200)
$K = 76$ for night-time periods (2200–0700)

and

$$L_{EPN} = 10 \log \frac{1}{T} \int_0^T 10^{0.1L(t)} \, dt \qquad (3.7)$$

where $L(t)$ = sound level in dB(A) or PNdb
$T = 20$ or 30 seconds to avoid including quiet periods between aircraft

The combined 24-hour NEF is computed using Eq. (3.8):

$$NEF_{day/night} = \log_{10}\left[\text{antilog}\left(\frac{NEF_{day}}{10}\right) + \text{antilog}\left(\frac{NEF_{night}}{10}\right) \right] \qquad (3.8)$$

A-Weighted Day-Night Average Sound Level (L_{DN} or DNL) and Day-Evening-Night Average Sound Level (L_{den})

In 1976, the Environmental Protection Agency (EPA) introduced the DNL for the measurement of community noise exposure. The FAA adopted the DNL for the measurement of cumulative aircraft noise under *Federal Aviation Regulations (FAR)*, Part 150, "Noise Compatibility Planning." This metric applies a 10-dB weighting penalty to aircraft movements in the nighttime period (i.e., 10:00 p.m.–7:00 a.m.) equivalent to multiplying such operations by 10.

In 2002, the European Parliament and Council issued Directive 2002/49/EC, which requires Member States to use L_{den} as the metric of overall noise levels at airports. In addition to the 10-dB penalty for night operations, L_{den} penalizes evening-hour movements during the period 6:00–10:00 p.m.

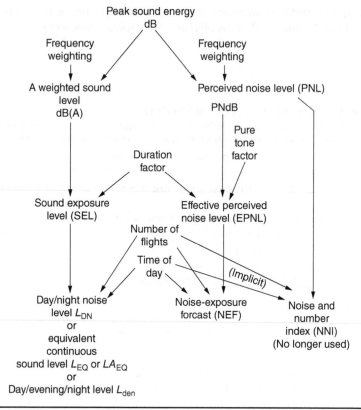

FIGURE 3.2   Relationships among noise measures.

L_{DN} and L_{den} describe the noise from the average exposure measured over a year. At a specific location, noise levels on a particular day can be higher or lower than the annual average.

It can be seen that NNI, NEF, DNL, and L_{den} are all very similar in basic form. There is much evidence to indicate that community response to noise impact can be correlated with any of these measures. The general relationships among noise measures are shown in Figure 3.2.

3.3   Community Response to Aircraft Noise

It should not be surprising that there is a wide range of individual responses to noise from aircraft operations in the vicinity of airports. Noise levels that are extremely annoying to some individuals cause little disturbance to others. The reasons for these differences are complex and largely socially based. Research has indicated that unlike individual reactions, community response is more predictable because of the large number of individuals involved. Figure 3.3 shows relationships that have been found to exist between levels of noise exposure

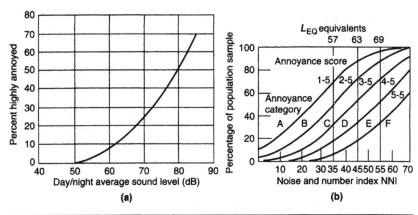

Annoyance Category	Feelings About Aircraft Noise
A	*Not annoyed.* Practically unaware of aircraft noise
B	*A liitle annoyed.* Occasionally disturbed
C	*Moderately annoyed.* Disturbed by vibration; interference with conversation and TV/radio sound, may be awakened at night
D	*Very annoyed.* Considers area poor because of aircarft noise; is sometimes startled and awakened at night
E	*Severly annoyed.* Finds rest and relaxation disturbed and is prevented from going to sleep; considers aircraft noise to be the major disadvantage to the area
F	*Finds noise difficult to tolerate.* Suffers severe disturbance; fells like moving away because of aircraft noise and is likely to complain

Figure 3.3   (*a*) Degree of annoyance from noise observed in social surveys. (*Schultz 1978*.) (*b*) Distribution of degrees of annoyance owing to aircraft noise exposure. (*Ollerhead 1973*.)

and community disturbance in terms of the percentage of persons annoyed (Ollerhead 1973; Schultz 1978; Hooper et al. 2009). It can be seen that below exposure levels of 55 L_{DN}, 57 L_{EQ}, and 35 NNI, the percentage of affected individuals who are highly annoyed by aircraft noise is very low. At exposure levels of 65 NNI, 69 L_{EQ}, and 80 L_{DN}, more than half the community is highly annoyed. Figure 3.3*b* is interesting in that it indicates that even at near-intolerable levels of noise exposure, about 10 percent of the population is either unaware of the noise or only occasionally disturbed.

3.4   Noise-Control Strategies

There are many ways in which operations on and in the vicinity of airports can be modified to control noise and to decrease its impact on airport neighbors (ICAO 2007). As early as 1986, 37 categories of noise-control actions had been identified as being in use at over 400 U.S. airports (Cline 1986). Several of these are discussed briefly.

Quieter Aircraft

Although considerable noise is generated by aerodynamic flow over the aircraft frame, most of the noise from modern transport aircraft has the engine as its source. The two principal component of engine noise are high-velocity exhaust-gas flows and air flows in the compressor fan system. The early turbojet engine was extremely noisy owing to the high velocity of the compressor tip speeds and the jet exhaust. Subsequent generations of high-bypass-ratio engines have included a number of achieved and proposed improvements, including

- Low-noise fans with swept stators
- Quieter intake liners, bypass and core stream liners
- Improved nozzle-jet noise suppressors
- Active noise-control fans
- Reduction in airframe noise
- Low-noise inlets
- Low-noise flaps, slats, and gear

Both the National Aeronautics and Space Administration (NASA) and the European Commission (EC) had research and development programs that sought to reduce aircraft noise through improved low-noise design resulting in a 6- to 7-dB reduction from the levels of aircraft in the mid-1990s.

Noise-Preferential Runways

Modern transport aircraft are not particularly sensitive to the crosswind component on landing and takeoff. Consequently, these operations can be conveniently carried out on a less than optimally oriented runway if that facility will reduce the noise nuisance to the community at large. Schiphol Amsterdam is an example of an airport that might well have abandoned the use of a particular runway were it not for the fact that this runway is well suited to direct noise nuisance away from the heavily populated suburbs of Amsterdam. At Los Angeles, an over-ocean operational procedure provides some relief from arriving aircraft noise to the close-in communities to the east of the airport between the hours of 0000 and 0630. During this period, aircraft approach the airport from over the ocean toward the east and depart over the ocean toward the west unless air traffic control (ATC) determines that the weather conditions are unsafe for such operations. Very much related to the noise-preferential runway concept is that of *minimum-noise routings* (MNRs) or *preferred-noise routings* (PNRs), which are designed to direct departing aircraft to follow routes over areas of predominantly low population density. Although the size of the noise footprint is not altered significantly, the impact in terms of disturbed population is much decreased. The use of MNRs and PNRs has been hotly contested by those adversely affected in terms of the social justice of the few bearing high noise exposure levels in order to protect the many. In the United Kingdom, the Noise Advisory Council has examined the practice of using MNRs and has recommended its continuation as being the best course of action for the community as a whole.

Operational Noise-Abatement Procedures

Several operating techniques are available that can bring about significant and worthwhile reduction in aircraft noise in the takeoff and approach phases in the vicinity of an airport as well as during operations while on the ground (ICAO 2010b).

Takeoff

To reduce noise over a community under the takeoff flight path, power can be cut back once the aircraft has attained a safe operating altitude. Flight continues at reduced power until reaching a depopulated area, when the full-power climb is resumed. At the point of cutback, noise levels can be reduced. Of course, there will be reduced benefits further down route, where noise levels are likely to be higher than those produced by a full-power climb. This is called *noise displacement*. However, by carefully planning the noise-abatement procedure (NAP) on takeoff, the level of noise exposure on the total community can be reduced.

Figure 3.4 shows the easily calculated theoretical effect of a power-cutback procedure on maximum flyover noise levels for points directly below the flight path. Staged-climb NAPs are common at many airports around the world. Figure 3.5 shows the results of tests carried out on a B747-100 by the Civil Aviation Authority (CAA)

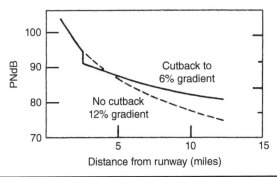

Figure 3.4 Effect of power cutback on ground noise levels.

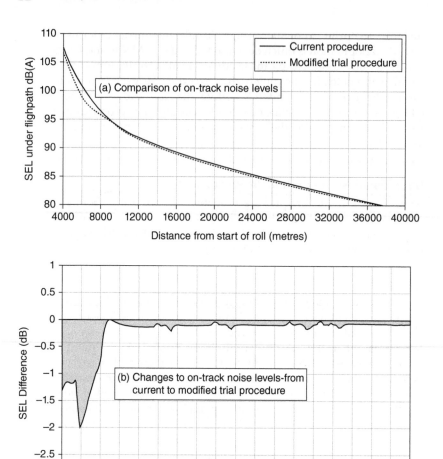

FIGURE 3.5 (a) Comparison of on-track noise levels for a Boeing 747 with and without power cutback. (b) Differences in on-track noise levels for a Boeing 747 with and without power cutback. (*Source: Ollerhead et al., 1989.*)

in Britain, in which noise improvements of 2 dB were widespread in the areas of high noise exposure in the initial climb paths, but the displacement of noise to downstream areas was negligible and considered not discernible. Noise experts point out that a more significant factor in noise of departing aircraft is the operator's practice of minimizing takeoff thrust to minimize fuel burn. Many aircraft thus are substantially lower when they pass the 6.5 km from rollout point than they would be with full-thrust takeoffs (Ollerhead et al. 1989). Later work is documented by ICAO (2008b).

Approach

Noise on approach can be reduced by adopting operational NAPs that keep aircraft at increased heights above the ground. Some of the following have been used at various airports:

- Interception of glide slope at higher altitudes when interception is from below the slope. Manchester Airport prohibits descent below 2000 feet (610 m) until the glide slope has been intercepted and requires that aircraft making visual approaches use the visual approach slope indicator system (VASIS) to avoid unnecessary low flying.

- Performing the final descent at a steeper than normal angle. Descents of 4 degrees have been used, but 3 degrees is a more normal angle.

- Two-segment approaches with the initial descent at 5 or 6 degrees of flaring to 3 degrees for the final approach and touchdown.

- Low-grade approaches with reduced flap settings and lower engine power settings demonstrate some reduction of noise. Reducing flap settings on the B737 from the normal 30 degrees reduces the noise by 2 EPNdB.

- Use of continuous-descent approaches employing secondary surveillance radar for height information. This prevents the use of power in a stepped descent and consequently reduces noise under some parts of the descent path. A combination of low-power and low-drag approach procedures has been used in the past with considerable success at Frankfurt Airport, which has severe environmental noise problems owing to its position within an urban area.

Runway Operations

The most significant improvement in noise impact that can be achieved when aircraft are on runways is control of the use of thrust reversal. Although thrust reversal is usually about 10-dB below take-off noise, it is an abrupt noise that occurs with little warning. Aircraft operations should be restrained from the use of thrust reversal on noise-nuisance grounds, except in cases where no other adequate means of necessary deceleration is available or where the airport setting does not require noise-control strategies.

Insulation and Land Purchase

Some relief to noise nuisance can be attained by the use of sound insulation. In some countries, those adversely affected by defined levels of noise nuisance are eligible for governmental or airport authority grants that must be used for double glazing or other sound-insulation procedures. Schemes of this nature operate, for example, in the noise-impacted areas around London Heathrow and Schiphol Amsterdam.

A more direct, although more expensive, method of reducing noise nuisance was adopted at LAX, where many homes and businesses in the immediate vicinity of the airport were purchased by the airport through mandatory purchasing procedures (eminent domain). In some cases, this type of action is the only recourse open to an airport when continued operation means intolerable living and working conditions for the neighboring population.

3.5   Noise Certification

In the spirit of its resolution of September 1968, ICAO established international specifications recommending the noise certification of aircraft that have reached acceptable performance limits with respect to noise emissions. Individual countries have developed their own parallel standards. Perhaps the most notable national requirements are the set of standards developed by the United States through the office of the FAA; these are published in the *Federal Aviation Regulations (FAR)* (FAA 2012).

The first generation of jet-powered commercial aircraft predated Annex 16 and early versions of the *FAR*. Such aircraft therefore were not covered by ICAO and the *FAR*, and airplanes such as the Boeing 707 and the DC-8 were designated as "non-noise-certificated (NNC)." The first standards for aircraft designed before 1977 were included in Chapter 2 of Annex 16, and airplanes such as the Boeing 727 and the Douglas DC-9 were designated as "Chapter 2 aircraft." In the parallel *FAR*, these were Stage 2 aircraft. Newer equipment was required to meet the improved standards on noise emission embodied in the later Chapter 3 (Stage 3) regulations. Examples of Chapter 3 aircraft are the Boeing 767 and the Airbus 319 (ICAO 2008a).

In order to reduce the noise nuisance from aircraft, the ICAO regulations have become more stringent over time, ensuring that fleets of older, noisier aircraft have been phased out and replaced with new air transport aircraft that are considerably quieter than similarly sized equipment of 50 years ago. Over time, aircraft have been required to conform to the increasingly more severe requirements by the periodic issuance of Chapter 2, Chapter 3, and Chapter 4 regulations; these have been promulgated by ICAO after international discussions and agreements (ICAO 2008a). Parallel regulations are the Stage 2, Stage 3, and Stage 4 modifications to the *FAR* in the United States (FAA 2012).

ICAO and FAA certification standards principally relate to the noise generated by an aircraft on approach and while on the runway and on flyover. The form of these standards is shown in Figures 3.6 and 3.7. Three noise-measurement points are defined

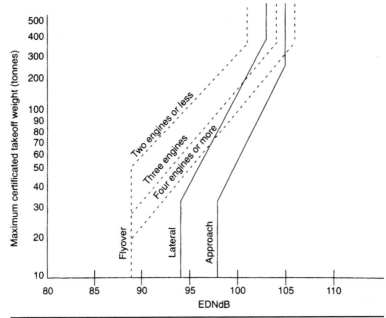

Figure 3.6 Aircraft noise certification limits. (*Sources: FAA, ICAO.*)

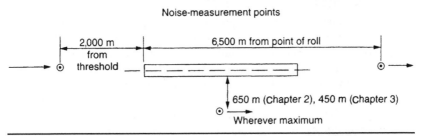

Figure 3.7 Location of noise-measurement points. (*Sources: FAA, ICAO.*)

under the approach and takeoff paths and laterally to the side of the runway. Maximum noise levels are set at these reference noise-measurement points; permitted noise levels are set to be dependent on maximum certificated takeoff weights, with the rationale that small, very noisy aircraft are socially undesirable and therefore should not be certificated. There are absolute maximum noise limits for even the largest aircraft. When examining Figure 3.6, it can be seen that an aircraft with noise characteristics plotting above or to the left of the curves is acceptable; those with plots to the right or below the curve fail to meet certification standards. The *FAR* are now identical to ICAO regulations. Figure 3.7 shows the location of the noise-measurement points.

In October 2001, the ICAO plenary meeting adopted a new and stricter Chapter 4 standard. This banned Chapter 2 aircraft from operating in major aviation states after April 1, 2002. Starting on January 1, 2006, newly certificated airplanes were required to meet Chapter 4 noise standards, as were Chapter 3 airplanes for which recertification to Chapter 4 was requested. The principal differences in requirements include

- A cumulative 10-dB over Chapter 3 levels
- The sum of the improvements at two measurement points to be at least 2 dB
- No tradeoffs permitted
- Standards for certification only not for new operational restrictions such as phaseouts
- Specific exemptions for new operating restrictions for developing countries

For general reference purposes, Table 3.1 shows the noise levels generated by a number of certificated aircraft in general usage in 2012 (FAA 2010).

3.6   Noise-Monitoring Procedures

In the whole area of airport noise problems, the airport authority probably can make no greater contribution than by establishing and operating an effective airport noise-monitoring program. The most beneficial programs have been those which have honestly and conscientiously monitored the status of noise pollution and encouraged an open exchange of information among the airport operator, the airlines, airline crews, the public, other airport authorities, and researchers in the field. The effectiveness of any such program is measured in terms of the computed reduction owing to implementation of monitoring procedures.

Noise monitoring at Manchester International Airport is a good example of a highly interactive program that has resulted in significant reduction in noise nuisance and aims to obtain continuing improvement. The airport is one of the world leaders in noise monitoring and noise-control procedures. Figure 3.8 shows the location of the runway in reference to surrounding urban development, designated preferred noise routings (PNRs), and the fixed noise-monitoring terminals. For special purposes, mobile noise-monitoring units are also used. These are moved to various locations according to need. Following United Kingdom practice, only departing operations are monitored. The microphones are located

- To conform as closely as possible to *FAR*, Part 36/ICAO, Annex 16 requirements of being 3.5 nautical miles (6.5 km) from start of roll and under the flight path.

Manufacturer	Airplane	Engine	Takeoff Gross Weight (lb)	Maximum Landing Weight (lb)	Takeoff dB(A)	Approach dB(A)
Airbus	A300B	CF6-50C	346,500	281,100	79.1	90.7
Airbus	A310-322	JT9D-7R4E1	330,690	267,850	79	89.2
Airbus	A319-131	V2522A5	158,730	149,910	73.2	83.5
Airbus	A320-211	CFM56-5A1	149,900	142,200	70.7	85.6
Airbus	A321-231	V2533-A5	165,340	165,440	68.1	85.5
Boeing	737-700	CFM56-7B20	154,500	142,500	74.3	86.1
Boeing	737-800	CFM56-7B27/2B1DAC	174,200	146,300	73.9	88.7
Boeing	737-900	CFM56-7B26	164,000	147,300	73.0	87.5
Boeing	747-400F	RB211-524H	875,000	666,000	89.4	92.5
Boeing	757-300	RB21-535 E4C	275,000	198,000	75.1	90.0
Boeing	767-300/300ER	RB211-524H	349,000	320,000	76.4	88.7
Boeing	777-300	PW 4098	550,000	524,000	74.4	89.6
BAE systems	146-300A	LF507	95,000	83,000	73.4	87.2
BAE systems	748 Series 2B	RR-DART-MK535	46,500	43,000	78.3	88.9
BAE systems	Jetstream 31	TBE331-10U-501H	15,200	14,600	61.7	74.7
Fokker	F-28 MK4000	Spey MK555-15H	73,000	64,000	75.5	86.3
Fokker	F70	RR Tay MK620-15	92,000	81,000	69.2	79.0
Embraer	EMB110-P2	PT6A-34	12,500	12,500	71.0	76.0
Embraer	EMB120-Brasilia	PW115	21,200	21,200	63.2	81.8
Embraer	EMB145ER	AE3007A1/1	45,410	42,540	65.9	82.5

Source: FAA AC36-3H.

TABLE 3.1 Measured Aircraft Noise Levels

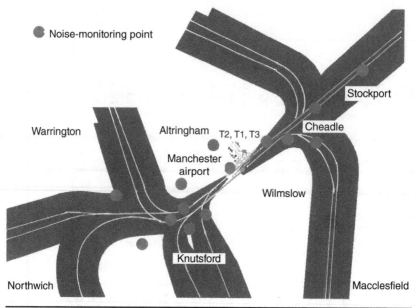

Figure 3.8   PNRs and noise-measuring points at Manchester Airport. (*Courtesy: Manchester International Airport.*)

- To provide measurement in the vicinity of the nearest possible built-up area where an NAP can be safely carried out.
- To be easily accessible for maintenance.
- To provide one sideline monitor that allows correlation of ATC records with those of the monitoring system, thus making it possible to identify individual flights.

In common with other U.K. airports, departure noise levels at Manchester Airport are limited to 103 PNdB by day (0700–2300) and 96 PNdB by night. Monitoring is geared toward detecting and assigning responsibility for violations of these limits by combining the techniques of noise monitoring with aircraft tracking that determines deviations from the acceptable swathes either side of the PNRs.

The automatic system in use at Manchester records each departure from the ATC radar track and records the noise level for each at the fixed monitoring points. Deviations from the PNRs are flagged. Each aircraft using the airport is required, after takeoff or go-around, to be operated in the quietest possible manner. Aircraft exceeding 90 dB(A) (103 PNdB) by day (0700–2300) and 83 dB(A) (96 PNdB) at night (2300–0700) at the monitoring points are subject to a penalty of £750 by day and £750 by night plus an additional £150 for each additional decibel.

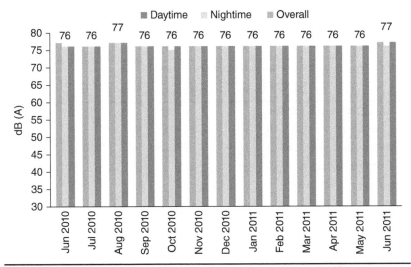

Figure 3.9 Average monthly noise levels: 13-month trend analysis. (*Courtesy: Manchester International Airport.*)

Each month, the airport publishes a report on monitored noise levels, the number of daytime and nighttime infringements, and the deviations from the PNRs. Figure 3.9 is an example of the reported daytime and nighttime noise levels over a year. Table 3.2 shows the published deviations from PNRs and the breakdown of noise infringements by airline for a particular month. Although associated with penalties to airlines for noisy operations, the chief purpose of the Manchester noise-monitoring control is positive rather than negative. The system is used as a continuous feedback to the airlines, to airport management, and to the airport's consultative committee, which includes membership from the environmentally affected communities around the airport. Based on the results documented in the Manchester Airport Noise and Tracking Information System (MANTIS), the airport consultative committee publishes a monthly summary report of the noise-monitoring system. Manchester Airport maintains a close working relationship with its surrounding communities to achieve an operation that minimizes avoidable negative environmental impacts.

Similar systems of noise monitoring exist widely in countries where noise impact creates severe problems with communities that surround busy urban airports. Not surprisingly, Chicago's O'Hare Airport has a noise-compatibility program that closely monitors noise impact according to *FAR*, Part 150 requirements. Using the flight tracks for the multirunway airport, the associated noise contours are generated using the FAA integrated noise model. The validity of the noise contours is verified using mobile noise-recording units, unlike the static recorders used at Manchester. Between 1979 and

Monthly MANTIS Summary Report
June 2011
Total movements: 15,766

Movements monitored: 15,752   **Detection rate (%): 100%**

	2011	2010
Daytime average peak noise level	77 dB(A)	77 dB(A)
Nighttime average peak noise level	76 dB(A)	76 dB(A)
Overall average peak noise level	77 dB(A)	76 dB(A)
Daytime noise infringements	3	2
Nighttime noise infringements	1	0
Total noise infringements	4	2

Track Infringements

	Runway 05	Runway 23
Total standard instrument departures (SIDs)	843	6967
Total MANTIS-correlated SIDs	813	6841
Total extreme deviations	0	0
Total overall deviations	21	77
Percentage deviation	2.6	1.1

Noise Infringements

Operator	No. of Infringements	Aircraft Type	Surcharge
Kalitta Air	3	Boeing 747-400 Freighter	£3,600
Monarch	1*	Airbus A330-200	£750

*Nighttime infringement.
Courtesy: Manchester International Airport.

TABLE 3.2   Monthly Infringement Report at Manchester Airport

1993, the number of homes adversely affected by noise in excess of 65 L_{DN} had decreased from 87,000 to 45,000, largely owing to the introduction of less noisy Chapter 3 (Stage 3) aircraft. This was a reduction of 48 percent. By the turn of the century, the reduction from 1979 levels was over 80 percent. The shrinkage of the noise-impact contours between 1979 and 2003 is shown in Figure 3.10. Under the O'Hare modernization program, in 2013 the airport was in the process of closing some of its existing runways and building new runways to end with a configuration of six east-west runways and two crosswind runways. Even with the new configuration and increased traffic, the noise levels at the airport will be very much lower than before the banning of Stage 2 aircraft.

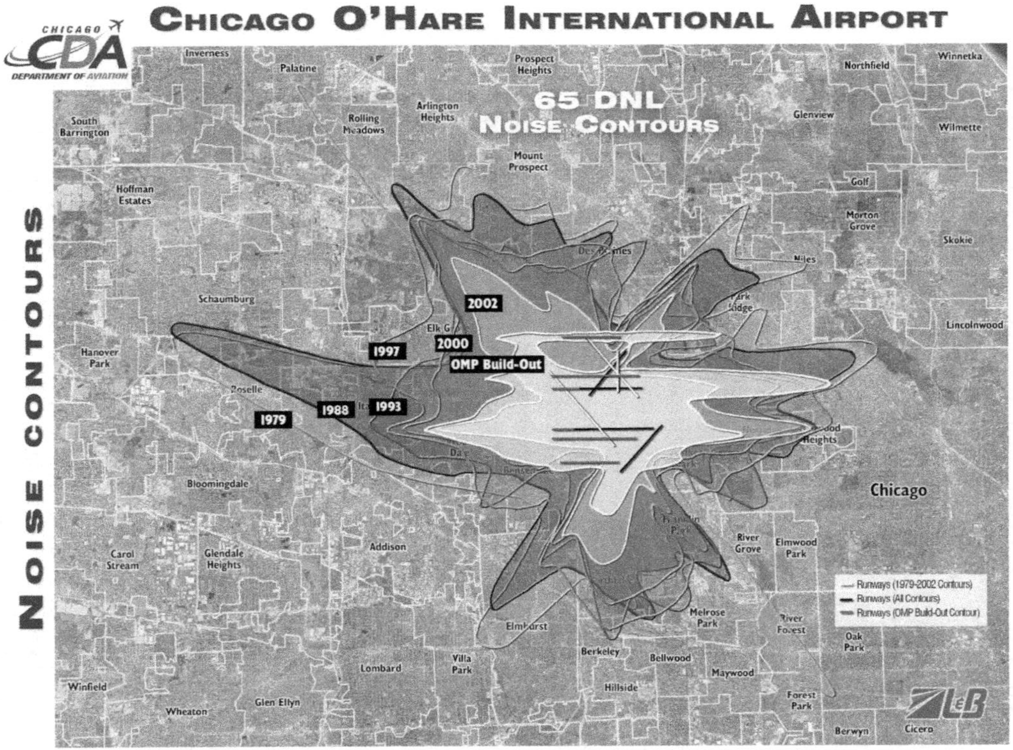

FIGURE 3.10 Historic and predicted future noise contours at Chicago O'Hare International Airport. (*Courtesy: Chicago O'Hare International Airport.*)

3.7   Night Curfews

The operation of any aircraft at night, especially large air transport aircraft, can be the cause of considerable noise annoyance. The factor of sleep disturbance is so troublesome that the standard measures of cumulative noise events such as L_{DN}, L_{EQ}, composite noise rating (CNR), and NNI weight each night operation by a factor between 10 and 15 in an attempt to reflect noise's real importance in nuisance value. Many governments and authorities claim that even such a heavy weighting does not lead to adequate safeguards with respect to night operations. Even very low levels of activity by noisy aircraft (e.g., one operation every half hour) could cause intolerable night conditions for many neighboring residents.

Consequently, night curfews on aircraft operations exist at many airports throughout the world (e.g., Zurich and Sydney). The nature of the curfews varies substantially among airports. At some facilities there is a complete ban on all operations, and the runways are effectively closed. Other airports permit the operation of some propeller aircraft that have low noise characteristics. Frequently, night freight movements are made by such "quiet" aircraft. Some airports such as London Heathrow allow for a quota of night movements that permits a heavily reduced level of operation. In Amsterdam, London, and Frankfurt, some curfew exemptions are granted under certain operational and scheduling circumstances to permit operations by noise-certificated aircraft. London, Tokyo, and Paris bend the curfew rules to allow delayed flights to land, but some airports operate a very restrictive curfew period that allows no exemptions. Sydney has a curfew from 11:00 p.m. to 6:00 a.m. During these hours, general passenger jets are excluded. Small propeller-driven aircraft, low-noise jets, and a limited number of freight movements are permitted, but during curfew hours, aircraft must operate over Botany Bay. The airport can get dispensation for unavoidable and unforeseeable circumstances that cannot be solved by alternative arrangements. Breaching the curfew at Sydney can incur a fine of up to A$550,000.

The nature of the curfew depends greatly on the local political atmosphere, the location and physical climate of the city involved, and the nature and volume of air transport through the airport. Curfews can be very effective in limiting nighttime disturbance. However, before activating a curfew, an airport must examine very carefully the effect that this constraint will have on airlines. Curfews increase the problem of peaking, and stringent curfews that accept no delayed aircraft, such as that in operation in Sydney, can produce alarming scheduling-window constraints when located at the end of long flight sectors (see Chapter 2).

3.8   Noise Compatibility and Land Use

Set in an environment such as farming land or forests, airports present few noise problems. It is the interaction of the noise from aircraft operations and land used for residential, commercial/industrial, and other urban uses that creates the undesirable noise impact of airports that is so familiar to airport operators and planners. Because airports are workplaces and terminal points for a mode of transport, left to themselves, they will generate urban development in their vicinities. This is likely to be in the form of residential areas for those working at the airport. Additionally, commercial and industrial development tends to be attracted to the airport because of commercial linkages with the aviation activities or convenient access to air transport. These directly associated land-use changes themselves generate secondary growth in the form of residences for the industrial and commercial workers, shops, schools, and a variety of other developments necessary for an expanding community. Because it is a large employer and consequently a generator of urban activity, without land-use control, an airport will very rapidly find that it has developments in its immediate vicinity that are incompatible with its own function (Ashford et al. 2011).

However, not all types of land use are equally incompatible with airports. Residential areas are recognized as being highly sensitive to aircraft noise and, therefore, every effort must be made to discourage the development of residential land use in the vicinity of airports. Some types of commercial and industrial uses are less sensitive; uses such as manufacturing and resource extraction, where internally generated noise levels can be very high, are usually reasonably compatible with a large, modern airport.

Recognizing the peculiar ability of airports to choke themselves environmentally, many governments around the world have developed land-use planning controls that apply specifically to airports to minimize the possibilities of incompatibility with the developing surrounding land uses. In the United States, the FAA has set out standards of airport land-use compatibility planning for use in the development of U.S. airports. Depending on its location within the airport's noise contour map (FAA 1977; ICAO 2008c), land surrounding airports is classified into four categories of noise exposure: minimal, moderate, significant, and severe. The locations of these classifications relative to a typical airport configuration are shown in the schematic in Figure 3.11. Each category is defined by a range of one of four noise-exposure indices: day/night average sound level (L_{DN}), *noise-exposure forecast* (NEF), composite noise rating (CNR), and community noise-equivalent level (CNEL). The most commonly used metric is the L_{DN}. Table 3.3 shows that areas within zone A are considered to be minimally affected by

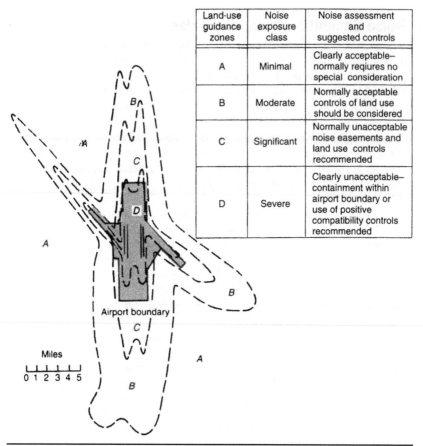

Land-use guidance zones	Noise exposure class	Noise assessment and suggested controls
A	Minimal	Clearly acceptable–normally reqiures no special consideration
B	Moderate	Normally acceptable controls of land use should be considered
C	Significant	Normally unacceptable noise easements and land use controls recommended
D	Severe	Clearly unacceptable–containment within airport boundary or use of positive compatibility controls recommended

Figure 3.11   Typical airport noise patterns. (*Source: FAA.*)

noise. Therefore, no special consideration of airport noise need enter into the designation of land use permitted in that zone. At the other extreme, zone D is severely affected, and land in this zone either should fall within the airport boundary or must be subject to positive compatibility controls. In the United Kingdom, the Department of the Energy in 1973 drew up a similar guideline for use originally with the NNI metric (HMSO 1973). This has been reinterpreted in terms of L_{EQ}, and listed in Table 3.4 are the current guidelines for development around London Heathrow Airport.

Many airports that originally were put down on greenfield sites have found themselves severely constrained within 20 to 30 years of operation. The airport administration therefore has a strong and legitimate interest in ensuring that future viable operation of the facility is not constrained by piecemeal development of incompatible neighboring land uses. Adoption of standards such as those set out by the FAA in the

Land Use Guidance Chart I: Airport Noise Interpolation

Land Use Guidance Zone (LUG)	Noise Exposure Class	Inputs: Aircraft Noise Estimating Methodologies				HUD Noise Assessment Guidelines	Suggested Noise Controls
		LDN	NEF	CNR	CNEL		
A	Minimal exposure	0–55	0–20	0–90	0–55	"Clearly acceptable"	Normally requires no special considerations
B	Moderate exposure	55–65	20–30	90–100	55–65	"Normally acceptable"	Land use controls should be considered
C	Significant exposure	65–75	30–40	100–115	65–75	"Normally unacceptable"	Noise easements, land use, and other compatibility controls recommended
D	Severe exposure	75 and higher	40 and higher	115 and higher	75 and higher	"Clearly unacceptable"	Containment within an airport boundary or use of positive compatibility controls recommended

Source: FAA.

TABLE 3.3 Major Land-Use Guidance Zone Classifications

	72 L_{EQ} and Above (60 NNI and above)	66–72 L_{EQ} (50–60 NNI)	60–66 L_{EQ} (40–50 NNI)	57–60 L_{EQ} (35–40 NNI)
Schools	Refuse	Most undesirable. When exceptionally it is necessary to give permission, e.g., for a replacement school, sound insulation should be required to a standard consistent with DfEE guidelines	Undesirable	Permission not to be refused on noise grounds alone
			Sound insulation to be required to Department of Education and employment guidelines	
Hospitals	Refuse	Undesirable	Each case to be considered on its merits	Permission not to be refused on noise grounds alone
		Appropriate sound insulation to be required		
Offices	Undesirable	Permit	Permit but advise insulation of conference rooms depending on position, aspect, etc.	
	Full insulation to be required	Full insulation to be required		
Factories, warehouses, etc.	Permit (It will be for the occupier to take necessary precautions in particular parts of the factory depending on the processes and occupancy expected. But see PPG24 for control of new factories, etc. in relation to their noise emissions.)			
New dwellings	66 L_{EQ} and Above (55 NNI and Above) Development will be refused with the exception of the one-to-one replacement of dwellings that must be constructed so as to provide a minimum sound attenuation in all habitable rooms of 35 dB		63–66 L_{EQ} (45–50 NNI) Permission not to be refused on noise grounds; sound attenuation of 25 dB in all habitable rooms in buildings	Up to 63 L_{EQ} Permission not to be refused on noise grounds.

Source: DfEE and Spelthorne Local Council, UK.

TABLE 3.4 Recommended Criteria for the Control of Development in Areas Affected by Aircraft Noise

United States and the planning authorities in the United Kingdom, if adhered to, will provide a reasonable basis for the continued compatible operation of the airport within its environment.

References

Ashford, N. J., S. Mumayiz, and P. H. Wright. 2011. *Airport Engineering: Planning, Design, and Engineering of 21st Century Airports*, 4th ed. New York: Wiley.

Cline, Patricia A. 1986. *Airport Noise Control Strategies*. Washington, DC: Federal Aviation Administration, Department of Transportation.

Federal Aviation Administration. 1977. *Airport Land Use Compatibility Planning* (AC150/5050-6). , Washington, DC: FAA, Department of Transportation.

Federal Aviation Administration. 2010. *Estimated Airplane Noise Levels in A-Weighted Decibels* (Advisory Circular AC 36-3H), April 25, 2002, updated to February 3, 2010. Washington, DC: FAA, Department of Transportation.

Federal Aviation Administration. 2012 (as updated). *Federal Aviation Regulations*, Part 36, "Noise Standards, Aircraft Type and Air Worthiness Certification." Washington, DC: FAA, Department of Transportation.

Her Majesty's Stationery Office. 1973. Department of the Environment Circular 10/73. London: HMSO.

Hooper, P., J. Maughan, I. Flindell, and K. Hume. 2009 (January). *Indices to Enhance Understanding and Management of Community Responses to Aircraft Noise Exposure*. Manchester, UK: Manchester Metropolitan University/University of Southampton.

International Civil Aviation Organization (ICAO). 2007. *Guidance on the Balanced Approach to Aircraft Noise Management*, 1st ed. (Document 9884). Montreal, Canada: ICAO.

International Civil Aviation Organization (ICAO). 2008a. Annex 16, *Environmental Protection*, Vol. 1: *Aircraft Noise*, 5th ed. Montreal, Canada: ICAO.

International Civil Aviation Organization (ICAO). 2008b (September). *Effect of PANS_OPS Noise Abatement Departure Procedures on Noise and Gaseous Emissions* (Doc. Cir. 317). Montreal, Canada: ICAO.

International Civil Aviation Organization (ICAO). 2008c. *Recommended Method for Computing Noise Contours around Airports*, 1st ed. (Document 9911). Montreal, Canada: ICAO.

International Civil Aviation Organization (ICAO). 2010a. *Noise Abatement Procedures: Review of Research, Development and Implementation Projects—Discussion of Survey Results*, 1st ed. (Document 9888). Montreal, Canada: ICAO.

International Civil Aviation Organization (ICAO). 2010b. *ICAO Environmental Report 2010* (Document ENVREP). Montreal, Canada: ICAO.

Ollerhead, J. B. 1973 (September). "Noise: How Can It Be Controlled?" *Applied Ergonomics*, Vol. 4, Issue 4 pp. 130–138.

Ollerhead, J. B., D. P. Rhodes, and D. J. Markman. 1989 (March). *Review of the Departure Noise Limits at Heathrow, Gatwick and Stansted Airports: Effects of Takeoff Weight and Operating Procedure on Noise Displacement* (R&D Report 9841). London: DORA, National Air Traffic Services.

Schultz, T. J. 1978 (August). "Synthesis of Social Surveys on Noise Annoyance," *Journal of Acoustical Society of America* 64. pp.377–405.

Further Reading

Horonjeff, R., F. X. McKelvey, W. J. Sproule, and Seth B. Young. 2010. *Planning and Design of Airports*, 5th ed. New York: McGraw-Hill.

CHAPTER 4

Airport Influences on Aircraft Performance Characteristics[1]

4.1 Introduction

The operations of airports are closely linked to the aircraft they serve. The linkage is ultimately of an economic nature based on the premise that public transport safety norms must never be degraded. Therefore, the function of the design and operation of runways and their approaches must be to allow safe transition between flight and ground maneuvering over the complete spectrum of air transport operations.

This chapter deals with the aircraft, in terms of their intrinsic performance, and the impact of an airport's facilities, surroundings, and weather conditions on performance and handling. The chapter also considers the legislative requirements determining what an airport must provide when certain levels of service are pledged.

Aircraft and airport operations personnel address the same circumstance—provision of safe operations on/off any runway—but they use dissimilar methodologies.

- The airport approach is to determine and publish *declared distances*, and the specification and derivation of these data through obstacle-limitation-surface (OLS) criteria are covered in detail in *Airport Engineering* by Ashford, Mumayiz, and

---

[1]The author of this chapter in the first two editions of this book was Robert Caves. In this third edition, this chapter has been extensively rewritten by Mike Hirst.

Wright (2011). This chapter will refer to the *declared distances* for runways, as presented for specific airports in *Air Information Publications* (AIPs) or stored in *Electronic Aviation Publication* (EAP) databases.

- An aircraft crew, throughout takeoff, approach, and landing, refers to speed (always indicated airspeed) and spatial information. They have much less precise knowledge of their position on or relative to a runway than is usually appreciated. In this chapter, relevant speed definitions that influence take-off and landing procedures are introduced and their definitions acknowledged. It is unlikely that most airport operations personnel will be conversant with these parameters.

4.2 Aircraft

This section considers largely the operation of commercial airliners, but the principles are similar for all types of aircraft. An *aircraft* is a heavier-than-air-machine that depends on the movement of air, either by the engine(s) or through the influence of the airframe's shape, and most significantly by the wing, to attain and sustain normal flight. An overriding parameter for an aircraft operator is that the aircraft must be able to perform services cost-effectively in order for operators to sustain viable service conditions.

The main consideration is the carriage of a useful load (the *payload*) over a declared range. The payload can comprise many elements—principally passengers, baggage, cargo or freight, and consumables (i.e., food stuffs and water/fluids for toilets, etc.). All-freight operations are also affected by the same considerations.

The nominal mass of the crew (flight crew and cabin crew) is usually included in the *operational empty weight* (OEW), which is the basic aircraft (with a nominal cabin configuration) without fuel or payload. Consumables, passengers, and their baggage (and/or, as appropriate, freight or cargo) are classified as payload. While much of these can be weighed prior to a flight, passengers are assessed using nominal mass values. The data in use vary from country to country, but typically 200 pounds (90 kg) is used for a crew member—this includes an approximate 20-pound (9-kg) baggage allowance—and 220 pounds (100 kg) is used for a passenger, including a typical 40-pound (18-kg) baggage allowance. In the United States and Europe, there are periodic assessments of passenger mass values (around once in 20 to 25 years), and the certifying authority may choose to revise the nominal values that certified operators must use on load sheets. Invariably, it rises over time. For very small aircraft, individual passenger mass values will be used, whereas for large aircraft there are mass categories—adult, child, and infant. While these tend to have only a minor impact on the majority of payload assessments, they can be significant for certain operations.

Many manufacturers report, in addition to OEW, an *aircraft-prepared-for-service* (APS) mass. This is greater than OEW because it adds (or subtracts—but this is rare) cabin variations and consumable allowances. Certified aircraft APS mass can be expected to include allowances that are specific to the airline and even the particular aircraft's seating configuration.

There will be a *maximum structural payload limit*, determined by loading criteria such as floor strength and the maximum allowable payload in sections of the fuselage. These criteria also affect where the aircraft center of gravity (CG) will be, and while CG location is an overwhelmingly important flight safety issue, it is of no direct consequence to an airport operator other than in contributing support pertinent to the loading of an aircraft. The aircraft operator's flight dispatcher holds the ultimate responsibility for assuring that an aircraft load is of the correct mass and that it is distributed appropriately within the aircraft. The operator will regard airport ground staff as reliable in terms of their observation of a loading operator, in that they are in a position to sense when anything out of the ordinary and that may have escaped their attention is taking place.

Fuel is loaded at the request of an operator. The operator will have knowledge of the prefueling content of an aircraft's fuel tanks, as well as the fuel load that is designated for an operation. The actual fuel load for a specific operation is determined at the time of the operation because it will take into account the sector distance, the actual route to be flown, and meteorological conditions. The operator may be able to load an aircraft with extra fuel, say, to conduct an outward and return flight, or to minimize the pickup volume at an airport where fuel is expensive—a practice often referred to as *tankering*. There may be requirements for the distribution of the fuel mass among the tanks and the sequence in which the tanks are filled, but these issues are either controlled by specialist staff or handled automatically by systems on the aircraft. The most important criteria that will be observed in determining a fuel load are that the requested load causes the aircraft neither to exceed its *maximum takeoff weight* (MTOW)—with its estimated payload—nor to exceed the declared fuel capacity of the aircraft type. Most fuel delivery is conducted in terms of volume (liters or gallons), but the crucial attribute is the mass of fuel on board. This depends on temperature and will be an issue handled by the airline and the fueling agent.

	Pounds (lb)	Kilograms (kg)
Maximum design taxi weight	777,000	352,441
Maximum design takeoff weight (MTOW)	775,000	351,533
Maximum design landing weight (MLW)	554,000	251,290
Maximum design zero-fuel weight (MZFW)	524,000	237,562
Operating-empty weight (OEW)	370,000	167,529
Maximum structural payload	154,000	69,853
Typical seating capacity		
Two classes	339,56 first class and 283 economy	
Three class	370,12 first class, 42 business class, and 316 economy	
Maximum cargo—lower deck	7,562 ft³	231.9 m³
Usable fuel		
U.S. gallons	47,890	
Liters	181,254	
Pounds	320,853	
Kilograms	145,541	

TABLE 4.1   Basic Mass Data for the Highest Gross Weight Variant of the Boeing 777-300ER

As indicated earlier, the maximum takeoff weight[2] (MTOW) must not be exceeded, although an aircraft may depart the stand with a designated taxiing fuel load that will allow a small (in relative terms) additional increase in weight. Table 4.1 presents overall mass data for the Boeing 777-300ER, and the data are extracted from a company publication that presents aircraft characteristics for airport planning (ACAP).

It is rare to find that adding OEW, maximum payload, and maximum fuel is a summation that equals MTOW. It is usually considerably greater. Because MTOW becomes the limiting attribute, if the aircraft carries its full payload, it cannot carry its greatest fuel load and thus attain its maximum range. As payload mass is reduced, equivalent fuel mass can be added, maintaining MTOW. Thus, when carrying sufficient fuel to achieve its maximum range, an aircraft can

---

[2]*Weight* is a term still used widely in aviation operations and is implied in abbreviations, but where metric/SI units are used, the word *mass* is preferred and now is insisted on by many regulatory authorities. In general, weight is expressed in pounds (lb) and mass in kilograms (kg). In some instances flight manuals now refer to MTOW as MTOM to stress that it is a mass, not a weight, value. The widely used convention of referring to all values, in abbreviation, as weight is used in this chapter.

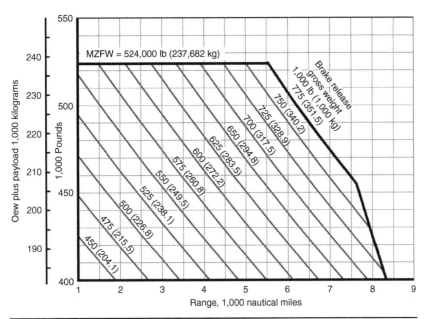

FIGURE 4.1   Boeing 777-300ER payload-range diagram.

carry only a proportion of its maximum payload. Eventually, no more fuel can be added, and as payload is reduced further, the additional range benefit is relatively small.

This can be plotted as a payload-range diagram, as shown at Figure 4.1. This example is again the Boeing 777-300ER. The plot has load plotted vertically and range plotted horizontally. The overall shape, with a flat maximum payload line, starting from the maximum zero-fuel weight (MZFW) on the vertical axis reaches the maximum takeoff weight (MTOW) at the point where it then begins to slope downward. Along this section, payload and fuel masses are exchanged, and the aircraft is at MTOW. At the next change of slope, the maximum fuel load has been reached and thereafter is a steeper drop to the maximum range with no payload, often referred to as the *ferry range*.

There are many ramifications associated with the point at which payload and range intersect on this plot. The only combinations that are allowable are within the envelope (enclosed by the bold lines and the axes). In general, the further to the right an operation is on this plot, the lower is the aircraft's operating cost per seat. This is a major issue for operators.

What is important to the airport is to appreciate that any aspect of its runway that limits the aircraft's takeoff weight will cause the payload-range attainable from the airport to be reduced and any reduction in allowable takeoff weight [sometimes referred to as *regulated,* or *restricted, takeoff weight* (RTOW)] will cause the steepest-sloping sector to commence at a shorter range and to maintain a similar gradient as the plotted lines at different aircraft weights shown in Figure 4.1. In fact, these are not straight lines but are slightly concave, but the significance is not of consequence in the applications referred to here. Note too that the takeoff-weight line that corresponds to the MZFW is a higher value: about 30,000 pounds (13,600 kg) higher in the example. This is attributable to allowances for reserves and is a variable that will not necessarily be constant across all operators of the type.

Figure 4.2 shows a diagrammatic payload-range plot and shows the major points that relate to aircraft mass values quoted in Table 4.1. The plot, because it shows the payload-range attributable to lower aircraft mass values, additionally provides an illustration of the effect of RTOW usually caused by limited takeoff distance on the attainable payload-range combinations that can be accommodated.

The *Aircraft Characteristics for Airport Planning* (ACAP) publications already quoted are produced to a format agreed on among manufacturers. They are specifically for planning and do not substitute for a flight crew operations manual (FCOM). In specific circumstances the FCOM is the best source of actual data.

Airport operations staff should be able to access FCOM-derived data through aircraft operations staff, and it is necessary that they do so if an operation is regarded as critical, for example, when assessing

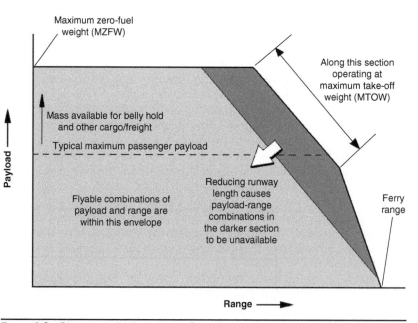

Figure 4.2 Diagrammatic illustration of a typical payload-range diagram.

the capability of a specific aircraft type to operate at or close to its performance limitations. (This is usually in planning but is a task deferred to operations staff.)

The ACAP is available from most major aircraft manufacturers through their website, or smaller manufacturers usually will provide the information on request. Companies often do not release performance data other than in response to a request from an airport. All information within these documents is generic in that the information pertains to a model specification that will be modified by the choices made by an operator with respect to fixtures and fittings in the cabin (e.g., number of galleys and toilets and even individual seat specifications), and the user has to appreciate how significant or not this might be. These data can be used by an airport planning team if they are content to refer to a general aircraft type in planning future operations.

In general, if current-day operations are being considered, the airport team is recommended to discuss requirements with the flight operations team of incumbent or target operators. The aircraft manufacturers will provide specific data to an airport when the circumstances are critical to safety and they are the consulted in confidence.

Figures 4.3 and 4.4 show the takeoff plots for an aircraft type and are diagrams again extracted from the example ACAP manual.

These plots provide an indication on how critical a runway distance value is to an aircraft operator and the significance of the effects air temperature and altitude on aircraft performance.

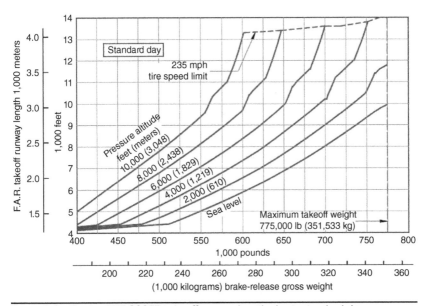

FIGURE 4.3   Boeing 777-300ER takeoff runway length chart, standard day.

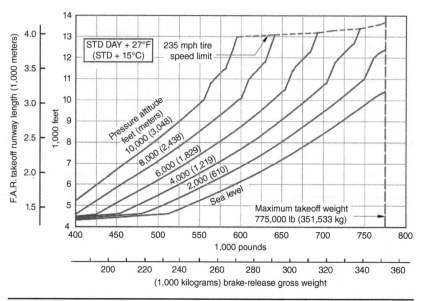

Figure 4.4   Boeing 777-300ER takeoff runway length chart, standard day + 27°F (15°C).

Entering the plots with the takeoff runway length [ACAP definitions are not specific about this but generally use declared takeoff distance available (TODA)], the user can read across to intercept a line (often this has to be interpolated) at the airfield elevation (i.e., pressure altitude) and then drop a line vertically onto the horizontal scale that will reveal RTOW. Referring back to the payload-range chart, the best payload and range combinations possible can be obtained. Charts often show *international standard atmosphere* (ISA) performance [with 15°C (27°F) ambient at sea level] and a plot of an ISA + 15°C (or ISA + 27°F). These can be interpolated to roughly to assess likely performance. They should not be extrapolated into lower or higher temperatures because the relationship of takeoff performance and air temperature is not linear. The general performance relationships to note are that, all other parameters being unchanged,

- As air temperature rises, the attainable takeoff weight decreases.
- As airport elevation increases, the attainable takeoff weight decreases.

Table 4.2 shows a sample of aircraft data indicating the airfield design requirements for each category of operation and gives some pertinent data for a representative set of aircraft. The weights and dimensions are from published data, but care must be taken when using specific values.

	Weight (lb 1000)				Limiting Field Length (m)			
	Max TO	Empty	Max Fuel	Typ Seats	Span (m)	Length (m)	ISA, sl	Track (m)
Long range								
A380-800	1,235	619	248	650	79.8	70.4	2,750	12.5
747-400	850	396	174	421	64.5	70.7	3,050	11
A340-600	811	386	153	375	63.4	74.7	3,050	10.7
777-300ER	775	370	146	370	63	73.4	3,000	10.7
MD-11	620	266	250	360	52	61.6	3,200	10.7
A340-300	559	302	118	315	60.4	63.6	3,050	10.7
DC10-30	555	238	235	255	50.4	55.6	3,500	10.7
787-9	484	253	102	300	60.1	55.9	2,270	8.7
IL-96-300	476	262	253	235	60.1	55.4	2,600	10.4
Medium range								
A330-300	459	264	170	315	60.4	63.7	3,050	10.7
B767-300ER	412	198	162	218	47.6	54.9	2,620	9.3
A310-300	346	175	108	218	43.9	46.7	1,675	9.6
757-200	250	137	34	201	38	47.3	2,350	7.3
Tu-204-200	244	130	72	180	41.8	46.1	2,250	7.3
Tu-154M	220	130	73	180	37.6	47.9	2,250	11.5
Short-medium range								
A321	183	105	18.7	186	34.1	44.5	1,750?	7.6
737-800	174	91.3	20.9	170	34.3	39.5	2,290	5.7
A320-200	170	89.3	18.5	164	34.1	37.6	2,134	7.6
737-400	155	74.2	19.1	150	28.8	35.2	2,560	5.2
MD82	149.5	78	29.2	146	28.9	36.5	2,300	5.1
C-series	128	73.5	n.a.	130	35.1	34.9	1,509	n.a.
ERJ-195	112	63.9	12.7	118	28.7	38.7	1,309	5.9
Fokker 100	103.6	54.1	8.6	80	29	37.1	1,500	5.1
146-200	97.5	54.4	9.1	103	26.4	31.6	1,050	4.7
CRJ-1000	91.8	50.7	6.8	102	26.2	39.1	2,079	4.1
ERJ-170	79	46.9	5.4	70	26	29.9	1,309	5.2
CRJ-200	51	30.5	6.5	50	21.2	26.8	1,918	3.4
ERJ-145	48.3	26.7	5.2	50	20	29.9	1,930	4.1

Sources: Largely manufacturer's data.

TABLE 4.2 Aircraft Data

There are several principles involved in matching aircraft to infrastructure. These reflect

- Demonstrated performance of the aircraft
- Application of assessed acceptable probability of any relevant failure
- The safety regulation of operations

Aircraft performance, as demonstrated for aircraft certification, is referred to as the *gross performance*. For the purpose of dimensioning the geometry of the environment within which it is considered safe to operate, the gross performance is factored, becoming *net performance* to take account of in-service variables. The variation is predicated to allow for variations caused by such influences as pilot skill, instrument inaccuracies, weight growth, and engine thrust reduction between overhauls.

Thus, for example, the demonstrated landing distance is factored by 1.67 under some regulations, including those of the United States, to derive the schedule landing-field length, and the gross climb performance is reduced by 0.9 percent in order to derive the performance that can be guaranteed. This information is published in the aircraft flight manual (and in the generic data in ACAP publications).

Regulations require that each new aircraft type demonstrates the distance required to land and take off under closely controlled conditions, with defined limitations on the pilot's reactions and carefully constructed safeguards to obviate any actions that might be inherently unsafe. Similarly, all other certificatory performance measures must be demonstrated for all applicable configurations of power and geometry, with all engines operating, and with the critical engine inoperative.

These result in the aircraft performance requirements being presented as

- Takeoff run required (TORR)
- Takeoff distance required (TODR)
- Accelerate-stop distance required (ASDR)
- Landing distance required (LDR)

Takeoff run required (TORR) is the net performance-assessed distance that the aircraft might need to travel while still in contact with the ground. This clearly sets a minimum runway length, but on its own this is not adequate and for safe airport operations much more is required.

Takeoff distance required (TODR) is the net performance measurement to reach a screen height, the distance being measured from the point at which the takeoff run commences. The screen heights used in the *Federal Aviation Regulations* (FAR; United States) and *Joint Aviation*

Regulations (*JAR*; Europe) certification requirements are similar, but vary among aircraft types and can introduce different circumstances depending on the aircraft's susceptibility to critical events. For most multiengine commercial jet airliners, a screen height of 35 feet (10 m) is used.

Accelerate-stop distance required (ASDR) is the distance required to accelerate to V_1[3] with all engines at takeoff power, experience an engine failure at V_1, and abort the takeoff and bring the airplane to a stop using only braking action without the use of reverse thrust. This should not exceed the paved runway length available at an airport.

Landing distance required (LDR) is measured from the threshold on the runway in use and includes the distance to the touchdown point and the landing run itself. The approach will be assumed to be conducted at the normal approach speed but that only brakes will be available after touchdown. As noted, the certified net performance typically will be the demonstrated gross performance increased by 1.67. Additional factors will be applied, typically 1.15 times, to account for a wet runway.

So that operators can match field-length requirements with the distances available, the airfield is required to publish for each runway the following so-called declared distances, which have been established on the basis of paved length, runway category, and local obstacles:

- Takeoff run available (TORA)
- Takeoff distance available (TODA)
- Accelerate-stop distance available (ASDA)
- Landing distance available (LDA)

When the operator has determined that the aircraft scheduled performance (corrected to the appropriate aerodrome elevation, air temperature, runway slope, wind, and runway surface condition at the required takeoff weight) results in a required distance for takeoff run and takeoff distance that are less than the declared distances available, the operation is deemed acceptable.

The operator also must conduct analyses of performance requirements and relate them to the declared distances available at any alternative airfield(s).

Declared distances are promulgated in the nation's *Aeronautical Information Publication* (AIP) or through the *Electronic Aeronautical Publication* (EAP) databases that are more common nowadays (see Chapter 11). They need to be accompanied by the airfield reference

---

[3]V_1 is the *critical engine failure recognition speed* or *takeoff decision speed*. It is the decision speed nominated by the pilot that satisfies all safety rules and above which the takeoff will continue even if an engine fails. It will vary greatly according to local and time-related, conditions.

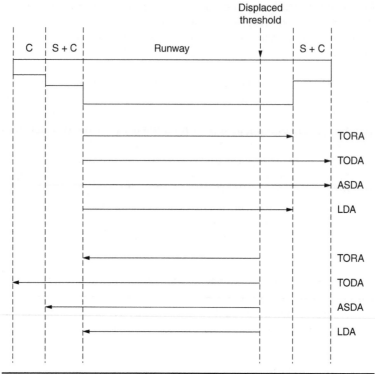

Figure 4.5 Runway declared distances.

temperature, the runway elevation, and the runway slope. It is the responsibility of the airport to notify, by means of *Notices to Airmen* (NOTAMs), any changes in these data caused by, for example, work in progress or accidents.

Declared distances take into account displaced thresholds, stopways, clearways, and *starter strips*, as shown in Figure 4.5.

An excess of TORA and TODA may allow a reduced-thrust takeoff to preserve engine life or an increased speed at the screen to improve climb performance. An excess of LDA may allow flexibility in planning for bad weather at the destination or may allow tanking of fuel. A more detailed discussion of field-length requirements is given in Section 3.2 of Ashford, Mumayiz, and Wright 2011.

4.3 Departure Performance

The takeoff portion of the departure procedure until the screen height is achieved is dealt with in detail in Section 3.2 of Ashford, Mumayiz, and Moore 2011. The essentials are restated here in order to introduce the operational choices.

The aircraft flight manual contains the following required distances:

- To 35 feet (10 m) with all engines
- To 35 feet (10 m) with the loss of one engine at the critical speed (V_1)
- To stop after loss of one engine at the critical speed

These distances will have been demonstrated in conformance with constraints on minimum speeds for rotation and for crossing the screen (the takeoff safety speed) and within the criteria for reacting to engine failure. The all-engine case then is factored by 1.5 in order to bring the probability of exceeding the resulting distance into the *remote-risk* category.

TODR is determined directly by adopting the greater of either the factored distance to the screen with all engines or the unfactored distance with the engine out. TORR is determined as the point equidistant between liftoff and the screen either factored for all engines or unfactored with an engine out. The accelerate-stop distance required (ASDR) is the unfactored rejected takeoff distance. These distances then may be corrected to specific conditions and compared with the TORA, TODA, and ASDA.

There is still a certain amount of disagreement about the adequacy of the regulations for the rejected-takeoff case, stemming from the number of accidents that have either overrun the runway after aborting or where the continued takeoff has been unsuccessful. Frequently these have been caused by substandard acceleration rather than a hard and noticeable engine failure. The onus is moving to the airport operator for the provision of runway-end safety areas (RESAs; see Section 4.5) in addition to the strip and prepared stopway, but this is by no means the only or best palliative.

An appropriate technical solution appears to be a ground-speed indication inside the cockpit, but the certification of rolling takeoffs makes it difficult to use this accurately, as does the trend toward using reduced thrust takeoffs to conserve engine life. The coalescing of Global Positioning System (GPS) position and speed measurement with surveyed airfield critical-distance locations could be used to monitor whether critical takeoff conditions are experienced, but no practical application has been described to date.

The situation is potentially much worse for aircraft of less than 12,500 pounds (6,700 kg) all-up weight because in general-aviation applications there is no engine-failure accountability for these aircraft below 200 feet (61 m). This is compensated to some extent by the all-engine requirement to reach a 50-foot (15-m) rather than a 35-foot (10-m) screen by a 1.25 factor on the demonstrated distance and by the TODR being not greater than the ASDR. These requirements do not always guarantee an acceptable operational solution, however.

Climb Segment	1	2	4
Twin engines	0	2.4	1.2
Three engines	0.3	2.7	1.5
Four engines	0.5	3.0	1.7

TABLE 4.3   Gross Climb-Gradient Requirement (%)

Aircraft used for commercial purposes (multiengine) and all aircraft with an MTOW in excess of 12,500 pounds (6,700 kg) are certified in accordance with a multistage certification process, whereby it must demonstrate performance equal to in excess of a minimum gross climb gradient throughout a four-segment climb. The requirements (*FAR 25* in the United States) are summarized at Table 4.3.

The segments begin or terminate at points that are defined by procedures (i.e., speed-related) and aircraft configuration. They are illustrated at Figure 4.6 and can be described as follows:

First segment. Critical engine inoperative, remaining engine(s) at takeoff thrust, landing gear extended, flaps in takeoff position, and aircraft at the minimum safety speed $V_{2,\text{min}}$.[4]

Second segment. Same as first segment, with gear retracted, speed increased by 20 percent, and proceeding to an altitude of 400 feet (122 m).

Third segment. Maintaining level altitude, retracting flaps and slats (fully or to appropriate settings), increasing speed by a further 5 percent, and engine power setting reduced to maximum continuous thrust.

Final (fourth) segment. Maintaining configuration at end of third segment, climb to altitude of 1,500 feet (457 m).

The segment capability is measured in flight testing, but in reality a commercial crew will fly to the specific points while at the same time maintaining the most favorable flight conditions to achieve the best possible climb and flying with due regard for the minimum speed conditions. For instance, it would be rare for a crew to maintain a level flight path in the third segment. This criterion is applied in the certification case to replicate a "worst case" scenario. The assurance that is critical in operations is that if the obstacle-limitation surfaces applied to runways are applied with rigor, they should be adequate to ensure that there is acceptable protection.

---

[4] V_2 is a referenced airspeed obtained after liftoff at which the required one-engine-inoperative climb performance (see Table 4.3) can be achieved.

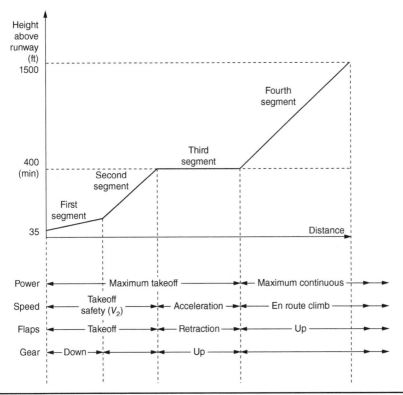

Figure 4.6   Climb-path segments.

In normal operations, climb-out technique is frequently also modified by considerations of noise abatement or fuel economy. Efforts to reduce the noise impact of departing aircraft include

- Flying the highest gradient possible (a technique used less as pure jets have given way to high-bypass turbofan aircraft engines)

- Using specific combinations of thrust and heading to avoid noise-sensitive areas

- Using thrust-cutback techniques that will balance the reduced climb rate with a lower perceived noise on the ground and without undue operational penalty

These performance requirements show how the maximum allowable takeoff weight may be limited by field length or by climb performance to meet either weight and temperature (WAT) limits or dominant-obstacle clearances. It is clear that frequently there is a margin available even at maximum structural weight over all these requirements. There are, in fact, other requirements that a dispatcher must check associated with en-route climb performance, landing performance, and limits on tires and brakes, but these are seldom critical. The pilot therefore has the discretion to perform the takeoff with less than takeoff thrust, provided that not less than 90 percent of the available power is selected. The amount of thrust required to stay within field-length and climb requirements normally can be selected accurately for specific airfield conditions by using an onboard computer, with consequent advantages to engine life and fuel consumption. On the other hand, not only must increased tire wear be considered but also the possible overall increase in risk compared with the historical statistical base for accidents per takeoff, whereby those with a significant margin between actual and allowable takeoff weight are predominantly a higher percentage of safe takeoffs.

Because the required takeoff weight is a function of the payload and fuel requirements, ultimately the payload will be tailored to the takeoff weight available by off-loading that part of the payload which will produce the least revenue. The primary fuel requirement may be reduced by fuel-management techniques, but a considerable proportion of the fuel uplift is to allow reserves for en-route winds, holding, and diversion to alternative landing fields. Thus, on short-haul flights, the reserves can exceed the primary fuel requirements, and this leads to an increase in the takeoff-weight requirement, which is particularly significant when it is realized that in an extreme diversion case, a long-haul jet might burn a quantity of fuel equal to a quarter of the reserve fuel simply to carry the reserves. It is open to question whether aircraft really need to carry the reserves traditionally required, with modern improvements in fuel flow-management, navigational accuracy, and weather forecasting. On short-haul flights, with these improvements and with excellent destination weather at departure time, the need to carry reserves for an alternative destination is particularly questionable if the destination has two independent runways. Operators can minimize fuel use and/or maximize payload by filing for closer destinations and then refiling in flight for the original destination (technically this can be referred to as an *en-route diversion*), but as air-traffic-management (ATM) stringencies increase, this is becoming a less usable option.

In summary, departure performance is dominated by the allowable takeoff weight, which is determined as the lowest of

- Maximum structural takeoff weight
- Climb performance limited by the WAT curve
- Takeoff field lengths: TORA, TODA, and ASDA
- Obstacle clearance
- En-route climb requirements
- Maximum structural landing weight

- Landing-field length, WAT, and diversion requirements
- Tire and brake limits

Any resulting margins between these limits and the required takeoff weight then may be used for tankering, to ease other limits, or to alleviate economic or environmental considerations.

4.4   Approach and Landing Performance

Approach performance is not necessarily in the aerodrome operator's scope of influence. At most airports, aircraft approach along a straight-in approach and at a steady descent angle. This is the glide-path angle, and 3 degrees is applied widely. This approximates a height loss of 300 feet per nautical mile and will mean that an aircraft maneuvered to 1,500 feet above the aerodrome elevation will need to conduct a straight-in approach from about 5 nautical miles. At 120 to 150 knots ground speed, the 3-degree approach causes an aircraft to maintain a steady rate of descent of between 600 and 750 feet/minute.

Where terrain or noise is a significant influence, a higher glide-path angle can be used, and 4.5 degrees is possible by the majority of aircraft types. The limitation is that the aircraft is now descending more rapidly, and on very clean designs, the low-drag airframe makes the attainment of a stabilized speed on approach more time-consuming. Glide-path angles as high as 7 degrees are flown at some terrain-congested airports, but this requires a "high drag" aircraft that can maintain stability at a low airspeed. Typically, this would be a propeller aircraft with large flaps. At 90 knots ground speed, the rate of descent is around 1,000 to 1,100 feet/minute.

Aircraft aim for the touchdown point, and this will be 1,000 feet (330 m) or so beyond the threshold marks on a typical commercial operations runway. Aircraft are expected to be around 50 feet (15 m) above ground at the threshold on a 3-degree approach and to flare, losing speed and reducing rate of descent, to land on or beyond the aiming point. After touchdown, the aircraft will be decelerated with brakes and any other mechanical means (reverse thrust on a jet or reverse pitch or flight idle on a propeller). There is considerable scope for variations in the performance of the landing, and these can include meteorological and runway surface conditions, including runway slope, that can lead to the distance from the landing reference—the threshold—to the end of the landing roll being very different on the same runway and with the same aircraft type. It is for this reason that the LDR requirement used in a flight manual has some onerous factoring applied to calculated and demonstrated distances.

Most regulations allow the manufacturer a choice of demonstrating the landing performance. One option is to use the most long-standing method, landing on a dry, hard runway with conservative assumptions as to height at the threshold, a large factor of safety, and no credit for reverse thrust. The second option is to demonstrate landing on a wet, hard surface from a lower height and higher speed at threshold and using all forms of retardation for which a practical procedure has been evolved. The latter case uses a much smaller factor of safety.

Specific flight plan calculations must take account of forecast runway conditions and wind, with the limiting field length being the lower of the no-wind and forecast-wind cases. The diversion airport also must be taken into account, but in this case the wet-runway factor is allowed to be 0.95. Although the regulations are easy to state, it should not be inferred that the operation is similarly easy to carry out. There are all the problems of accurate alignment and speed control on the approach; adjustment of speed and heading for crosswinds, gusts, and wind shear; and maintenance of direction on the runway, as well as the primary problem of arresting the descent rate without inducing an extended float. The aircraft design is severely tested in this phase of the flight, yet the pilot seldom can compromise in favor of sparing the structure. Indeed, the latest short-haul aircraft specifications are very concerned that brake cooling should not affect turnaround times, that crosswinds should not affect regularity, and that the autopilot should be able to cope with nonlinear and decelerating approaches.

Performance and handling on the approach are just as important as the ground phase of landing in producing a safe completion of a flight. The most vital consideration is accurate achievement of the correct conditions at the threshold (i.e., height, speed, descent rate, track, and power). In order to attain these conditions consistently, the ground aids must be satisfactory, and the aircraft must have adequate performance on the approach to correct discrepancies in the flight path and to respond to emergencies.

The high vertical momentum of modern jets combined with wind gradients, gusts, and wind shear make it essential to provide slope guidance in the form of *visual approach slope indicators* (VASIs) or the most common *precision approach path indicators* (PAPIs). A full-precision approach, of course, is the best aid to accurate flying. In the limit, the approach can be flown completely automatically, the various categories being tabulated as in Table 4.5. The original purpose in developing automatic landings was to increase regularity and save the costs of diversions.

One of the most popular aircraft in production is the Boeing 737-800, which has good operating economics, but it is a design that illustrates the ultimate mass that current technology can accommodate on a single-axle main gear that will fit in the space available on this particular airframe. Therefore, it is an unusual aircraft in that during short-haul applications it can require a longer LDA than TORA. This example is quoted as a reminder that generalizations can be dangerous and that generally accepting that aircraft need a greater takeoff

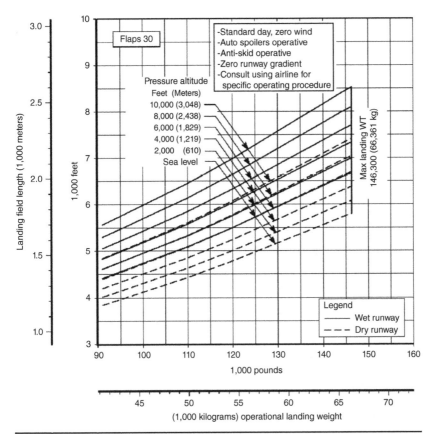

FIGURE 4.7 Boeing 737-800 landing-distance performance.

distance than landing distance is not always true. Bear in mind, too, that Boeing has remedied the susceptibility of the type by offering advancements that have reduced the LDR of appropriately equipped models of the aircraft marque. Figure 4.7 shows the ACAP declaration (for airport design) of the Boeing 737-800 landing-distance requirement. Detail on the plot shows the influence of airport elevation and the allowance for a wet runway. The latter is guidance, and flight crews will have to make judgments on surface conditions. They depend on air traffic services to report relevant information, such as measured braking efficiency or the depth of water on a runway.

4.5 Safety Considerations

The preceding description has highlighted why there is so much greater variation in the distance taken to land an aircraft than to attain takeoff. Given these tendencies, it is not surprising that this phase of flight still attracts more than its share of accidents.

Flight Phase	Duration (%)	Accidents (%)	Relative Exposure
Taxi	n.a.	15	—
Takeoff	1	10	10.0
Initial climb	1	7	7.0
Climb (flaps up)	14	5	0.357
Cruise	57	11	0.193
Descent	11	3	0.273
Initial approach	12	13	1.083
Final approach	3	14	4.667
Landing	1	22	22.0

Source: Boeing, Jet Fleet Statistical Summary (2001–2010).

TABLE 4.4   Phase of Flight Data

Table 4.4 is extracted from the regularly produced Boeing statistical analysis of commercial jet airliner operations and shows 17 percent of all accidents occurring in takeoff and initial climb (roughly to the end of segment three in certification terms) and 36 percent in final approach and landing. Note that 15 percent of accidents (the source analysis considered accidents that resulted in fatalities) additionally were in the taxi phase or within the airport between runway and gate for arrivals and departures. The latter are not related to aircraft performance, but landing and takeoff accidents are and they total over 50 percent of all accidents in the recorded data. The statistics have been factored by duration of flight time to relate the *relative exposure* to an accident in a time-related manner, and it is information such as this, showing enormous relative exposure in the runway-related phases of flight, that leads to considerable effort to force all airport operators to consider the areas around a runway and to reduce accidents there as much as possible.

Long-standing requirements have been to maintain a relatively level and protected runway strip with protected areas at both ends, or RESAs. The latter are required largely to accommodate undershoots and overshoots within areas adjacent to the runway ends that are clear of hazardous obstacles. ICAO–recommended RESA provision for a commercial-operation runway is an area that is at least twice the runway width and 240 m long (ICAO Annex 14). Some lesser requirements are predicated for small (Code 1 and 2) and nonprecision or visual-use-only runways. The RESA should be relatively flat, clear of significant obstacles, and have a surface of adequate strength to accommodate emergency and rescue vehicles. Since the

RESA is beyond the runway strip, this requirement demands strict control over land up to 970 feet (300 m) beyond the end of the paved area, and it is not always easy for all airports to accommodate.

Local terrain (where it dips below threshold height), the presence of transport infrastructure, especially on embankments or in cuttings, and so on might not allow some airports to meet these requirements without a significant reduction in the declared distances they promulgate. It is essential for all aerodromes, therefore, to justify the provisions they use by presenting safety assessments that show that an acceptable level of risk is attained in expected operations. This is being mandated on an annual basis by some national authorities; for example, U.K. Civil Aviation Authority (CAA) Safety Notice SN-2012/004 (March 30, 2012) states: "The annual requirement for Aerodrome Licence Holders to review and determine the RESA distance, even if there were no actual changes to the operations at the aerodrome, is now withdrawn. Instead, the risk of a runway excursion should now be assessed on a regular basis as circumstances change, as determined by the aerodrome as part of its normal Safety Management System (SMS) process."

A regulator may issue a permitted dispensation if the traffic movement level is low, if the traffic mix is biased toward smaller aircraft, and/or if the typical LDR of operations is much less than the LDA. The regulator still may require mitigation, for example, requiring that the aerodrome install an emergency material arresting system (EMAS) within the available RESA. This issue is addressed further in Chapter 11.

The CAA–mandated procedure is typical of what is regarded as "best practice" in safety management and has become more commonly required to have been shown to be taken into account in the most recent decade or so. The attainment of economic performance will be taken into account by the regulator, but it must not be assumed that mitigation will be permitted automatically on these grounds.

A critical example is the requirement to climb after an engine failure, especially at an airport with a mountain off the end of the runway. In such a situation, the required category risk is achieved by invoking progressively severe operational limitations as the intrinsic risk increases. This concept of protection is illustrated in Figure 4.8. In the case of mountains being close, the regulations would restrict conditions where landing and takeoff operations could be conducted safely and might require that takeoff is in one runway direction, subject to tailwind criteria. An example of this limitation is Salzburg, Austria, where runway 16 is the preferred landing direction (and the only one in instrument conditions), and runway 34 is the mandated takeoff direction for commercial operations, with operations restricted by operators if tailwind values exceed limits they agreed with the regulator.

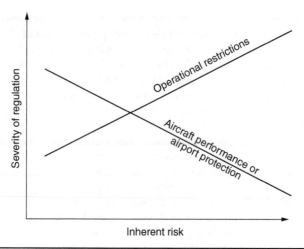

FIGURE 4.8 Operational and performance regulation tradeoff.

The general philosophy of matching the safe operation of aircraft and airports has been reviewed in this section.

4.6   Automatic Landing

The original purpose in developing automatic landings was to increase regularity and save the costs of diversions. There are, however, many other advantages in the form of pilot workload, control of touchdown dispersion, softer landings, and the maintenance of flight path even in difficult wind-shear conditions. Experience with the equipment is showing increased reliability and a greater harmony between pilots and autopilots. It appears that the monitoring and takeover functions of the pilot are sufficiently undemanding for fail-passive equipment to be used even for Category IIIA. Most of the benefits therefore accrue to the airline, whereas only full use of the system can justify the expense of the ground equipment, particularly for Category III. Also, different regulations must be brought into effect; the equipped aircraft must be allowed to bypass other stacked aircraft, and it must be provided with positive control on the ground in minimum visibility.

Safe automatic landing operations depend on the same factors that have been considered for visual landings, but with much greater emphasis on the concept of decision height. This depends on the method of operation, the specification of the equipment, and the runway visual range (RVR), as shown in Table 4.5, as well as on the obstacle-clearance criteria. The latter must take account of the demonstrated height loss during a missed approach.

Category	DH (ft)	RVR (m)
I	200	800
II	100	400
IIIA	—*	200
IIIB	—*	50

*No decision height applicable.

TABLE 4.5 Decision Heights (DHs) and Runway Visual Ranges (RVRs) for Precision-Approach Runways

The required RVR is a function of the pilot's angle-of-view cutoff, the intensity and beam spread of the lighting system, the vertical and horizontal structure of the fog, and the location of the pilot's eyes relative to the aids and his or her intrinsic visual reference needs. It must be measured at least at three positions along the runway. The touch-down-zone reading must be passed to the pilot within 15 seconds of reading, followed by the other readings if they are lower than the first and less than 2,625 feet (800 m).

Any proposal to operate automatic landings must show the feasibility of the proposed minima for each runway, including the adequacy of the facilities and the obstacle-clearance capability. An airport wishing to declare a runway as suitable to receive automatic landings must consider

- Obstacle clearances
- Glide-path angles
- Terrain on approach [It should be essentially flat for 1,000 feet (300 m) before threshold over a 200-foot-wide (60-m-wide) strip.]
- Runway length, width, and profile
- Conformity and integrity of the instrument landing system (ILS)
- The visual aids and their integrity
- The level of air traffic control (ATC) equipment and its meteorological monitoring requirements

The most critical aspect of performance is the ability to climb after a missed approach has been declared. It must be possible to demonstrate adequate climb performance in each of the three following flight conditions:

1. Positive net gradient at 1,500 feet (457 m) above the airfield in the cruise configuration with one engine out

2. A gross gradient not less than 3.2 percent at the airfield altitude with all engines at maximum takeoff power in the final landing configuration [This is to allow a safe overshoot (balked landing).]

3. A minimum gross climb gradient at the airfield altitude with the critical engine out and all others at maximum takeoff power in the final landing configuration but with the gear up (The gradients should be not less than 2.1 percent for twin-engine aircraft, 2.4 percent for three-engine aircraft, and 2.7 percent for four-engine aircraft.)

Landing WAT charts indicate the maximum landing weights at which these gradients can be achieved as functions of altitude and temperature. These performance criteria will allow safe operation in the vicinity of airfields only if used in conjunction with minimum descent altitudes and taking into account the handling of each aircraft type, the ground aids available, and the local terrain conditions. In setting minimum descent altitudes and associated decision heights, it is important to realize that there is an inevitable height loss between the decision to declare a missed approach and the establishment of a positive climb gradient, even with the demonstrated performance quoted earlier. This is due to delays in responding to pitch and power inputs and to the initial downward momentum of the aircraft.

The protected-surface funnel shown in Figure 4.9 includes surfaces to protect in the missed-approach situation. These surfaces are designed to give clearance below the climb paths guaranteed by the landing WAT limits to the remote-risk level.

The approach path is governed by the need to maintain clearance over obstacles both on the expected flight path and in the general vicinity of the airfield. The former are protected by the surfaces in

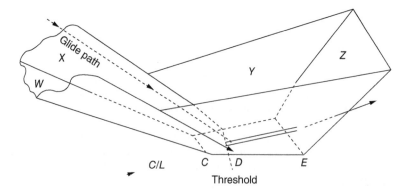

Figure 4.9 Obstacle-assessment surfaces. W and X are approach surfaces; CDE is the footprint; Y is a transitional surface; and Z is the missed-approach surface.

Figure 4.9 or by those in Annex 14, whereas the latter (and any obstacles that break the protected surfaces) are cleared by the imposition of margins over the declared obstacle-clearance altitudes (OCAs) or obstacle-clearance heights (OCHs). The derivation of these is fully explained in ICAO PAN-OPS (ICAO 2006), and there may be variations in national legislation.

In essence, reductions in noise and fuel use are obtained by avoiding until the latest time the drag and hence the thrust that come with the selection of full landing flaps; the time at which the speed is stabilized, thus, is also delayed so as to maintain a safe margin over the stall speed. Lufthansa first developed the technique to decrease noise over a sensitive area 8 miles (13 km) out on the approach to Frankfurt. A more general version of the procedure was later adopted by the International Air Transport Association (IATA), and it is now used widely in the continuous-descent approach (CDA) procedure favored for fuel and noise minimization in air traffic management (ATM) procedures.

Studies in Europe and the United States in the 1990s, using research aircraft with advanced avionics showed that precise four-dimensional (3D position and time) flight paths were feasible using control parameters compatible with ATC processes. This led to the development of CDA procedures for many major airports. In ATM terms, the arrival rate into the terminal area (TMA) is matched to the acceptance rate of the airport, thus minimizing radar vectors and path stretching. This should allow the use of a near-idle-thrust descent from cruise altitude and minimized maneuvering time at low altitudes.

There are some disadvantages; for example, it is not possible to operate in this way without affecting decision heights. ATC also must restrict the acceptable speed range in the early stages of the approach, and either all operations must adopt the technique or separations on the approach must increase. This can limit the proportion or even the acceptance of all turboprop aircraft at busy airports. Current ATM research programs are considering the impact of these limitations and the possibility of having segregated traffic flows based on aircraft categories. This is not based on aircraft performance but influenced by performance, and the motivation for studies is the preservation of runway movement capacity.

A final consideration in approach performance is the problem of wake turbulence. It is necessary to pay attention to the order in which aircraft of different weights are allowed to approach or take off so that smaller aircraft do not follow larger aircraft when runway capacity must be maximized. At Heathrow, the arrival runway capacity was found to vary from 34.7 to 31.8 movements per hour as the percentage of heavy aircraft varied from 10 to 50 percent. This has not improved with the advent of new large aircraft, such as the Airbus A380, and the topic is addressed in Section 4.8.

4.7   Operations in Inclement Weather

As with vortices generated from preceding aircraft, as well as with the use of automatic landings, one of the prime consequences of operations in inclement weather is reduced runway capacity, in this case owing to increased runway occupancy time. Not only will average times increase, but the standard deviation on occupancy time also will increase. This is a function of the decreasing value of high-speed exits as well as the increased braking distances, in that a turning radius of 1,000 feet (305 m), which would be acceptable for 50 miles per hour (88 km/h) in good weather, will be usable at only 20 miles per hour (32 km/h) in slippery conditions. On the other hand, the aiming point for touchdown is a function of the distance from threshold to the most likely exit, but it does not vary significantly with weather conditions. Since runway occupancy is likely to be the critical capacity parameter when inclement weather forces the occupancy time up toward 90 seconds, a case can be made for designing exit locations and angles in relation to an aircraft's poor-weather landing performance.

Wet runways do not really qualify as inclement weather in that, as described earlier, regulations exist to cover this rather normal situation, and the condition normally can be controlled by good runway grooving and drainage. The real difficulty comes when the runway is contaminated with standing water, slush, snow, or ice. There are some regulations to cover these cases. The *JARs* require that data be established for aborting or continuing takeoff on runways with very low friction and with significant precipitation, as well as simply wet conditions. They also require that 150 percent of the average depth of precipitation be used in subsequent calculations based on these data.

Flight or operations manuals contain data on performance in slush and ice and on aquaplaning. Slush is perhaps the worst offender in that it affects both acceleration and braking. Thus, it can increase the emergency distance required on takeoff by 50 percent and can be particularly dangerous on an exploratory touch-and-go. When the friction coefficient is less than 0.05, flight manuals must advise a V_{stop}. This is similar to the normal decision speed for abandoning takeoff, but it is concerned only with the speed from which the aircraft at a given weight may be stopped within the runway length available. It might be less than the minimum control speed and does not imply any ability to continue takeoff if an engine fails.

Aquaplaning is a function of speed and the depth of the standing water, the friction reducing to levels equivalent to an icy runway as the water fails to disperse beneath the tire. These contaminated runway conditions give rise to two specific problems. First, there is the difficulty, already discussed, of the consistent evaluation of the contamination on aircraft performance. The variation is not confined to variations in contamination but includes such factors as tire wear and pressure, pilot technique, and the extent to which the dynamic aquaplaning gives way to a sustained aquaplaning, and at much-reduced depths of precipitation.

The second group of problems concerns the method of informing the pilot correctly of the prevailing runway conditions. For some years, the two methods in use have been feedback from pilots who just used the runway and measurement of the depth of precipitation (the depth and type of precipitation being calibrated to give a percentage increase in takeoff and landing distance required). It is preferable to measure the runway friction directly by towing a device down the runway, and this will essentially mimic a loaded aircraft tire. Whatever the technique used (every manufacturer has its own lists of pros and cons), there are still many variables concerning the tire and runway surface to bear in mind when relating friction readouts to generate predicted rollout distance.

There are clearly doubts about the reporting of contamination and performance in known contamination. When considered together with other problems, such as "soft" failures (e.g., tire bursts and partial loss of power) in contrast to the "hard" complete engine failures around which the regulations are drawn up, there will always be occasions when the pilot is in doubt in the region of takeoff safety speed. In these circumstances, it always should be safer to reject rather than to risk a nonflying takeoff. It is safer to overrun in a straight line than to risk losing directional control because the gear is stronger in a fore-and-aft sense. This points to a real need for RESA because the implied safety in these pilot decisions is false if the terrain at the end of the runway is difficult or nonexistent.

A remaining item of inclement weather is wind shear. Prior to its interpretation, this phenomenon was attributed to have caused several accidents that were provisionally attributed to pilot error. The most serious form of wind shear is associated with cold air creating a gust front preceding a thunderstorm. It comprises horizontal shears, produced by turbulence in the cold sublayer, and the reversed direction of the warm inflow moving up and over the cold sublayer that causes the worst problems. The associated changes in airspeed (many tens of knots in a period of less than 5 seconds) can produce very strong tailwinds very soon after the pilot has taken remedial action for a gusting headwind, and it thus may cause an aircraft to stall. Systems of low-level anemometers have been developed that compare horizontal wind strength and direction around the periphery of an airport with the normal central reading. Wind-shear warning is given when the discrepancy exceeds a preset tolerance. These systems have been superseded by laser radar (*ladar*) devices in experimental installations that have performed well. It seems best to detect circumstances, because they are short term, and to suspend operations for a limited period, accepting that delay, congestion, and residual costs are incurred but safety is preserved. Aware of the business implications, and with the greater fidelity available in modern flight-control systems, aircraft-based wind-shear alleviation systems are still being considered.

4.8 Specific Implications of the Airbus A380 (New Large Aircraft)

Over the last decade, the Airbus A380 has evolved to match the ICAO definition for a category F aircraft [maximum span 262 feet (80 m), maximum length 262 feet (80 m), and maximum height 80 ft (24.4 m)]. The aircraft entered service in 2007 and is the only aircraft in service or planned in the foreseeable future in the category. Meanwhile, airports do occasionally have to accommodate the even-larger-span Antonov An-225 freighter, but this is treated as a one-off aircraft and is operated under dispensation.

Issues of airport compatibility related to category F aircraft include

Stands. 262 feet (80 m) wide and 262 feet (80 m) deep with 23 feet (7-m) minimum clearance between adjacent stand and with a perpendicular track behind the stand (aircraft stand to taxi lane centerline) of 165 feet (50.5 m)

Taxiways. 82 feet (25 m) wide [maximum aircraft landing gear track is 52.5 feet (16 m)]; minimum separation between parallel taxiways is 320 feet (97.5 m)

Runways. New build to be 197 feet (60 m) wide with 21.3 feet (6.5-m) shoulders, but 148 feet (45 m) wide with 25 feet (7.5-m) shoulders is proving to be acceptable (especially because the A380 has reverse thrust on the inboard engines only)

Runway-taxiway. Minimum separation of 624 feet (190 m) (instrument runway)

The aircraft is compliant with existing aircraft performance requirements and does not pose any additional burden in terms of takeoff and landing distance requirements. Indeed, it has been designed to need shorter runways than the Boeing 747 operating similar-range services. It is worthy of note that the Boeing 787 Dreamliner and the equivalent Airbus A350 XWB (although they are smaller aircraft) are also aiming to attain shorter field performance requirements.

In terms of environmental compatibility, engine technology has alleviated any significant noise problems, and while premanufacturing studies showed that wake vortices should not be a problem, increased approach-separation spacing is being applied, treating the aircraft as a "super" category and allowing a following "super" to be 4 nautical miles separated, a "heavy" 6 nautical miles separated, and jets and turboprops as much as 8 nautical miles separated. This is not influencing airport capacity greatly while A380 movements are small, but the aircraft impact on capacity at airports with considerable traffic mix will become more significant as the aircraft population rises. This seems inevitable because the fleet was approximately 70 aircraft in the beginning of 2012, and production is running on the order of 30 to 36 aircraft annually, against an outstanding commitment to 180 further aircraft (at mid-2012).

The most potentially pressing performance problem with the aircraft is the impact of wake-vortex turbulence when in cruise. Although the vortex intensity is much less in cruise than it is in the landing and takeoff configuration, the application of reduced vertical separation minima (RVSM), reducing separation between crossing or passing movements to 1,000 feet (305 m) from 2000 feet (610 m) when cruising has led to some reported incidents. Research had suggested that there should be no problem because, fundamentally, air viscosity absorbs vortex energy and breaks up the vortex over a period that should not exceed 2 minutes. There is concern that sections of vortices can form "loop," almost as if creating the smoke ring that a cigarette smoker can produce by exhaling smoke while holding his or her tongue in the center of the mouth. These loops could be stable enough to drift down from an aircraft's path for longer periods of time, but finding incidences of this phenomenon is elusive. It is believed that there should be no inherent risk in terms of structural loading or flight-path upset, but the aircraft in cruise has relatively small performance margins, and it is a flight phase where cabin activity can be considerable. Therefore, this does potentially involve a lot of risk. It is a reported limitation that has no bearing on airport performance, and alleviation conditions do seem to be understood and to have been applied successfully.

References

Ashford N.J., S. Mumayiz, and P.H. Wright. 2011. *Airport Engineering*, 4th ed. Hoboken, NJ: Wiley.

Boeing. 2004. B777-200LR/-300ER: *Airplane Characteristics for Airport Planning* (D6-583292, October 2004). Seattle: Boeing Airplane Company. Available on the company website. http://www.boeing.com/

International Civil Aviation Organization (ICAO). 2006. *Procedures for Air Navigation Services: Aircraft Operations*, Vol. II (Document 8168 Ops/611), 1st ed. Montreal, Canada: ICAO.

CHAPTER 5

Operational Readiness[1]

5.1 Introduction

One of the criteria for judging the efficiency of an airport is the availability of operational facilities: runways, instrument-approach aids, lighting, fire and rescue services, mechanical and electrical systems, people movers, baggage-handling systems, air bridges, and so on—in short, the "readiness" state of the airport to provide the operational facilities appropriate to the types of airlines and aircraft using the airport. All of this involves a considerable commitment to maintenance on the part of airport management. Increasingly, airport managers are documenting this aspect of their responsibilities, often with the help of control systems that monitor availability of critical facilities and equipment from a centralized operations control center. Table 5.1 shows a British Airports Authority (BAA) Heathrow management report on "passenger sensitive" equipment as an example of a detailed analysis of monthly performance for a year. The availability target for all five items of equipment in this particular year was 98.5 percent.

5.2 Aerodrome Certification

The principle of central-government responsibility for safety aspects of public transportation extends to the licensing of airports. The International Civil Aviation Organization (ICAO) uses the term *aerodrome certificate* and requires that all Member States put in place a system for granting certification for airports with international commercial service. The ICAO *Manual on Certification of Aerodromes* (ICAO 2001) lists the obligations of the airport operators and provides a guide on how to document compliance, in the form of an approved certification manual (see Chapter 17). The compliance

---

[1]This chapter was reedited and rewritten by William Fullerton.

Monthly	Jan	Feb	Mar	Apr	May	Jun	Jul	Aug	Sep	Oct	Nov	Dec
Loading bridges	99.03	98.89	98.36	97.95	98.54	97.14	96.54	96.35	97.02	96.85	97.98	97.21
Passenger conveyors	97.67	98.69	97.78	98.74	94.97	96.24	98.47	98.24	98.84	98.69	99.13	98.54
Baggage conveyors	98.85	99.20	98.80	98.64	98.64	94.07	98.76	98.24	99.06	98.59	99.17	99.09
Lifts	99.50	99.25	99.24	98.95	99.17	98.33	98.73	98.34	99.30	99.35	99.10	98.36
Escalators	98.97	99.09	99.45	99.48	99.25	99.26	98.55	99.62	99.48	98.98	97.52	97.65
Aggregate result	98.25	99.07	98.73	98.64	98.41	97.03	97.95	97.63	98.46	98.27	98.53	98.03
Calculative												
Loading bridges	98.73	98.74	98.71	97.95	98.23	97.97	97.52	97.29	97.25	97.19	97.28	97.27
Passenger conveyors	98.00	98.10	98.07	98.74	96.86	96.66	97.10	97.05	97.35	91.53	97.73	97.82
Baggage conveyors	99.05	99.07	99.04	98.87	98.75	97.19	97.58	97.71	97.94	98.03	98.17	98.27
Lifts	99.16	99.17	99.17	98.95	99.06	98.78	98.77	98.68	98.79	98.88	98.90	98.84
Escalators	98.80	98.83	98.88	99.48	99.37	99.33	99.14	99.23	99.28	99.24	99.08	98.86
Aggregate result	98.82	98.85	98.84	98.64	98.52	98.01	97.99	97.92	98.01	98.05	98.11	98.10
Moving annual												
Loading bridges	98.78	98.76	98.71	98.63	98.61	98.47	98.30	98.11	97.98	97.83	97.78	97.56
Passenger conveyors	98.22	98.18	98.07	98.03	97.72	97.60	97.61	97.47	97.47	97.51	97.72	97.90
Baggage conveyors	99.09	99.08	99.04	99.03	98.99	98.65	98.53	98.47	98.48	98.44	98.44	98.44
Lifts	99.20	99.20	99.17	99.14	99.15	99.05	99.03	98.97	98.99	99.02	99.02	98.96
Escalators	98.89	98.89	98.88	98.97	98.95	99.85	99.85	98.98	98.98	98.93	98.93	98.94
Aggregate result	98.89	98.88	98.84	98.80	98.74	98.58	98.51	98.42	98.39	98.33	98.34	98.30

Source: BAA (1989).

TABLE 5.1 Passenger Sensitive Equipment Performance Results

requirements themselves are contained in Appendix 14 to the Chicago Convention of 1944, which is published as a separate manual (Annex 14) with worldwide distribution. Annex 14 contains the standards and recommended practices for safe operations, including required facilities, services, and operating conditions, for international airports of all sizes.

In common with most ICAO Member States, the United States makes certification compulsory for airports used by air carriers, both scheduled and charter. In Great Britain, the Civil Aviation Authority (CAA) requires, in addition, that any airport being used for flight training also must be licensed. The CAA issues two classes of license, *public use* and *ordinary*. The essential difference between them is that a public-use airport must be available to all would-be operators without discrimination, whereas the ordinary category may be restricted if the owner so desires. Typically, an aircraft manufacturer's airfield falls into the latter category.

The requirements for the licensing or certification of an airport are set out in national regulations. In the case of the United States, they are to be found in Part 139 of the *Code of Federal Regulations* (*CFR 2004*), which specifies certain criteria that must be met in relation to pavement areas (i.e., runways, taxiways, and apron) and safety areas (i.e., overrun areas); marking and lighting of runways, thresholds, and taxiways; airport fire and rescue services; handling and storage of hazardous articles and materials; emergency plan; self-inspection program; ground vehicles; obstructions; protection of navaids; public protection; bird-hazard reduction; and assessment and reporting of airport conditions, including areas where work is in progress and other unserviceable areas.

In the case of Great Britain, the legal requirements are set out in the *Air Navigation Order and Regulations,* CAP 393, Articles 76 to 79 (CAA 2010a). Although the requirements are very similar in the United States and Great Britain, U.S. regulations contain additional rules dealing with maintaining the operational readiness of the airport (e.g., pavement repairs; clearance of snow, ice, etc.; and lighting maintenance). In addition, the Federal Aviation Administration (FAA) issues very comprehensive guidelines regarding implementation in the 150 and 139 series of advisory circulars. In general, the certificate holder has to satisfy the regulating authority that

1. Airport operating areas on the airport and in its immediate vicinity are safe.

2. Airport facilities are appropriate to the types of operations taking place.

3. The management organization and key staff are competent and suitably qualified to manage the aircraft flight-safety aspects of the airport.

Conditions that fall short of safety requirements for a given airport may result in reduced or halted air operations. For this reason, virtually all airport programs related to operational readiness feature a strong safety component. Safety-relevant programs include Aeronautical Information Services (AIS), apron management, vehicle/pedestrian control, access control, foreign-object-debris (FOD) control, wildlife-hazard management, hazardous-material handling (including fuel spills), construction safety, and maintenance, among others. Emergency operations—actions designed to control the negative effects of an unplanned disruption or accident—also can be considered relevant to safety.

These programs—and the airside facilities they control—can be monitored systematically through a safety management system, as covered in Chapter 15.

5.3 Operating Constraints

Visibility

Air traffic moves under either visual flight rules (VFRs) or instrument flight rules (IFRs) depending on weather conditions and prevailing traffic densities. VFR operations are possible where weather conditions are good enough for the aircraft to operate by the pilot's visual reference to the ground and to other aircraft. Operational runways are classified according to the weather conditions in which they can operate. The worse the condition in which a runway is to operate, the greater is the amount of visual and instrument navigational equipment that must be provided. Runways can be classified according to their ability to accept aircraft at different degrees of visibility (ICAO 2010).

Noninstrument runway. This is a runway intended for the operation of aircraft using visual approach procedures only.

Instrument-approach runway. This is a runway served by visual aids and a nonvisual aid providing at least directional guidance for a straight-in approach.

Precision-approach runway, Category I. This is an instrument runway served by an instrument landing system (ILS) and visual aids, intended for use in operations down to a decision height of 200 feet (60 m) and visibility of no less than 2,600 feet (800 m) or a runway visual range (RVR) of 1,800 feet (550 m).

Precision-approach runway, Category II. This is an instrument runway served by ILS and visual aids, intended for use in operations down to a decision height of 100 feet (30 m) and an RVR of 1,000 feet (300 m).

Precision-approach runway, Category III. This is an instrument runway served by ILS to and along the runway with further subcategories.

Category IIIA. This is intended for operations down to an RVR of 575 feet (175 m) and zero decision height using visual aids during the final phase of landing.

Category IIIB. This is intended for operations down to an RVR of 160 feet (50 m) and zero decision height using visual aids for taxiing.

Category IIIC. This is intended for operations without reliance on visual reference for landing or taxiing.

Runway visual range (RVR) is defined as the distance over which the pilot of an aircraft on the centerline of the runway can see the runway surface markings or the lights delineating the runway or its centerline. This range is now frequently determined automatically by RVR sensors, such as those shown in Figure 5.1, which are set just off the runway shoulders. *Decision height* is defined as the minimum height at which the pilot will make the decision either to land or to abort the attempt to land.

Low visibility and ceilings can, in addition to presenting safety challenges, result in reduced airfield capacity. Figure 5.2 shows the sort of record that should be available to help determine the economic viability of high-category operations—"Potential regularity" refers to the percentage of flights that could operate on a runway under various approach categories. Simply recording the number of hours that the RVR is below the limit for Category I is not particularly helpful. At airports where low RVRs occur at night or in the very early morning when there is little traffic, the number of hours of poor visibility overestimates the level of traffic disruption poor visibility would cause. At other airports, however, severe and prolonged morning mist or haze could affect peak-hour operations, and without Category II or III capability, the development of the airport might be made difficult.

Note that each major category requires significant improvements in runway visual aids, secondary power backup specifications, aircraft equipment, pilot training, and airport procedures. In other words, moving from one category to another represents a significant investment that affects various elements of the airport system. The investment decision will be based on financial analysis of the number

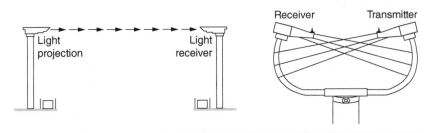

FIGURE 5.1 Runway visual range (RVR) measurement equipment types.

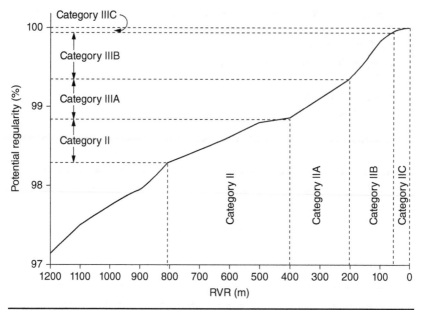

FIGURE 5.2   Impact of reduced visibility on potential regularity at London Heathrow.
(*BAA.*)

of weather days per year that would close the airport if precision-approach operations were not possible.

The figure shows graphically the results of a computation done by the BAA on the effect of reduced visibility in terms of potential regularity (i.e., operational impact). It can be seen that the proportion of operations requiring Category II and III operations is less than 2 percent. Category IIIc conditions affect less than 0.05 percent of operations. Nevertheless, as far back as the 1980s, the principal operator at Heathrow, British Airways (BA), decided that a "blind landing" capability was economically justifiable because the airline finds itself able to operate when its competitors are grounded. Completely automatic "hands off" landings of BA aircraft at London Heathrow are now routine, as are operations of many other carriers.

Crosswind Effects

Regulating bodies such as ICAO and the FAA require that an airport has sufficient runways, both in number and orientation, to permit use by the aircraft for which it is designed, with a usability factor of at least 95 percent with reference to wind conditions. Modern heavy transport aircraft are able to operate in crosswind components of up to 30 knots without too much difficulty, but for operational purposes runway layouts are designed more conservatively. Annex 14 requires an orientation of runways that permits operations at least 95 percent of the time with crosswind components of 20 knots (37 km/h) for Category A and

B runways, 15 knots (27.8 km/h) for Category C runways, and 10 knots (18 km/h) for Category D and E runways (ICAO 2010). FAA regulations differ slightly (FAA 1989). Runways must be oriented so that aircraft can be landed at least 95 percent of the time with crosswind components not exceeding 15 miles per hour (24 km/h) for all but utility airports and 11.5 miles per hour (15.5 km/h) for utility airports.

The usability factor should be based on reliable wind-distribution statistics collected over as long a period as possible, preferably not less than 10 consecutive years. As aircraft have become heavier, the provision of crosswind runways has become less important at large hubs, where there is a generally prevailing direction of wind. However, crosswind runways are still operated at many airports when winds vary strongly from the prevailing direction or where light aircraft are operated.

The usability of a runway or a combination of runways is most easily determined by the use of a *wind rose,*[2] which is compiled from a tabular record of the percentage incidence of wind by direction and strength, as shown in Table 5.2. For clarity of presentation, the table shows a record of the percentage of time the wind falls within certain speed ranges (in knots), with the direction recorded to the nearest of 36 compass directions.

A wind rose is drawn to scale with rings at 0 to 10, 11 to 16, 17 to 21, 22 to 27, and 28 and over knots, as shown in Figure 5.3. The percentage of time that a crosswind component occurs in excess of 13 knots[3] can be determined using the following example with runway 10–28 (oriented 105 to 285 degrees). (For the purposes of this example, it is assumed that true north and magnetic north are identical; in practice, runway bearings are magnetic, and wind data are referred to true north. Therefore, runway bearings must be corrected prior to plotting.) The direction of the main runway 10–28 is plotted through the center of the rose, and 13-knot crosswind-component lines are plotted to scale parallel on either side of this centerline. The sum of percentages of wind components falling outside the parallel component lines is the total amount of time that there is a crosswind component in excess of 13 knots. Table 5.3 indicates that this occurs for a total of 2.72 percent of the time for this particular runway direction. Therefore, this runway would conform to FAA standards if proposed as the only runway of a U.S. airport. The reader is invited to check that it also would meet ICAO standards. Note that estimates of part areas of the rose had to be made in compiling Table 5.3.

---

[2]The subsequent calculation method is now performed routinely by computer programs. The manual example is shown for instructional purposes only.
[3]For Airport Reference Codes A-II and B-II, the FAA mandates that the maximum allowable crosswind component is 13 knots. This is also the ICAO standard for similar airport designs.

Hourly Observations of Wind Speed

Direction	0–3 / 0–3	4–6 / 4–7	7–10 / 8–12	11–16 / 13–18	17–21 / 19–24	22–27 / 25–31	28–33 / 32–38	34–40 / 39–46	≥41 / ≥47	Total	Average Speed Knots	Average Speed Mi/h
01	469	842	568	212						2,091	6.2	**7.1**
02	568	1263	820	169						2,820	6.0	**6.9**
03	294	775	519	73	9					1,670	5.7	**6.6**
04	317	872	509	62	11					1,771	5.7	**6.6**
05	263	861	437	106						1,672	5.6	**6.4**
06	357	534	151	42	8					1,092	4.9	**5.6**
07	369	403	273	84	36	10				1,175	6.6	**7.6**
08	158	261	138	69	73	52	41	22		814	7.6	**8.8**
09	167	352	176	128	68	59	21			971	7.5	**8.6**
10	119	303	127	180	98	41	9			877	9.3	**10.7**
11	323	586	268	312	111	23	28			1,651	7.9	**9.1**
12	618	1397	624	779	271	69	21			3,779	8.3	**9.6**
13	472	1375	674	531	452	67				3,571	8.4	**9.7**
14	647	1377	574	781	129					3,008	6.2	**7.1**
15	338	1093	348	135	27					1,941	5.6	**6.4**
16	560	1399	523	121	19					2,622	5.5	**6.3**
17	587	883	469	128	12					2,079	5.4	**6.2**
18	1046	1984	1068	297	83	18				4,496	5.8	**6.7**

19	499	793	586	241	92						2,211	6.2	**7.1**
20	371	946	615	243	64						2,239	6.6	**7.6**
21	340	732	528	323	147	8					2,078	7.6	**8.8**
22	479	768	603	231	115	38	19				2,253	7.7	**8.9**
23	187	1008	915	413	192						2,715	7.9	**9.1**
24	458	943	800	453	96	11	18				2,779	7.2	**8.2**
25	351	899	752	297	102	21	9				2,431	7.2	**8.2**
26	368	731	379	208	53						1,739	6.3	**7.1**
27	411	748	469	232	118	19					1,997	6.7	**7.7**
28	191	554	276	287	118						1,426	7.3	**8.4**
29	271	642	548	479	143	17					2,100	8.0	**9.3**
30	379	873	526	543	208	34					2,563	8.0	**9.3**
31	299	643	597	618	222	19					2,398	8.5	**9.8**
32	397	852	521	559	158	23					2,510	7.9	**9.1**
33	236	721	324	238	48						1,567	6.7	**7.7**
34	280	916	845	307	24						2,372	6.9	**7.9**
35	252	931	918	487	23						2,611	6.9	**7.9**
36	501	1,568	1,381	569	27						4,046	7.0	**8.0**
00	7729										7,729	0.0	**0.0**
Total	21,676	31,828	19,849	10,437	3,357	529	166	22			87,864	6.9	**7.9**

Source: FAA (1989).

TABLE 5.2 Wind Table: Wind Direction and Number of Recorded Hours over a 10-Year Period

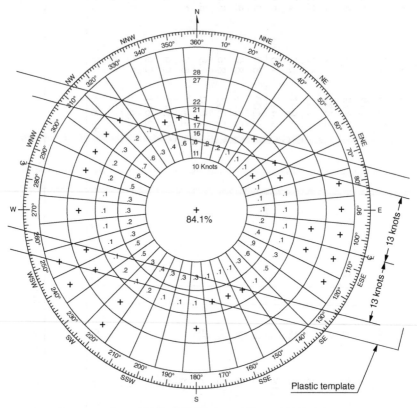

A runway oriented 105°–285° (true) would have 2.72% of the winds exceeding the design crosswind/component of 13 knots.

FIGURE 5.3 Wind rose. (*FAA.*)

Bird-Strike Control

Since the beginning of aviation, birds have been recognized as a hazard to aviation. In the early days, damages tended to be minor, such as cracked windshields, dented wing edges, and minor fuselage damage. However, fatal accidents owing to bird strikes occurred as far back as 1912, when Cal Rogers, the first man to fly coast to coast across the United States, was subsequently killed in a bird strike. As aircraft have become faster, birds have become less able to maneuver out of the way, and the relative speed at impact has increased. Damage also increased when turbine-engine aircraft were introduced. Ingestion of birds into the engine can cause a blocking or distortion of airflow into the engine, severe damage to the compressor or turbine, and an uncontrollable loss of power. On January 15, 2009, USAirways Flight 1549 lost thrust in both A-320 engines after striking a flock of Canada geese shortly after takeoff from New York's LaGuardia airport. In what

Direction	Hourly Observations of Wind Speed Knots / Mi/h										Average Speed	
	0–3 / 0–3	4–6 / 4–7	7–10 / 8–12	11–16 / 13–18	17–21 / 19–24	22–27 / 25–31	28–33 / 32–38	34–40 / 39–46	≥41 / ≥47	Total	Knots	Mi/h
01				0.12								
02				0.12								
03				0.05								
04				0.04								
05				0.01								
06												
07												
08						0.01						
09												
10												
11												
12												
13						0.01						
14					0.01							
15												
16				0.01								
17				0.04								
18				0.14	0.10							
19				0.16	0.10							
20				0.16	0.10							

TABLE 5.3 Wind Table: Estimated Area Not Included as Being Below 13-Knot (24-km/h) Crosswind Component

Direction	Hourly Observations of Wind Speed Knots Mi/h										Average Speed	
	0–3 0–3	4–6 4–7	7–10 8–12	11–16 13–18	17–21 19–24	22–27 25–31	28–33 32–38	34–40 39–46	≥41 ≥47	Total	Knots	Mi/h
21				0.20	0.20							
22				0.11	0.10							
23				0.03	0.19							
24					0.05							
25					0.01							
26												
27												
28												
29												
30												
31												
32					0.01							
33					0.05							
34				0.04								
35				0.25								
36				0.30								
00												
Total				1.78	0.92	0.02				2.72		

Source: FAA (1989).

TABLE 5.3 Wind Table: Estimated Area Not Included as Being Below 13-Knot (x-km/h) Crosswind Component (Continued)

has come to be called the "Miracle on the Hudson," the crew successfully landed the aircraft in the Hudson with no loss of life.

Loss of life through bird strike may be unusual, but airport operators must be aware that the potential for a disaster can exist in the vicinity of an airport where aircraft are operating at the low altitudes at which they are likely to come in contact with birds. International and national aviation regulating bodies therefore have prepared advisory documents that guide airport operators in methods of reducing the risk of bird hazards through programs of bird-strike control (ICAO 2011).

ICAO recommends that the airport operators assign bird-control duties to a wildlife officer, preferably someone with training in biology or related sciences. As a first step, the wildlife officer should commission a study of the bird species that present a hazard at the airport, including size and weight, flying and flocking patterns, density of population, and nesting locations relative to the approaches. From this information and from a history of reports of bird strikes and evidence of bird impacts (i.e., bird remains found during daily inspections), the degree of risk presented by bird hazard can be assessed.

Any bird, if present in sufficient quantities, can present a hazard to aviation at an airport. However, because different birds exhibit remarkably different behavior patterns, only a few are likely to create hazards. Past accidents involving large passenger aircraft on the U.S. East Coast indicate the particular hazard associated with gulls and Canada geese. Birds present on the airport are there because the facility provides a desirable environment for such natural requirements as food, shelter, safety, nesting, rest, and passage for migratory routes. Successful bird-strike control largely depends not on driving birds off but on creating an environment on the airport and in its immediate vicinity that is not attractive to birds in the first place.

Typical countermeasures will

1. Control garbage, especially the location of garbage dumps near the airport. Garbage-disposal dumps should not be located within 8 miles (13 km) of an airport.

2. Control of other food sources, such as insects, earthworms, and small mammals, by a variety of measures, such as poisons, insecticides, cultivation, and hunting to ensure that the open space at and near the airport does not encourage a food supply likely to be attractive to the troublesome species.

3. Eliminate as much as possible the occurrence of surface water that can form suitable habitat for water birds. Control measures include filling or draining areas or netting open-water areas.

4. Where possible, control farming in close proximity to operations. If open areas held by the airport for future expansion are leased as farmland, a ban on cereal cropping, for example, should be written into the leases.

5. Promote vegetation that discourages the presence of birds and avoid vegetation, such as trees, hedgerows, and berry-bearing shrubbery, that attracts birds.

6. Ensure that buildings in the airport area do not provide suitable nesting places for birds such as swallows, starlings, and sparrows that have become used to living in human-made environments.

Even where habitat-control measures have been taken to discourage some birds from being attracted to an airport, other birds might appear in significant numbers. It might become necessary to disperse and drive off birds using more active measures. These measures should not be used instead of habitat control because, given an attractive environment, when one bird is driven off, another is likely to take its place. Dispersing and expulsion techniques include

- Pyrotechnic devices (e.g., firecrackers, rockets, flares, shell crackers, live ammunition, gas cannons)
- Recorded distress calls
- Dead or model birds
- Model aircraft and kites
- Light and sounds of a disturbing nature
- Trapping
- Falcons
- Narcotics and poisons

If the presence of birds is a serious problem that threatens to disrupt the safe operation of the airport, the operator has no choice but to initiate a control program that will reduce the hazard to an acceptable level.

However, effective action cannot always be taken by the airport operator alone. Indeed, an effective control program must include all airside tenants and possibly property owners adjacent to the airport because these areas could be a source of bird or animal activity hazardous to air traffic. ICAO recommends that the larger community be organized to form a wildlife hazard committee that jointly considers and approves measures to reduce bird and other wildlife risk to an acceptable level.

5.4 Operational Areas

Pavement Surface Conditions

It is essential that the surfaces of pavements, especially runways, be kept as free as possible of contaminants and debris to ensure safe aircraft operations. A *contaminant* is defined as "a deposit (such as snow, ice, standing water, mud, dust, sand, oil, and rubber) on an airport pavement, the effect of which is detrimental to pavement braking conditions" (FAA 2010). *Debris*, on the other hand, refers to loose material such as sand, stone, paper, wood, metal, and pavement fragments that could be detrimental to operation by damaging aircraft structures or engines or by interfering with the operation of aircraft systems.

Potentially hazardous pavement conditions include deterioration of pavement, such as holes, depressions, cracking, and failed seals, which can produce debris with the potential to cause a fatal accident. For this reason, the FAA requires that pavement holes of more than 3 inches (7.6 cm) in depth be repaired immediately.

Especially since the introduction of jet aircraft, airport operators find that they must pay increasing attention to the available friction between runways and tires on landing and to the precipitant drag effect on takeoff. The danger of damage from debris also has increased owing to the higher speeds at takeoff and landing, the nature of the jet engine, and the danger of ingestion, especially for underslung engines. The seriousness of the problems involved is recognized by national airworthiness authorities, which routinely recommend that landing distances on wet runways be increased over those for dry conditions. Jet aircraft are also highly susceptible to the effect of precipitant drag, which occurs on slush- or water-covered runways, seriously affecting the ability of aircraft to obtain flying speed safely on takeoff. ICAO publishes special recommendations on operational measures dealing with the problem of taking off from slush- or water-covered runways. Ideally, the airport operator, although unable to keep a runway dry, would like to be in a position to keep the runway clear of contaminants and debris. However, snow, slush, and blowing sand might present conditions where operation will continue with less than optimal pavement surface conditions during a continuous clearing process. Therefore, procedures are set up to measure runway surface friction and the precipitant drag effect so that the pilots can adjust their techniques to existing conditions. In summary, the occasions under which assessment of the runway surface condition might be required include

- *The dry runway case*—infrequent measurement to monitor texture and wear and tear through the normal life of the runway

- *The wet runway case*—taking care to note the dramatic interaction of wet conditions with rubber deposits, which can result in a serious deterioration of the friction coefficient

- The presence of a significant depth of water and the possibility of aquaplaning

- The slippery runway case owing to the presence of ice, dry snow, wet snow, compacted snow, or slush, which reduce the coefficient of surface friction

- A significant depth and extent of slush, wet snow, or dry snow that can produce a significant level of precipitant drag

Because of the potential for runway excursions in extreme weather, especially overruns, the FAA and ICAO have recently advocated the extension of runway end-safety areas (RESA)—graded, obstacle-free zones to safely stop aircraft that have exited the runway end. Where available land is insufficient, specially designed systems can be installed to "capture" exiting aircraft with minimal damage and no loss of life.

At a very busy airport that frequently experiences conditions where braking might be impaired by contaminants, an adequate level of runway cleaning equipment must be maintained. Equipment also must be available to check the results of cleaning by measuring friction and drag. Rubber accumulation is inevitable on all active runways. Appendix 2 of Part 2 of the ICAO *Airport Services Manual* contains an inspection guide for the visual estimate of rubber deposits accumulated on the runway (ICAO 2002). The time interval between rubber buildup assessments depends on factors such as air traffic volumes (frequency and type of aircraft), climatic conditions, pavement type, and the pavement's service and maintenance requirements. ICAO recommends that airports with more than 210 landings per week should conduct detailed inspections weekly. If rubber contamination is excessive, the rubber can be removed by a variety of methods. These include high-pressure water cleaning, chemical removal, and even sand blasting (with vacuum containment to prevent debris). Each method has advantages and disadvantages; the best practice for each airport will depend on such factors as the condition of the pavement, the ability to control runoff (if water is used), and the environmental effects of the chemicals used.

At a less busy airport, where conditions of impaired braking are experienced only infrequently, but where operations must continue despite inadequate cleaning equipment, assessment of runway friction is essential, and equipment for measuring these effects must be available to enable pilots to adjust their operations to the existing conditions. At an even less important airport where operations can be suspended, it is essential to have equipment to assess runway friction to be able to make a decision on when conditions have reached the point where suspension of activities is necessary. It is important to remember that even where the removal of snow and ice is given high priority, there is frequently a significant loss of friction on an apparently dry, cleared runway. At airports that regularly experience heavy snowfalls, for example, in northern Europe and North America, clearance might have to be discontinued for a short while during a storm to permit some operations to continue. Runways are unlikely, in such conditions, to be completely clean. There are also likely to be local slippery patches. The airport authority will need to measure and assess surface conditions to inform pilots of the overall condition and to determine the areas requiring more cleaning treatment.

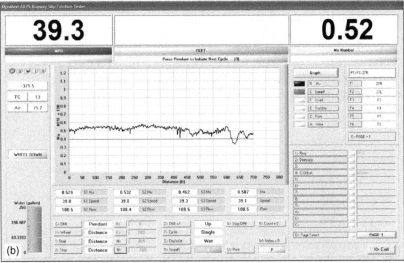

FIGURE 5.4   (a) Instrumented pickup truck with retractable fifth wheel. (*Dynatest.*)
(b) PC readout of Dynatest runway friction test. (*Dynatest.*)

Various types of friction testing equipment are available. Several
versions are small trailers with a measuring device (Mu Meter) that is
attached behind a towing vehicle. Figure 5.4*a* shows a truck with a
retractable fifth wheel that performs the friction test. An illuminated
control panel by the driver gives readings at the same time as a record

Measured/Calculated Coefficient	Estimated Braking Action
0.40 and above	Good
0.39–0.36	Medium to good
0.35–0.30	Medium
0.29–0.26	Medium to poor
0.25 and below	Poor

Source: FAA (2008).

TABLE 5.4   Relationship Between Coefficient of Friction and Braking Efficiency

is made of the coefficient of friction with a corresponding estimated braking action. Figure 5.4*b* shows the format of the visual display of the record of the runway surface condition. For further descriptions of such equipment, the reader is referred to FAA (1991, 2008) and CAA (2010b).

Table 5.4 indicates the relationship between the coefficient of friction and the subjective estimate of braking efficiency. It is quite possible for a thin film of ice to reduce the coefficient of friction on an aircraft pavement from 0.50 to 0.15, reducing the braking efficiency to less than a third of that in the dry condition.

Recognizing that an airport authority must be in a position to evaluate the level of runway friction, assessment of pavement condition should never take precedence over the clearance operations themselves. Within the operational areas, safety and efficiency require observing the following clearance priorities for the various areas involved:

- Runways
- Taxiways
- Aprons
- Holding bays
- Other areas

Snow clearance is frequently coordinated through the operation of the *snow committee,* consisting of members from the airline operators, meteorology, air traffic system (ATS) services, and airport administration. Clearance is laid down in a snow plan that ensures that agreed-on procedures exist for the provision and maintenance of equipment; for clearance according to stated priorities; for installation of runway markers, snow fencing, and obstruction marking; and for providing for maneuvering aircraft. For further details on snow clearance, the reader is referred to FAA (1991, 1992, 2008), which indicate the availability and uses of such specialized equipment as the snow blower shown in Figure 5.5. While snow conditions occur for only a limited period during the year, airports might turn to outside

Figure 5.5 Snow blower. (*Oshkosh.*)

contractors to provide snow clearance services, having the equipment moved onto site only prior to the snow season.

Debris presents a separate and different problem at airports. Jet turbine engines are extremely susceptible to damage from ingestion of solid particles of debris picked up from the pavement surfaces. Tire life is also reduced by wear and cuts induced by sharp objects on the pavements, deteriorating pavement surfaces and edges, and poor, untreated pavement joints. Damage also can occur to the skin of aircraft from objects thrown up from the pavement. This is precisely what occurred on July 25, 2000, to Air France Flight 4590, the Concorde flight from Paris Charles de Gaulle International Airport to JFK International Airport in New York. During the takeoff run, the Concorde suffered a ruptured tire from a piece of debris that had fallen off an aircraft that had just departed from the same runway. The tire debris struck the underside of the wing, eventually causing a rupture to a fuel tank, leading to a fire and a fatal crash.

Problems arising from debris can be reduced by regular inspections of the pavement surface condition of all operational areas and by establishing a sweeping and cleaning program that sets up priorities and frequencies. In order to help designate the particular location of debris, a plan of the paved areas should be divided into manageable paved segments of approximately 1,640 feet (500 m) (ICAO 1984). Descriptions of these pavement sections can be entered into a database that keeps a record of inspection results, current condition, and any operational restrictions in a pavement management system (PMS). A PMS also can perform cost-benefit analyses to recommend when and what kind of repairs/replacement should take place.

Runway inspections are almost always conducted in vehicles owing to their extensive area and evacuation considerations. These vehicles should travel at the lowest velocity possible in the opposite direction of takeoff or landing for air traffic safety. Increasingly, busy airports are using remote sensing to detect the presence of debris, wildlife, and defects on the operational airside pavements. The FAA has approved the use of certain technologies, including radar- and video-based systems. An example of such a system is the Singapore Changi iFerret automated FOD detection system using a vision-based system linked to air traffic control (ATC) and ground operations control. In apron areas, the cleanliness of the airline operators and other users determines the amount of litter and debris that is present. Cargo areas are particularly susceptible to the presence of fragments of strapping, nails, and container and pallet debris. The problem can be reduced by careful adherence to a disciplined program of maintaining litter-free pavements. Runways, taxiways, and holding bays are subject to debris eroded from the pavement shoulders. This kind of debris can be reduced by shoulder-sealing treatments, which might be necessary on highly trafficked facilities (FAA 2007). Removal of debris is achieved by powered mobile brooms, vacuum and compressed-air sweepers, and magnetic cleaners. As a further incentive, some airports provide brightly painted litter bins adjacent to aircraft gates/parking bays in which any litter or debris found in the aircraft parking areas can be deposited. The airport operator must strive to instill a culture of safety in all personnel who have access to airside facilities.

5.5   Airfield Inspections

First signs of airfield unserviceability might well arise as a result of routine daily inspections, which are usually the responsibility of the airport owner. ICAO (2001) requires that airport operators document several kinds of inspection and maintenance activities on the airside:

- Inspection of the movement area and obstacle-limitation surface by the airport operator, including runway friction and water-depth measurements on runways and taxiways

- Inspection procedures for aeronautical lights (including obstacle lighting), signs, markers, and the electrical system (including secondary power supplies)

- Procedures for monitoring obstacle-limitation surfaces, controlling obstacles under the authority of the operator, and notifying the civil aviation authority of any changes

- Protection of sites for radar and navigational aids [This includes control of activities (such as inspections and ground maintenance) near sensitive installations and placement of warning signs to ensure that the performance of these sites is not be degraded.]

All these inspections require immediate remedial action to address any serious deficiencies discovered.

ICAO provides some suggestions on the appropriate frequency of movement-area inspections (ICAO 1983):

- *Runways.* Four inspections per day

- *First light.* Detailed, full-width inspection of approximately 15 minutes per runway (two passes)

- *Morning.* Inspection of all runways and a detailed shoulder inspection

- *Afternoon.* Same as morning

- *Nightfall.* Complete runway surface inspection (fills gap between earlier inspections and lighting inspections)

- *Taxiways.* Daily inspections for those in use

- *Aprons.* Daily inspections

- *Infields.* Inspections to verify height of grass, cleanliness, and drainage according to maintenance schedule

It is absolutely vital that radiotelephone (RT) contact and liaison between ATC and the vehicle being used for inspection be fault-free because the inspection vehicle will be moving on the operating area, which includes runways and taxiways, where there might be a danger of collision with aircraft. The vehicle should be driven slowly enough for a thorough visual inspection to be made and, if necessary, stopped for a closer inspection of a particular area or for the removal of debris. During the inspection, particular attention should be given to

- Surface condition of runways, taxiways, runway holding points, stopways, and clearways

- Presence of standing water, snow, ice, sand, rubber deposits from aircraft tires, and oil or fuel spillage

- Presence of debris on any part of the movement area

- Status of any work in progress on the airport and the presence of any associated materials, obstructions, ditches, and so on, together with the correct ground-hazard warnings

- Condition of daylight movement/indicator boards and signs, including boundary markers

- Damage to light fittings, broken glass, and so on

- Growth of grass/plants or any other causes of lights being obscured

- Any congregation of birds or presence of animals or unauthorized persons likely to interfere with operations

In the event of any objects being discovered that are identified as coming from an aircraft, immediate steps would need to be taken to check recent departures with ATC, who will decide if it is necessary to send a message. Prior to darkness, an inspection to check the operation of all lighting systems should be carried out if night operations are to take place. This inspection will be particularly aimed at an examination of

- Runway/taxiway lighting
- Obstruction lights
- Airport rotating beacon/identification beacon
- Traffic lights guarding the operating/movement area
- Visual approach slope indicators (VASIS)
- Approach lights visible from within the airport boundaries

A full approach lighting inspection can be carried out only by means of a flight check. With larger lighting systems, some form of photographic record has to be made.

In view of the variety and complexity of all those aspects of an airport that need to be inspected, it is advisable to use a checklist, together with a map of the airport, so that a systematic record can be made of the results of the inspection. An example of a suggested checklist for an airfield inspection is given in Figure 5.6. Each airport will need to develop lists appropriate to its own particular circumstances. If any unserviceability is discovered as a result of an inspection and the condition cannot be rectified immediately, then suitable *Notices to Airmen* (NOTAM) action has to be taken (see Section 11.6 in Chapter 11).

5.6 Maintaining Readiness

Maintenance Management

The primary concern of operations management is to ensure the continuous availability of all operational services. To achieve this, a systematic approach to maintenance management is called for, the extent of which will depend on the types of operations at a particular airport. Clearly, a major air carrier hub will require a vastly more complex maintenance program than an airport dealing only with general aviation (GA) types of operations. There are two essential aspects, however, that will be common to any airport maintenance program:

- A documented schedule of routine maintenance
- A comprehensive system of maintenance records, including costs

Regularly Scheduled Inspection Checklist

| Date: _____ | Day: _____ | ✓ Satisfactory
× Unsatisfactory |
| Time: _____ | Inspector: _____ | |

Facilities	Conditions	✓	Remarks
Pavement areas	Pavement lip over 3"		
	Hole 5" diam. 3" deep		
	Cracks/spalling/bumps		
	FOD: Gravel/debris/etc.		
	Rudder deposits		
	Ponding/edge dams		
Safety areas	Ruts/humps/erosion		
	Drainage/construction		
	Objects/frangible bases		
Markings and signs	Visible/standard		
	Hold lines/signs		
	Frangible signs		
Lighting	Obscured/dirty/faded		
	Damaged/missing		
	Inoperative		
	Faulty aim/adjustment		
Navigational aids	Rotating beacon		
	Wind indicators		
	REILs/VGSI systems		
Obstructions	Obstruction lights		
	Cranes/trees		
Fueling operations	Fencing/gates/signs		
	Fuel marking/labeling		
	Fire extinguishers		
	Grounding clips		
	Fuel leaks/vegetation		
Snow & ice	Surface conditions		
	Snowbank clearance		
	Lights & signs obscured		
	NAVAIDS/fire access		
Construction	Barricades/lights		
	Equipment parking		
ARFF	Equipment/crew avail.		
	Communications/alarm		
Public protection	Fencing/gates		
	Signs		
Wildlife hazards	Dead birds		
	Flocks of birds/animals		

FIGURE 5.6 Regularly scheduled inspection checklist.

Many of the airport facilities, such as radio communications, radio and radar approach aids, and airfield lighting, are of such critical importance to flight safety that every effort has to be made to ensure that failures do not occur. Among the elements involved are

- Radio communications (air/ground) transmitters and receivers
- Aeronautical fixed telecommunications
- Telephones
- Approach and landing aids
- Radio/radar
- Lighting
- Fire and rescue services
- Aircraft movement areas
- Power plant and distribution system

Preventive Maintenance

The process of preventive maintenance is concerned mainly with regular inspection of a system and all its component parts with the objective of detecting anything likely to lead to a component or system failure and taking appropriate action to prevent that happening. Such action might involve cleaning or replacing parts on a predetermined schedule. Whatever the action called for, this cannot be determined in the first place unless a planned inspection schedule is established. An example of a preventive maintenance schedule for medium-intensity approach lighting is given in Table 5.5. Runway lighting is somewhat simpler to maintain, but the same systematic inspection is required (Table 5.6). Centerline and touchdown-zone lighting will, of course, be much more vulnerable to damage as a result of being run over by aircraft. The fact that they are located below ground level also makes them vulnerable to water infiltration.

Electrical Maintenance

There are few systems on an airport that do not depend in one way or another on electrical power. Indeed, the power requirements of modern airports are equal to those of small towns. Operational facilities, especially those concerned with aviation technical services, make heavy demands on the public power supply. Standby power must be available to provide a secondary power supply to these essential services in the event of breakdown of the main supply. The arrangements for a secondary power supply depend in part on the switchover-time requirements, that is, the time interval between loss of power and the availability of a secondary supply (FAA 1986). This could be critical in the case of precision-approach aids or lighting, as indicated in Tables 5.7 (FAA 2009) and 5.8 (ICAO 2006), respectively. The demands

Maintenance Requirement	Daily	Weekly	Monthly	Bimonthly	Semiannually	Annually	Unscheduled
1. Check for burned-out lamps	X						
2. Check system operation		X					
3. Replace burned-out lamps		X					
4. Check semiflush lights for cleanliness		X					
5. Record input and output voltages of control cabinet			X				
6. Clear any vegetation obstructing the lights			X				
7. Check angle of elevation of lights					X		
8. Check structures for integrity					X		
9. Check approach area for new obstructions					X		
10. Check photoelectric controls (if used)					X		
11. Check electrical distribution equipment						X	
12. Check insulation resistance of cable						X	
13. Check fuse holders, breakers, and contacts						X	
14. Replace all PAR 38 lamps after 1,800 hours on maximum intensity							X

Source: FAA (2009).

TABLE 5.5 Preventive Maintenance Inspection Schedule for Medium-Intensity Approach Lighting

Maintenance Requirement	Daily	Weekly	Monthly	Bimonthly	Semiannually	Annually	Unscheduled
1. Check for burned-out or dimly burning lights	X						
2. Repair or replace defective lights		X					
3. Clean lights with dirty lenses			X				
4. Check the intensity of selected lights			X				
5. Check the torque of mounting bolts				X			
6. Clean and service light; check electrical connections					X		
7. Check for water in the light base					X		
8. Remove lamps after 80 percent of service life							X
9. Remove snow from around fixtures							X
10. Check wires in saw cuts							X

Source: FAA (2009).

TABLE 5.6 Preventive Maintenance Inspection Schedule for Centerline and Touchdown-Zone Lighting Systems

Runway	Aids Requiring Power	Maximum Switchover Time
Noninstrument	Visual approach slope indicators	2 minutes
		2 minutes
	Runway edge	2 minutes
	Runway threshold	2 minutes
	Runway end	2 minutes
	Obstacle	
Instrument approach	Approach lighting system	15 seconds
	Visual-approach slope indicators	15 seconds
		15 seconds
	Runway edge	15 seconds
	Runway threshold	15 seconds
	Runway end	15 seconds
	Obstacle	
Precision approach Category I	Approach lighting system	15 seconds
	Runway edge	15 seconds
	Runway threshold	15 seconds
	Runway end	15 seconds
	Essential taxiway	15 seconds
	Obstacle	15 seconds
Precision approach Category II	Approach lighting system	1 second
	Runway edge	15 seconds
	Runway threshold	1 second
	Runway end	1 second
	Runway centerline	1 second
	Runway touchdown zone	1 second
	Essential taxiway	15 seconds
	Obstacle	15 seconds
Precision approach Category III	(Same as Category II)	(Same as Category II)

Source: ICAO (2006).

TABLE 5.7   Recommended Switchover Times in the Event of Power Failure: Lighting

range from a maximum permitted interruption of 15 seconds to zero or completely uninterrupted supply, as in the case of an ILS localizer and glide slope for Category II and III approaches. The source of secondary supply usually is one or more diesel-driven generators. In the case of a zero-switchover-time requirement, the arrangement is to have these facilities supplied by a generator with a coupled energy-storage flywheel.

Type of Runway	Aids Requiring Power	Maximum Switchover Times
Instrument approach	SRE	15 seconds
	VOR	15 seconds
	NDB	15 seconds
	D/F facility	15 seconds
Precision approach, Category I	ILS localizer	10 seconds
	ILS glide path	10 seconds
	ILS middle marker	10 seconds
	ILS outer marker	10 seconds
	PAR	10 seconds
Precision approach, Category II	ILS localizer	0 seconds
	ILS glide path	0 second
	ILS inner marker	1 second
	ILS middle marker	1 second
	ILS outer marker	10 seconds
Precision approach, Category III	(Same as Category II)	(Same as Category II)

Source: ICAO (2006).

TABLE 5.8   Recommended Switchover Times in the Event of Power Failure: Ground-Based Radio Aids

The generator is driven by an electric motor. In the event of a main power failure, the generator derives the required driving power from the flywheel until the coupled standby diesel generator takes over the full load. To further safeguard the remote possibility of the generator failing, a second generator is coupled in parallel with the first.

Such stringent requirements as these call for an appropriate level of maintenance and, alongside this, a suitable level of workforce. The requirements for maintenance personnel in a typical airport electrical shop (Category II airport) are listed in Table 5.9.

Operational Readiness: Aircraft Rescue and Firefighting

There is one essential element of the operating system—the rescue and firefighting service (RFFS)—where maintaining readiness applies as much to personnel as to machines. Opportunities for RFFS personnel to carry out their assigned tasks under "real" conditions fortunately are rare; as a result, they can maintain readiness to deal with an aircraft accident only by constant practice. It is especially difficult to maintain a peak of performance under these circumstances, and it is

Workshop/ Office	High-Voltage Plant	Secondary Power Supply	Lighting Systems	Low-Voltage 24-h Service
1 manager	1 foreman	1 foreman	1 foreman	1 foreman
3 clerks	5 electricians	5 fitters	5 electricians	41 electricians
			3 fitters	15 fitters
			1 driver	

Source: Frankfurt Airport.

TABLE **5.9**   Personnel Requirements in a Typical Electrical Shop of a Category II Airport

for this reason that facilities should be provided for "hot fire" practices and, also if possible, for practices in smoke-filled confined spaces, preferably simulated aircraft interiors.

Some airports employ independent aircraft and firefighting services. It will be important in these cases to carry out occasional tests of the ability of the personnel and equipment to meet the required performance criteria (see Chapter 12) and in particular to test their communications and coordination procedures.

Safety Aspects

Whether maintaining electrical or mechanical systems, the nature of maintenance work exposes those who carry it out to certain risks, including natural phenomena such as lightning strikes while working out on the airfield. A comprehensive set of guidelines on dealing with risks should be drawn up by management. The nature of these risks can be indicated by an examination of some of the common causes of accidents:

- Working on equipment without adequate coordination with equipment users
- Working on equipment without sufficient experience on that equipment
- Failure to follow instructions in equipment manuals
- Failure to follow safety precautions
- Using unsafe equipment
- Failure to use safety devices
- Working at unsafe speeds
- Poor housekeeping of work areas

The FAA issues guidelines on continual checking of the safety conditions at an airport through its program of Airport Safety Self-Inspection (FAA 2004). Where safety-related matters require action by the operations department of the airport company or authority, these actions will be coordinated through the airport operations control center (see Chapter 16).

Most air carrier airports will have a multiplicity of electrical and mechanical systems with large amounts of associated equipment requiring continuous checking and servicing by skilled maintenance personnel. The extent of this work will very much depend on the amount and type of aircraft operations, the weather categories in which the airport operates, and the number of passengers, visitors, and staff using the airport.

With the increasing sophistication (and complication) of the equipment used at airports, the use of automatic equipment/system monitoring to provide prompt warning of equipment failures, once limited to certain airport operational facilities (e.g., ILS, lighting), is now a commonplace application for all key installations. These systems also can provide comprehensive performance records and interface with other airport control systems, as well as logistics systems for maintenance work orders and spare-part availability. But a vital element for achieving the exacting state of readiness needed in air transport is for airports to have available the resources/workers and materials to enable a rapid and effective response to any deficiencies in the airport's operating infrastructure.

Airfield Construction

Despite good maintenance programs, the airport operator will find that major airside repairs will be necessary at some time, which inevitably will introduce construction activities to the movement area. Obviously, the visual guidance systems on the airside, which ensure the safe and expeditious flow of aircraft, vehicles, and personnel, are not designed to accommodate construction activities. Significant safety planning is an absolute must.

Construction on or near the movement area is inherently dangerous because it introduces several additional risk factors to an already challenging environment from the point of view of safety management. Factors that increase risk begin first simply with the transit of additional personnel and vehicles in and out of the movement area through temporary access points—with the concomitant risks of unauthorized access of personnel and additional vehicles with potential for runway incursions. Second, construction frequently changes the circulation of both vehicles and aircraft in the movement area; thus other operators, accustomed to the original pattern of movement, have to adapt to a change in the environment. Third, the introduction of untrained personnel to the movement area, unfamiliar with the rules and risks of their new environment, creates serious control challenges. Fourth, construction equipment and barriers create obstacles that require marking, lighting, and monitoring.

Fifth, construction activities themselves create dust, debris, and noise, which can adversely affect the safety of the movement area.

Adequate management of these risks begins with a good contract-procurement system that ensures that contractors who are awarded work on or near the movement area have good safety-management programs, proper equipment, and insurance policies appropriate to the risk of working airside. The next step is to implement a thorough airport permit system based on the satisfaction of certain safety requisites. No work should begin without the knowledge of ATC and apron management. In cases where airside construction work goes beyond what would be considered normal maintenance activities and affects air traffic, publication of a NOTAM likely will be required.

The permit application should be accompanied with a detailed work plan, with precise movement routes, communication procedures with ATC and apron control, evacuation procedures, scheduled briefings, construction inspections, turnover procedures, and control measures (including equipment inspections, cleaning, dust control, and obstacle marking). In addition, the work should be planned to coincide with periods of lower activity. Avoid peak periods, if possible.

The submitted work plan will be subjected to a risk analysis by the aerodrome operator's safety management team. Safety management systems cover the main elements of risk management (i.e., hazard identification and mitigation), including some obvious safety measures that should be taken into account (see also Chapter 16):

- Isolation of the work area with correctly marked barriers
- Use of reflective vests by construction personnel, in addition to normal personal protective gear
- Daily safety briefings of personnel to ensure that they are aware of work limits, authorized access routes, communication procedures, and inherent dangers on the apron (e.g., jet blast, FOD, and collisions)
- Assignment of a works coordinator, in constant communication with ATC and apron control
- Vehicle inspections to ensure serviceability, sufficient fuel for evacuation, proper markings, and absence of FOD in tires
- Protection of loose material from wind and rain erosion
- Strict adherence to authorized work hours
- Take into consideration security aspects linked to outside workforce

All these controls should be detailed in the work plan and implemented during construction.

The airport operator will have to inspect daily for compliance. Work progress should be monitored as well by the engineering or maintenance department to ensure that required changes to the work hours or extension of the work period is anticipated with enough advance warning to gain approval and properly inform ATC, apron control, and operators using the movement area.

As a final consideration, while all work in or near the maneuvering area presents a serious risk that must be managed, work within the runway strip is subject to special restrictions. Annex 14 divides the obstacle-free zone surrounding the runway strip into three zones. While the limits vary according to airport classification, the following are typical:

- *197 feet (60 m) from the runway centerline*—very limited work area [30 to 92 square feet (9–28 m²)], obstacle height restricted to 3.3 feet (1 m)

- *197 to 246 feet (60–75 m) from the runway centerline*—work area not restricted, obstacle height restricted to 6.6 feet (2 m)

- *Beyond 246 feet (75 m) from the runway centerline*—no work-area limits

If construction work that cannot comply with these restrictions must go forward, a runway closure likely will be required.

Conclusion

The most critical task of an airport operator is to ensure the safe, reliable, and expeditious movement of aircraft, passengers, and cargo through the airside to landside and vice versa. A systematic approach is necessary to ensure the operational readiness of the critical facilities that make such movement possible. ICAO, as well as several Member States, has developed certification criteria that describe the minimum standards for operation of these facilities. For further information, the reader is urged to refer directly to ICAO Document 9774, *Manual on Certification of Aerodromes*.

References

Civil Aviation Authority (CAA). 2010a. *Air Navigation: The Order and the Regulation* (CAP 393). London: CAA.

Civil Aviation Authority (CAA). 2010b. *The Assessment of Runway Surface Friction Characteristics* (CAP 683). London: CAA.

CFR 2004: *Airport Certification. Code of Federal Regulations*, Part 139. Washington, DC: Office of the Federal Register, National Archives, and Records Administration.

Federal Aviation Administration (FAA). 1986. *Standby Power for Non-FAA Airport Lighting Systems* (AC150/5340-17B). Washington, DC: FAA, Department of Transportation.

Federal Aviation Administration (FAA). 1989. *Airport Design* (AC150/5300-13, changes 1-13). Washington, DC: FAA, Department of Transportation.

Federal Aviation Administration (FAA). 1991. *Runway Surface Condition Sensor Specification Guide* (ACJ50/5220-13B). Washington, DC: FAA, Department of Transportation.

Federal Aviation Administration (FAA). 1992. *Airport Snow and Ice Control Equipment* (AC150/5220-20). Washington, DC: FAA, Department of Transportation.

Federal Aviation Administration (FAA). 2004. *Airport Safety Self-Inspection* (ACI5015200-18C). Washington, DC: FAA, Department of Transportation.

Federal Aviation Administration (FAA). 2007. *Guidelines and Procedures for Maintenance of Airport Pavements* (AC150/5380-6B). Washington, DC: FAA, Department of Transportation.

Federal Aviation Administration (FAA). 2008. *Airport Winter Safety and Operations* (ACI50/5200-30C). Washington, DC: FAA, Department of Transportation.

Federal Aviation Administration (FAA). 2010. *Airport Foreign Object Debris (FOD) Management* (AC150/5210-24). Washington, DC: FAA, Department of Transportation.

International Civil Aviation Organization (ICAO). 1983. *Airport Operational Services* (Document 9137) in *Airport Services Manual,* 1st ed. Montreal, Canada: ICAO.

International Civil Aviation Organization (ICAO). 1984. *Airport Maintenance Practices* (Document 9137) in *Airport Services Manual,* Part 9, 1st ed. Montreal, Canada: ICAO.

International Civil Aviation Organization (ICAO). 2001. *Manual on Certification of Aerodromes* (Document 9774) in *Airport Services Manual,* 1st ed. Montreal, Canada: ICAO.

International Civil Aviation Organization (ICAO). 2002. *Pavement Surface Conditions* (Document 9137) in *Airport Services Manual,* Part 2, 4th ed. Montreal, Canada: ICAO.

International Civil Aviation Organization (ICAO). 2010. Annex 14: *Aerodromes,* Vol. 1: *Aerodrome Design and Operations,* 5th ed. Montreal, Canada: ICAO.

International Civil Aviation Organization (ICAO). 2011. *Bird Control and Reduction* (Document 9137) in *Airport Services Manual,* Part 3, 4th ed. Montreal, Canada: ICAO.

CHAPTER 6

Ground Handling

6.1 Introduction

The passenger and cargo terminals have been described as interface points between the air and ground modes (Ashford et al. 2011). The position of the terminals within the general system has been shown conceptually in Figure 1.3, and the actual flows within the terminals in the more detailed system diagrams of passenger and cargo terminals are shown in Figures 8.1 and 10.7, respectively. Within the context of these diagrams, the movement of passengers, baggage, and cargo through the terminals and the turnaround of the aircraft on the apron are achieved with the help of those involved in the ground handling activities at the airport (IATA 2012). These activities are carried out by some mix of the airport authority, the airlines, and special handling agencies depending on the size of the airport and the operational philosophy adopted by the airport operating authority. For convenience of discussion, ground handling procedures can be classified as either terminal or airside operations. Such a division, however, is only a convention in that the staff and activities involved are not necessarily restricted to these particular functional areas. Table 6.1 lists the airport activities normally classified under ground handling operations. The remainder of this chapter deals with these activities, but for convenience, the major areas of baggage handling, cargo, security, and load control have been assigned to other chapters to permit a more extensive discussion of these items.

6.2 Passenger Handling

Passenger handling in the terminal is almost universally entirely an airline function or the function of a handling agent operating on behalf of the airline. In most countries of the world, certainly at the major air transport hubs, the airlines are in mutual competition. Especially in the terminal area, the airlines wish to project a corporate image, and passenger contact is almost entirely with the airline, with the obvious exceptions of the governmental controls of health, customs, and immigration. Airline influence is perhaps seen at its

Terminal

Baggage check
Baggage handling
Baggage claim
Ticketing and check-in
Passenger loading/unloading
Transit passenger handling
Elderly and disabled persons
Information systems
Government controls
Load control
Security
Cargo

Airside

Ramp services
Supervision
Marshaling
Startup
Moving/towing aircraft
Safety measures
On-ramp aircraft servicing
Repair of faults
Fueling
Wheel and tire check
Ground power supply
Deicing
Cooling/heating
Toilet servicing
Potable water
Demineralized water
Routine maintenance
Nonroutine maintenance
Cleaning of cockpit windows, wings, nacelles, and cabin windows
Onboard servicing cleaning
Catering
In-flight entertainment
Minor servicing of cabin fittings
Alteration of seat configuration
External ramp equipment
Passenger steps
Catering loaders
Cargo loaders
Mail and equipment loading
Crew steps on all freight aircraft

TABLE 6.1   The Scope of Ground Handling Operations

extreme in the United States, where individual airlines on occasion construct facilities (e.g., the old United terminal and the new Jet Blue terminals at New York JFK). In these circumstances, the airlines play a significant role in the planning and design of physical facilities that they will operate. Even where there is no direct ownership of facilities, industry practice involves the designation of various airport facilities that are leased to the individual airlines operating these areas. Long-term designation of particular areas to an individual airline results in a strong projection of airline corporate image, particularly in the ticketing and check-in areas and even in the individual gate lounges (Figure 6.1).

Figure 6.1   Airline-designated check-in area.

A more common arrangement worldwide is for airlines to lease designated areas in the terminal, but to have a large proportion of the ground handling in the ramp area carried out by the airport authority, a special handling agency, or another airline. At a number of international airports, the airline image is considerably reduced in the check-in area when common-user terminal equipment (CUTE) is used to connect the check-in clerk to the airline computers. Use of the CUTE system can substantially reduce the requirements for numbers of check-in desks, particularly where there is a large number of airlines and some airlines have very light service schedules or the airline presence is not necessary throughout the whole day. Desks are assigned by resource managers on a need basis. Check-in areas are vacated by one airline and taken up by another based on departure demand. The airline's presence at check-in desks is displayed on overhead logo panels that are activated when an airline logs onto the CUTE system (Figure 6.2). Common Use Self Service or CUSS is a shared kiosk offering check-in facilities to passengers without the need for ground staff. The CUSS kiosks can be used by several participating airlines in a single terminal.

The airside passenger-transfer steps (Figure 6.3) and loading bridges (Figure 6.4) might be operated by the airline on a long-term leasing arrangement or by the airport authority or handling agency at a defined hiring rate to the airlines. With the advent of very large aircraft (e.g., the A380), multiple loading bridges are required to cope

FIGURE 6.2   Computer-assigned CUTE passenger check-in desks at Munich Airport.

Figure 6.3   Airline passenger steps.

Figure 6.4   Elevating passenger air bridge.

Figure 6.5   Three-loading-bridge configuration serving an A380.

with passenger flows to and from a single aircraft (Figure 6.5). They require experienced handling, but even these are normally operated by the airlines.

Apron passenger-transfer vehicles are usually of the conventional bus type. Both airline and airport ownership and operation are common, airline operation being economically feasible only where the carrier has a large number of movements. Figure 6.6 shows a typical airport-owned apron bus. Where a more sophisticated transfer vehicle, such as the mobile lounges shown in Figure 6.7 are used, it is usual for the operation to be entirely in the hands of the airport authority.

6.3   Ramp Handling

During the period that an aircraft is on the ground, either in transit or on turnaround, the apron is a center of considerable activity (IATA 2004). Some overall supervision of activities is required (ICAO 2010) to ensure that there is sufficient coordination of operations to avoid unnecessary ramp delays. This is normally carried out by a ramp coordinator or dispatcher who monitors departure control. Marshaling is provided to guide the pilot for the initial and final maneuvering of the aircraft in the vicinity of its parking stand position. In the delicate

FIGURE 6.6   Apron passenger transport bus.

FIGURE 6.7   Mobile lounge for passenger transport across the apron.

process of positioning the aircraft, the pilot is guided by internationally recognized hand signals from a signalperson positioned on the
apron (Figure 6.8). Where nose-in docking is used next to a building,
self-docking guides such as the Aircraft Parking and Information System (APIS) using optical moiré technology or the Docking Guidance
System (DGS) using sensor loops in the apron pavement enabling the
pilot to bring the aircraft to a precise location to permit the use of
loading bridges (Ashford et al. 2011). Marshaling includes the positioning and removal of wheel chocks, landing-gear locks, engine
blanking covers, pitot covers, surface control locks, cockpit steps, and

SIGNALMAN'S
POSITION

DAYLIGHT   DARKNESS
TO ATTRACT
OPERATORS
ATTENTION

DAYLIGHT   DARKNESS
STRAIGHT
AHEAD

DAYLIGHT   DARKNESS
TURN TO
RIGHT

DAYLIGHT   DARKNESS
REMOVE
CHOCKS

FIGURE 6.8   Ground signalman marshalling an aircraft. (*Courtesy: IATA.*)

tail steadies. Headsets are provided to permit ground-to-cockpit communication, and all necessary electrical power for aircraft systems is provided from a ground power unit. When the aircraft is to spend an extended period on the ground, the marshaling procedure includes arranging for remote parking or hangar space.

The ramp handling process also includes the provision, positioning, and removal of the appropriate equipment for engine starting purposes. Figure 6.9 shows an engine air-start power unit suitable for providing for a large passenger aircraft.

Safety measures on the apron include the provision of suitable firefighting equipment and other necessary protective equipment, the provision of security personnel where required, and notification of the carrier of all damage to the aircraft that is noticed during the period that the aircraft is on the apron.

Frequently there is a necessity for moving an aircraft, requiring the provision and operation of suitable towing equipment. Tow tractors might be needed simply for pushing out an aircraft parked in a nose-in position or for more extensive tows to remote stands or maintenance areas. Figure 6.10 shows a tractor suitable for moving a large passenger aircraft. It is normal aircraft-design practice to ensure that undercarriages are sufficiently strong to sustain towing forces without structural damage. Tow tractors must be capable of moving aircraft at a reasonable speed [12 mi/h (20 km/h) approximately] over considerable taxiway distances. As airports grow larger and more decentralized in layout, high-speed towing vehicles capable of operating in excess of 30 mi/h (48 km/h) have been developed, although

Figure 6.9 Mobile apron engine air-start vehicle.

Figure 6.10 Aircraft tow tractor.

speeds of 20 mi/h (32 km/h) are more common. Usually aircraft that are being towed have taxiway priority once towing has started. Therefore, reasonable tow speeds are necessary to avoid general taxiing delays.

6.4   Aircraft Ramp Servicing

Most arriving or departing aircraft require some ramp services, a number of which are the responsibility of the airline station engineer. When extensive servicing is required, many of the activities must be carried out simultaneously.

Fault Servicing

Minor faults that have been reported in the technical log by the aircraft captain and that do not necessitate withdrawal of the aircraft from service are fixed under supervision of the station engineer.

Fueling

The engineer, who is responsible for the availability and provision of adequate fuel supplies, supervises the fueling of the aircraft, ensuring that the correct quantity of uncontaminated fuel is supplied in a safe manner. Supply is either by mobile truck (bowser; Figure 6.11) or

Figure 6.11   Mobile apron fuel tanker (bowser).

Figure 6.12   Mobile aircraft fuel dispenser for fueling from apron hydrant system.

from the apron hydrant system (Figure 6.12). Many airports use both systems to ensure competitive pricing from suppliers and to give maximum flexibility of apron operation. Oils and other necessary equipment fluids are replenished during the fueling process.

Wheels and Tires

A visual physical check of the aircraft wheels and tires is made to ensure that no damage has been incurred during the last takeoff/landing cycle and that the tires are still serviceable.

Ground Power Supply

Although many aircraft have auxiliary power units (APUs) that can provide power while the aircraft is on the ground, there is a tendency for airlines to prefer to use ground electrical supply to reduce fuel costs and to cut down apron noise. At some airports, the use of APUs is severely restricted on environmental grounds. Typically, ground power is supplied under the supervision of the station engineer by a mobile unit. Many airports also can supply power from central power supplies that connect to the aircraft either by apron cable (Figure 6.13) or by cable in the air-bridge structure.

FIGURE 6.13   Apron cable electrical supply.

Deicing and Washing

Figure 6.14 shows a typical multiuse vehicle suitable for spraying the fuselage and wings with deicing fluid and for washing the aircraft, especially the cockpit windows, wings, nacelles, and cabin windows. This self-propelled tanker unit provides a stable lift platform for spraying or for various maintenance tasks on conventional and wide-bodied aircraft. Apron drainage facilities must permit the recapture and recycling of deicing fluid (ICAO 2000).

Cooling/Heating

In many climates where an aircraft is on the apron for some time without operation of the APU, auxiliary mobile heating or cooling units are necessary to maintain a suitable internal temperature in the aircraft interior. The airline station engineer is responsible for ensuring the availability of such units.

With increasing fuel costs and environmental concern, much interest has been focused on centralized compressed-air units delivering air to the aircraft gate positions (usually called *fixed air supply* or *preconditioned air,* Figure 6.15) and to mobile compressors at the gates (known simply as *compressed-air systems*). Pneumatic systems can deliver high-pressure air for both heating and cooling and for air-starting the engines.

FIGURE **6.14**   Deicing/washer vehicle.

FIGURE **6.15**   Fixed ground cooling unit attached to an air bridge.

Where fixed air systems are used, cockpit controls can ensure either internal heating or cooling on an individual aircraft basis depending on the requirement. Studies indicate that the high cost of running aircraft APUs now means that fixed air systems can completely recover capital costs from the savings of two years of normal operation. Where airlines have infrequent flights to an airport, APUs are still used.

Other Servicing

Toilet holding tanks are serviced externally from the apron by special mobile pumping units. Demineralized water for the engines and potable water are also replenished during servicing.

Onboard Servicing

While external aircraft servicing is being carried out, there are simultaneous onboard servicing activities, principally cleaning and catering. Very high levels of cabin cleanliness are achieved by

- Exchange of blankets, pillow, and headrests
- Vacuuming and shampooing of carpets
- Clearing of ashtrays and removal of all litter
- Restocking of seatback pockets
- Cleaning and restocking of galleys and toilets
- Washing of all smooth areas, including armrests

Catering

Personnel clear the galley areas immediately after disembarkation of the incoming passengers. After the galley has been cleaned, it is restocked, and a secondary cleaning takes care of spillage during restocking. Internationally agreed standards of hygiene must be met in the handling of food and drink from their point of origin to the passenger. Where route stations are unable to meet either quality or hygiene standards, catering supplies are often brought from the main base. Figure 6.16 shows the loading operation of a catering truck. These trucks are usually constructed from a standard truck chassis with a closed-van body that can be lifted up by a hydraulic scissor lift powered by the truck engine. Two different types of catering trucks are available: low-lift vehicles suitable for servicing narrow-bodied aircraft up to 11.5 feet (3.5 m) doorsill height and high-lift vehicles for loading wide-bodied jets.

6.5   Ramp Layout

During the design phase of a commercial air transport aircraft, considerable thought is given to the matter of ramp ground handling. Modern aircraft are very large, complicated, and expensive.

Figure 6.16   Catering truck in loading position.

Therefore, the apron servicing operation is also complicated and consequently time-consuming. Unless the ramp servicing procedure can be performed efficiently, with many services being carried out concurrently, the aircraft will incur long apron turnaround times during which no productive revenue is earned. Inefficient ramp servicing can lead to low levels of aircraft and staff utilization and a generally low level of airline productivity. The complexity of the apron operation becomes obvious when Figure 6.17 is examined. This figure shows the apron positions typically designated for servicing and loading equipment for a Boeing 747. It can be seen that the aircraft door and servicing-point layout has been arranged to permit simultaneous operations during the short period that the vehicle is on the ground during turnaround service. The ramp coordinator is required to ensure that suitable equipment and staff numbers are available for the period the aircraft is likely to be on the ground. Complicated as Figure 6.17 is, it hardly shows the true complexity of the problem. Because the ground equipment is necessarily mobile, the neat static position shown is made less easy by problems of maneuvering equipment into place. Positioning errors can seriously affect the required free movement of cargo trains, transporters, and baggage trains. Over the last 25 years, the arrival of low-cost carriers (LCCs) has put considerable pressure

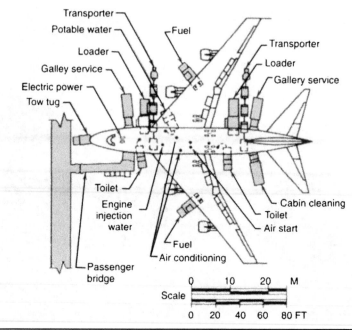

Figure 6.17   Ramp layout for servicing a B747SP. *Note:* Under normal conditions, external electrical power, air-start, and air conditioning are not required when the auxiliary power unit is used. (*Courtesy: Boeing Airplane Company.*)

on ramp efficiency with the demand for very short turnaround times. Some LCCs have negotiated contracts with the airport company that stipulate forfeiture of landing charges when a turnaround time of as low as 20 minutes is exceeded.

Particular attention must be paid to the compatibility of apron handling devices with the aircraft and other apron equipment. The sill height of the aircraft must be compatible with passenger and freight loading systems. In the case of freight, there is the additional directional compatibility requirement. Transporters must be able to load and unload at both the aircraft and the terminal onto beds and loading devices that are compatible with the vehicles' direction of handling. Many transporters can load or unload in the one direction only. The receiving devices must be oriented to accept this direction.

Most mobile equipment requires frequent maintenance. In addition to normal problems of wear, mobile apron equipment is subject to increased damage from minor collisions and misuse that do not occur in the same degree with static equipment. Successful apron handling might require a program of preventive maintenance on apron equipment and adequate backup in the inevitable case of equipment failure.

Safety in the ramp area is also a problem requiring constant attention. The ramps of the passenger and cargo terminal areas are high-activity locations with much heavy moving equipment in a high-noise environment. Audible safety cues, such as the noise of an approaching or backing vehicle, are frequently not available to the operating staff members, who are likely to be wearing ear protection. Very careful training of the operating staff is required, and strict adherence to designated safety procedures is necessary to prevent serious accidents (IATA 2012; CAA 2006).

6.6   Departure Control

The financial effects of aircraft delay fall almost entirely on the airline. The impact of delays in terms of added cost and lost revenue can be very high. Consequently, the functions of departure control, which monitors the conduct of ground handling operations on the ramp (not to be confused with ATC departure), are almost always kept under the control of the airline or its agent. Where many of the individual ground handling functions are under the control of the airport authority, there also will be general apron supervision by the airport authority staff to ensure efficient use of authority equipment.

The complexity of a ramp turnaround of an aircraft is indicated by the critical-path diagram shown in Figure 6.18. Even with the individual servicing functions shown in simple form, it is apparent that many activities occur simultaneously during the period the aircraft is

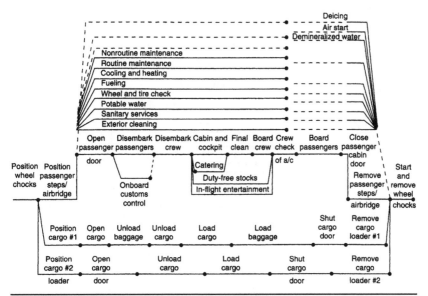

FIGURE 6.18 Critical path of turnaround ground handling for a passenger transport aircraft taking cargo.

on the ramp. This functional complication is a reflection of the physical complexity found on the ramp (see Figure 6.17).

The ramp coordinator in charge of departure control frequently must make decisions that trade off payload and punctuality. Figure 6.19 shows the effect of intervention by departure control in the case of cargo loading equipment breakdown. Figure 6.19a indicates satisfactory completion of a task within a scheduled turnaround time of 45 minutes. Figure 6.19b shows a 10-minute delay owing to equipment breakdown being reduced to a final ramp delay of only 5 minutes by the decision not to load nonrevenue airline company stores.

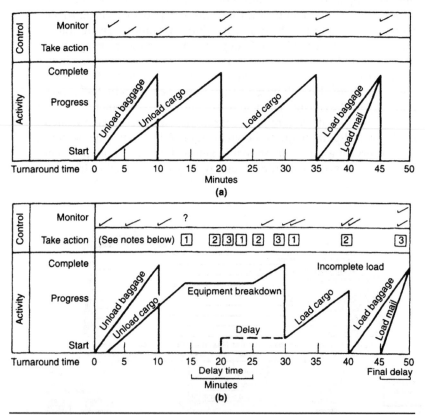

FIGURE 6.19   Effect of breakdown and delay on apron dispatch: (a) activity normal, no control action; (b) delay through breakdown, control required. Action 1: Assess the nature of the problem and how long the problem (breakdown of the cargo loader) will take to sort out. Action 2: Take corrective action immediately or call equipment base and ask the engineer to come to the aircraft immediately or call up a replacement loader. Action 3: Advise all other sections/activities that will be affected by the breakdown. Give them instructions as necessary (e.g., notify movement control of a delay, tell passenger service to delay boarding, etc.).

6.7   Division of Ground Handling Responsibilities

There is no hard-and-fast rule that can be applied to the division of responsibility for ground handling functions at airports. The responsibility varies not only from country to country but also among airports in the same country.

Prior to airline deregulation, handling activities were carried out mainly by airlines (acting on their own behalf or for another airline) or the airport authority. At many non-U.S. airports, all handling tasks were undertaken by the airport authority (e.g., Frankfurt, Hong Kong, and Singapore). The converse was almost universally true in the United States. Virtually all airport ground handling was carried out by the airlines. (In the old Soviet Union, all aviation activities were the responsibility of one organization, Aeroflot. This included the functions of the civil aviation authority, the airline, and the airports.)

Since deregulation, there has been a general movement toward liberalization and the introduction of competition in airport operations. In the mid-1990s, the European Union introduced regulations that required airports to use two or more ground handing operators where the scale of operation made this economic (EC 1996). This policy has been mirrored all around the world. Specialist companies are now providing some or all ground handling services at most large and medium-sized airports. In some facilities, the airlines still prefer to use their own staff where there is major contact between the company and the public. Ticketing, check-in, and lounge services are retained by the airline, but on the ramp, functions such as marshaling, steps, loading and unloading of baggage and cargo, and engine starts are carried out by the handling companies. Table 6.2 shows the results of a recent survey of how ground handling varies for a number of selected airports (see Acknowledgments).

In theory, some economic advantages of scale could be expected from centralized ground handling operations. One body operating airport-wide should be able to plan staff and equipment requirements with less relative peaking and probably with less duplication of facilities. Similar economies could be expected with standby equipment. Routine preventive maintenance also should be less expensive with a smaller proportion of the equipment out of service for repairs. However, the advantages that accrue from these areas are most likely to be countered by the disadvantages that accrue from too centralized an operation and lack of competition. Airports using only one ground handling organization are also vulnerable to severe industrial action from a relatively small group of workers. Gains in efficiency may be more than lost in unreasonable wage claims, and it might be difficult to introduce any level of competition once a monolithic agency has been set up, which is why the European Commission (EC) has introduced regulations that discourage or

Activity	Airport	Airlines	Airport Handling Company	Airline Handling Company	Not Applicable
Baggage check-in	10.19%	37.96%	11.11%	39.81%	0.93%
Baggage handling inbound	15.00%	31.00%	11.00%	41.00%	2.00%
Baggage handling outbound	15.69%	32.35%	10.78%	40.20%	0.98%
Baggage claim	20.41%	31.63%	7.14%	39.80%	1.02%
Passenger check-in	11.01%	38.53%	11.01%	38.53%	0.92%
Transit passenger handling	10.42%	31.25%	10.42%	34.38%	13.54%
Disabled passenger services	18.87%	30.19%	9.43%	40.57%	0.94%
Flight information systems	80.25%	8.64%	4.94%	3.70%	2.47%
Ground transportation systems	56.63%	3.61%	16.87%	12.05%	10.84%
Security	71.08%	7.23%	13.25%	6.02%	2.41%
Airside ramp services	26.32%	24.21%	8.42%	40.00%	1.05%
Airside supervision	67.82%	10.34%	3.45%	18.39%	0.00%
Airside marshalling	36.73%	24.49%	7.14%	30.61%	1.02%
Airside startup	22.68%	28.87%	6.19%	37.11%	5.15%
Airside ramp safety control	65.96%	17.02%	0.00%	15.96%	1.06%
Airside on-ramp aircraft servicing	15.05%	34.41%	4.30%	39.78%	6.45%
Airside aircraft fault repair	2.25%	55.06%	3.37%	34.83%	4.49%
Airside fueling	15.29%	14.12%	27.06%	41.18%	2.35%
Airside wheel and tire check	4.12%	46.39%	6.19%	41.24%	2.06%
Airside ground power supply	34.29%	22.86%	7.62%	34.29%	0.95%
Airside deicing	13.79%	16.09%	10.34%	19.54%	40.23%

172

Airside cooling/heating	26.60%	15.96%	8.51%	32.98%	15.96%
Airside toilet servicing	18.56%	26.80%	7.22%	42.27%	5.15%
Airside portable water	24.73%	22.58%	6.45%	38.71%	7.53%
Airside demineralized water	10.00%	17.50%	6.25%	30.00%	36.25%
Airside routine aircraft maintenance	0.00%	51.04%	6.25%	34.38%	8.33%
Airside nonroutine aircraft maintenance	0.00%	53.26%	4.35%	29.35%	13.04%
Airside exterior aircraft cleaning	6.32%	32.63%	7.37%	42.11%	11.58%
Onboard servicing cabin and cockpit cleaning	9.38%	31.25%	7.29%	51.04%	1.04%
Onboard servicing catering	8.05%	25.29%	11.49%	50.57%	4.60%
Onboard servicing in-flight entertainment	1.20%	55.42%	2.41%	27.71%	13.25%
Onboard minor servicing of cabin fittings	1.19%	54.76%	4.76%	27.38%	11.90%
Onboard changing seat configuration servicing	2.41%	54.22%	4.82%	22.89%	15.66%
Onboard external ramp equipment provision and manning	9.57%	38.30%	7.45%	38.30%	6.38%
Onboard passenger steps servicing	14.44%	30.00%	11.11%	43.33%	1.11%
Onboard catering loader servicing	8.14%	26.74%	9.30%	50.00%	5.81%

Table 6.2 Distribution of Responsibilities for Ground Handling Operations at 72 Selected Airports (for Participating Airports, see Acknowledgments)

prevent monopoly positions in ground handling (EC 1996). The very scale of large airports to a degree negates the idea of being able to operate ground equipment from one pool. Physically, the total provision probably will have to be broken into a number of relatively self-sufficient and semiautonomous organizations based on the various parts of a single large terminal or on the individual-unit terminals of a decentralized design.

In general, the ground handling function is not an area of considerable profit for an airport authority. Labor and equipment costs are high, and in general, either revenues barely cover attributable expenses or, as in many cases, are actually less than costs. These losses often are cross-subsidized using revenues from other traffic areas, such as landing fees or nontraffic concession revenues.

6.8 Control of Ground Handling Efficiency

The extreme complexity of the ground handling operation requires skilled and dexterous management to ensure that staff and equipment resources are used at a reasonable level of efficiency. As in most management areas, this is achieved by establishing a system of control that feeds back into the operation when inefficiencies appear. The method of control used at any individual airport depends on whether the handling is carried out by the airline itself, by a handling agency such as another airline, or by the airport authority.

Four major reporting tools help to determine whether reasonable efficiency is being maintained and permit the manager to discern favorable and unfavorable operational changes.

Monthly complaint report. Each month, a report is prepared that shows any complaints attributable to ground handling problems. The report contains the complaint, the reason behind any operational failure, and the response to the complainant.

Monthly punctuality report. Each month, the manager in charge of ground handling prepares a report of all delays attributable to the ground handling operation. In each case, the particular flight is identified, with its scheduled and actual time of departure. The reason for each delay is detailed. The monthly summary should indicate measures taken to preclude or reduce similar future delays. Typical aircraft servicing standards are 30 to 60 minutes for a transit operation and 90 minutes for a turnaround. Where LCC operations are involved, these times may be reduced considerably.

Cost analysis. The actual handling organization will, at least on a quarterly basis, analyze handling costs. These costs should include capital and operating costs. For airlines, airports, and handling agencies, this can be achieved fairly easily on a monthly basis by a computerized management reporting system that allocates expenditures and depreciation to management cost centers, even for relatively small operators. It is normal practice to have budgeted expenditures in a number of categories. Variances between budget and actual expenditure require explanation.

General operational standards. To ensure an overall level of operational acceptability, periodic inspections of operations and facilities must be made. This is especially important for airlines carrying out their own handling away from their main base or at airports where they are handled by other organizations. For the airport operation, it is equally important. Whether or not the handling is carried out by the airport, the general standards reflect on airport image. Inspections ensure that agreed standards are maintained and highlight areas where standards are less than desirable. Table 6.3 shows the form of checklist used by an international airline to ensure that handling at outstations conforms to company standards. The airport operator should maintain a similar checklist for all major airlines operating through the airport that do their own handling, omitting the areas related to administration and accounting. In all areas possible, the evaluation should be carried out using quantitative measures. Subjective measures should be avoided because they are not constant between evaluators and may not be constant over time even with a single evaluator.

6.9   General

Ground handling of a large passenger aircraft requires much specialized handling equipment, and the total handling task involves considerable staff and labor inputs. Good operational performance implies a high standard of equipment serviceability. In northern climates, it is usual to assume that equipment will be serviceable for 80 percent of the time during the winter and 85 percent during the summer. Backup equipment and maintenance staff must be planned for the periods of unserviceability. Most companies operate failure-maintenance procedures for ground handling equipment rather than expensive preventive-maintenance programs. Availability of equipment and staff becomes a problem where airports do not designate individual gates to specific airlines. Some airlines, operating with low frequencies into some European airports, find that there is no policy of preferred gates for them. This can mean considerable movement of airline equipment and staff around the airport. In the United States, a different handling problem can occur. U.S. ground handling, passenger handling, load control, and baggage-handling crews often have a system of bidding for shift choice based on seniority. Consequently, a non-U.S. airline

Passenger services: Check-in
  Operational adequacy of general check-in desks
  First-class check-in service
  Waiting time at check-in
  Seat selection procedure
  Information display
  Courtesy and ability of check-in staff
  Passenger acceptance control
  Standby control, late passengers, overbooking, rebates
  Acceptance of excess, special, and oversized baggage
  Baggage tagging, including transfer, first class
  Security of boarding passes/ticketing/cash and credit vouchers
  Minimum and average check-in times
  Preparation of passenger lists
  Control of catering orders
  Ticket issues and reservations

Passenger services: Security
  Personal search or scan efficiency
  Hand baggage search efficiency
  Inconvenience level and waiting times

Passenger services: Escort and boarding
  Effectiveness of directions and announcements
  Staff availability for inquiries at waiting and boarding points
  Assistance at governmental control points
  Control of boarding procedure
  Liaison level between check-in and cabin staff
  Service levels of special waiting lounges for premium ticket holders
  Special handling: Minors, handicapped

Passenger services: Arrivals
  Staff to meet flight
  Information for terminating and transfer passengers
  Transfer procedures
  Assistance through government control points
  Special passenger handling: Minors, handicapped
  Baggage delivery standards
  Assistance at baggage delivery

Passenger services: Delayed/diverted/canceled flights
  Procedures for information to passengers
  Procedures for greeters
  Messages including information to destination and en-route points
  Procedures for rerouting and surface transfers
  Meals, refreshments, and hotel accommodations

TABLE 6.3 Checklist for Monitoring the Efficiency of Ground Handling

Passenger services: Baggage facilities
    Compilation of loss or damage reports
    Baggage tracing procedures
    Claims and complaints procedures

Passenger services: Equipment
    Check security and condition of all equipment: Scales, reservations
      printer, seat plan stand, ticket printer, credit-card imprinter, calculators, etc.
    Condition and serviceability of ramp vehicles
    Serviceability and appearance of ramp equipment
    Maintenance of ramp equipment and vehicles
    Control of ramp equipment and vehicles
    Driving standards and safety procedures
    Communications: Telephones, ground-air radio, ground-ground radio

Ramp handling: Aircraft loading/unloading
    Care of aircraft exteriors, interiors, and unit load devices
    Adequacy of loading instructions and training
    Ramp equipment planning and availability
    Positioning of equipment to aircraft
    Loading and unloading supervision
    Securing, restraining, and spreading loads
    Operation of load equipment
    Operation of aircraft onboard systems
    Securing partial loads
    Ramp security
    Ramp safety
    Pilferage and theft

Ramp handling: Cleaning/catering
    Standard of cockpit and cabin cleaning/dressing
    Toilet/potable-water servicing
    Catering loading/unloading
    Availability of ground air
    Air-jetty operations

Ramp handling: Load control (for airline only)
    Load sheet accuracy and adequacy of presentation
    Load planning
    Advance zero-fuel calculation and flight preparation

Ramp handling: Aircraft dispatch
    Punctuality record
    Turnaround/transit supervision
    Passenger release from aircraft
    Passenger waiting time at boarding point
    Logs and message files
    Accuracy of records of actual departure times
    Flight plan, dispatch meteorological information

TABLE 6.3 Checklist for Monitoring the Efficiency of Ground Handling (*Continued*)

Ramp handling: Postdeparture
  Accuracy and time of dispatch of postdeparture records and messages

Cargo handling: Export
  Acceptance procedures
  Documentation: Procedures and accuracy
  Reservations: Procedures and performance
  Storage: Procedures and performance
  Makeup of loads: Procedures and performance
  Check weighing
  Palletization and containerization: Procedures and performance

Cargo handling: Import
  Breakdown of pallets/containers: Procedures and performance
  Customs clearance of documents
  Notification of consignees
  Dwell time of cargo
  Lost/damaged cargo procedures
  Proof of delivery procedures
  Handling of dangerous goods procedures
  Handling of restricted goods procedures
  Handling of valuable consignments procedures
  Handling of live animals procedures
  Handling of mail

Administration of ground handling
  Office appearance
  Furniture and equipment condition
  Inventory records: Ramp equipment/vehicles/office equipment/furniture
  Budgeting: Preparation and monitoring
  Control of cash/invoices/tickets/accounting/sales returns/strongboxes/keys/
    airport records/stationery
  Complaints register
  Staff appearance
  Condition of manuals/local instructions/emergency procedures/standing
    orders/general office file

TABLE 6.3   Checklist for Monitoring the Efficiency of Ground Handling (*Continued*)

being handled by a U.S. carrier may well find that there is a seemingly continuous training problem as ground handling crews change. It is even possible on a long stopover to have two different crews that each need instruction on equipment operation. This U.S. problem has diminished with the growth of new third-party independent ground handling companies that no longer operate to these old union practices.

References

Ashford, N. J., S. Mumayiz, and P. H Wright. 2011. *Airport Engineering: Planning, Design, and Development of 21st Century Airports.* Hoboken, NJ: Wiley.

Civil Aviation Authority (CAA). 2006. *Airside Safety Management* (CAP 642). London: CAA, Her Majesty's Stationary Office.

European Commission (EC). 1996. *Access to the Groundhandling Market at Community Airports* (Council Directive 96/67/EC). Brussels: EC, October 15.

International Air Transport Association (IATA). 2004. *Airport Development Reference Manual,* 9th ed. Montreal, Canada: IATA.

International Air Transport Association (IATA). 2012. *Airport Handling Manual,* 32nd ed. Montreal, Canada: IATA.

International Civil Aviation Organization (ICAO). 2000. *Manual of Aircraft Ground Deicing/Anti-icing Operations,* 2nd ed. (Document 9460-AN/940). Montreal, Canada: ICAO.

International Civil Aviation Organization (ICAO). 2010. *Annex 14, Aerodromes,* 5th ed. Montreal, Canada: ICAO.

CHAPTER 7

Baggage Handling

7.1 Introduction

This chapter deals with baggage handling at airports from process, system, and organizational perspectives. The chapter itself is divided into the following topics:

- Context, history, and trends
- Baggage-handling processes
- Baggage-handling equipment, systems, and technology
- Process and system design drivers
- Organization
- Management and performance metrics

7.2 Context, History, and Trends

Baggage handling is an essential element of airport operations, but as with other utility functions, it is often remarked on only when it goes wrong. The effects of failure can range from a few passengers not receiving their bags when they arrive at their destination to the widespread disruption of airport operations, including flight cancellations, along with all that such events entail for airlines and passengers.

Historically, baggage appeared near the top of passengers' list of complaints, but this is no longer the case. An analysis of customer complaints over the period 2009–2012 (Figure 7.1) shows that baggage-related issues accounted for less than 5 percent of all complaints. A total of 3.8 percent of complaints are attributable to third parties—airlines and their handlers—and only 0.3 percent are attributable to terminal operations—the baggage-handling systems themselves.

This improvement has been the result of an industry-wide appreciation of the costs associated with poor baggage-handling performance combined with investments in advanced, automated baggage systems around the world. Even so, the cost to the airport and airline

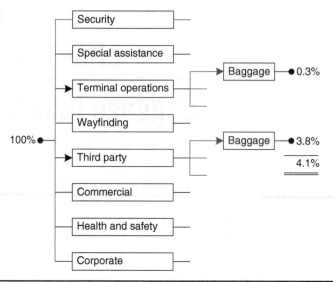

Security

Special assistance

Baggage —●0.3%

Terminal operations

Wayfinding

100%● —

Baggage —●3.8%

Third party

4.1%

Commercial

Health and safety

Corporate

FIGURE 7.1 Baggage component of customer complaints.

community (and hence the traveling public) is still large—the International Air Transport Association's (IATA's) director general, Giovanni Bisignani, remarked at The Wings Club in February 2009 that the global costs of mishandled bags were US$3.8 billion.

Even though baggage handling is usually carried out by an airline or its appointed handler, this distinction is often unclear to passengers. Thus, if they suffer problems or delays with baggage, passengers will assume that it is a failing of the airport, thereby risking its reputation. In practice, both airports and airlines have important roles to play, and a collaborative approach to managing baggage handling leads to a better outcome for all parties.

The scale and complexity of baggage handling have changed over the course of the last few decades, and this has led to a spectrum of baggage-handling solutions that range from the simple to the very sophisticated, based on the needs of the airline customers.

The most obvious trend is the progressive introduction into service of ever larger aircraft. This has led to the introduction of *unit-load devices* (ULDs, or containers) as a means of expediting the loading and unloading of both baggage and cargo, which otherwise would be slow or even unmanageable if dealt with as loose items.

As the cost of air travel has declined, it has become accessible to a wider range of passenger types. With this comes a wider range of items that passengers want to carry, thereby putting additional stress on out-of-gauge (oversize) processes.

In addition, as routes to newly developed markets increase, using baggage as a means of transporting trade items becomes ever more popular, offering, in some cases, a faster, cheaper, and more secure way of moving high-value items than as freight.

The requirement to screen passengers and their baggage has been introduced by most jurisdictions. For modest passenger and baggage flows, this can be dealt with without separating passenger and bag. For larger airports with large quantities of baggage to screen, however, in-line systems offer screening rates of approximately 1,200 bags per hour per machine—an order of magnitude greater than that which can be achieved by a metal-detecting arch and accompanying baggage-screening equipment—become necessary.

There are also trends that tend to reduce the quantity and size of hold baggage. Budget airlines, for instance, usually will charge extra for any hold luggage, and even full-service carriers will make charges for extra bags or excessive weight. Health and safety rules in some regions also exert a downward pressure on the weight of baggage that can be handled.

What history indicates, above all, is that changes in technology, legislation, competition, markets, and even short-term events such as the Olympics can have a profound and rapid effect on the types and volume of baggage and processing needs.

7.3   Baggage-Handling Processes

Overview

A typical set of baggage processes is shown in Figure 7.2. While all commercial airports will have check-in, reclaim, and flight build facilities (also called *makeup*), only hub airports will have any significant transfer-baggage facilities. Hub airports with multiple terminals also may have a significant interterminal transfer process connecting passengers and their bags arriving at one terminal with their departure flights in a different terminal.

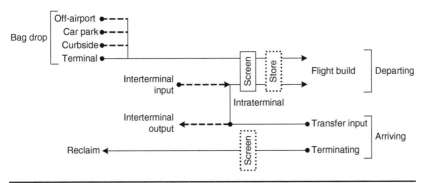

FIGURE 7.2   Typical baggage processes.

Bags entering the system via a bag drop generally will be screened in the terminal of departure. Once in the baggage system, optionally, they may be stored and then delivered to a flight build output. From there they are taken to the departing aircraft and loaded.

Terminating bags arriving at a terminal will be delivered to reclaim for collection by passengers. In some circumstances and jurisdictions, terminating bags are screened for illicit items.

Transfer bags arriving at a terminal will be input into the baggage system and routed to the terminal of departure. Once there, the process follows that for locally checked-in baggage. The major elements in this process are described in turn in the following sections.

Bag Drop

Off-airport check-in can be offered in a number of ways including in-town airline offices, check-in counters at downtown train stations, and services supporting check-in and bag drop at hotels. For example, in Hong Kong, most airlines have check-in counters at both Hong Kong and Kowloon Stations. Airport Express passengers can check in and leave baggage at these facilities so that they are free to visit the city for the rest of day before leaving for the airport without having to carry their baggage around with them.

Car-park and curbside check-ins are convenient ways to check in for a flight and to drop bags without having to take them through a crowded airport building. They typically operate as follows:

- Pull up to a booth in a car park or the curb adjacent to the departure terminal, and present a photo ID along with a confirmation number, destination, flight number, or e-ticket number to an agent.
- Hand checked bags to the agent, collect the baggage receipt and boarding pass, and proceed straight to security.

In 2012, American Airlines offered curbside check-in at 66 U.S. airports, whereas Delta offered the service at approximately 100 U.S. airports.

Terminal check-in is ubiquitous. This, historically, has been carried out at staffed combined check-in and bag-drop desks. The passenger presents travel and identification documents to the check-in agent, who assigns a seat number. If the passenger has baggage to check in too, then the agent typically will print and attach bar-coded tags to each bag and issue baggage receipts to the passenger. The agent then will dispatch the bags into the baggage system.

Since more and more check-in functions can be performed online (e.g., seat selection, boarding-card printing), the traditional check-in arrangement at airports is becoming increasingly deconstructed to allow passengers to make use of the functions they require while

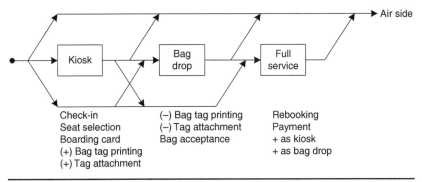

FIGURE **7.3** Flexible check-in options.

bypassing others. Figure 7.3 shows how passengers can access the required combination of functions for their needs.

Typically, there will be three waves, each wave being encountered by passengers as they makes their way through the departure concourse. The first wave is provided by self-service kiosks that support check-in, seat selection, and boarding-card printing (Figure 7.4). Optionally, these kiosks may support bag tag printing and attachment. The kiosks are less space-consuming, can be more flexibly located, and are cheaper than conventional desks. Because these

FIGURE **7.4** Self-service kiosk.

kiosks allow for more units and keep transaction times down, passengers benefit from fewer queues than otherwise would be the case with conventional check-in. And because one member of staff can host and support a group of kiosks, operational costs for airlines and handlers are reduced for a given level of service.

The second wave is a bag drop where passengers can deposit hold baggage. Often these bag drops are physically indistinguishable from a conventional check-in desk and are staffed in the same way—it is simply that they are used purely for baggage acceptance. A typical example of such facility is shown in Figure 7.5.

There is growing interest in self-service bag drops, where passengers can deposit baggage without the need for a member of staff. Qantas is an early adopter of this approach for domestic traffic. In this arrangement, bag tags are printed and attached at a check-in kiosk [or permanent radiofrequency ID (RFID) tags are used for frequent flyers] so that when the passenger reaches the bag drop, there is little more to do than put the bag onto the receiving conveyor (Figure 7.6). The average process time is in the range of 20 to 30 seconds per bag. This short process time (compared with 1 to 2 minutes or more for conventional check-in and bag drop), coupled with multiple bag drops, means that there are rarely queues of passengers waiting to deposit bags.

Figure 7.5 Staffed bag drop.

Figure 7.6   Self-service bag drop.

The third wave has full-service desks. Here, any of the functions performed in waves one and two also can be performed, and they can deal with additional functions such as taking payment for excess baggage or rebooking.

Passengers still may reach the departure gate with baggage that an airline may not choose or be able to carry in the cabin. Therefore, there usually will be the opportunity, at the gate, for an agent to tag a bag and then have the bag loaded into the aircraft hold. Since this is a time-consuming activity that would slow up aircraft boarding if left to the last moment, airports and airlines often will employ a series of measures to minimize the number of last-minute gate bags. This usually will involve one or more of the following:

- Agents inspecting all baggage at check-in for size to capture all non-cabin-compatible items.
- Limits placed on the size of baggage at passenger screening, necessitating the prior check-in of items that cannot be carried in the cabin.
- Agents spotting passengers waiting in and around gate areas with unsuitable baggage so that the items can be tagged and loaded before boarding begins.

Usually gate bags do not need to be rescreened because they will have been checked along with the passenger through the processes needed to reach the gate.

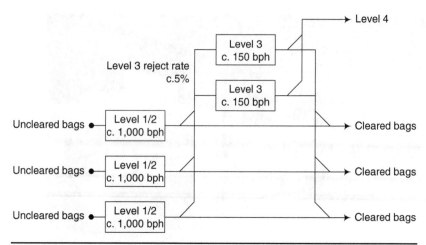

FIGURE 7.7   Multilevel screening protocol.

Hold Baggage Screening

Once bags have entered the baggage system, generally they will be screened using in-line x-ray machines [also known as *explosive-detection systems* (EDS)] to ensure that dangerous or prohibited items are not present. A typical European screening process is shown in Figure 7.7. Uncleared bags are examined by a level 1 hold-baggage-screening (HBS) machine. These machines typically can process bags at rate of more than 1,000 per hour. If the machine and its image-processing algorithm is able to determine that there is no threat present, the machine will clear the bag. For perhaps 30 percent of bags, the image-processing algorithm will not be able to clear the bag confidently, so the image will be passed to a human operator for a level 2 decision. In most cases the bags then will be cleared, but typically 5 percent of all incoming bags will still be unresolved and will require a more detailed examination. These bags will be sent to a level 3 HBS machine, which uses computed tomography to give a three-dimensional image, allowing a more thorough examination by an operator. Level 3 machines typically have a throughput of 150 bags per hour. In the vast majority of cases, no threat will be present, and the operator will clear the bag. In a very small fraction of cases, the images taken at level 3 still will be inconclusive, and the bags will be sent to level 4, where a physical examination of the bag will be carried out.

The multilevel protocol adopted in the United States is as follows:

Level 1 screening is performed with EDS units. All bags that can physically fit in an EDS unit are directed to level 1 screening and scanned using an EDS. All bags that automatically alarm at level 1 are subject to level 2 screening.

During *level 2* screening, Transportation Security Administration (TSA) personnel view alarm bag images captured during the level 1 EDS scan and clear any bags whose status can be resolved visually. All bags that cannot be resolved at level 2 and all bags that cannot be directed to level 1 because of size restrictions are sent to level 3 screening.

Level 3 screening is performed manually and involves opening the bag and the use of explosive-trace-detection (ETD) technology. Bags that do not pass level 3 screening (typically, a small percentage of total bags) are either resolved or disposed of by a local law enforcement officer.

The TSA has published guidelines and design standards for hold-baggage screening that provide an excellent introduction to the U.S. implementation of hold-baggage screening (TSA 2011).

Bag Storage

Originally, baggage-handling systems had no need to provide bag storage—bags for a flight were accepted at check-in only when the flight makeup positions were available for use, typically two to three hours before the scheduled departure time. Over time, the need for additional bag storage has increased. One factor is the growth in transfer traffic, which can mean that an inbound flight and its connecting bags arrive well before the planned flight makeup positions for the departing flight are open. Another reason is the desire to allow passengers to check in bags when they choose. And increasingly, bag stores can be used to manage and buffer the flow of bags to flight makeup positions, thereby enabling more efficient use of staff and infrastructure or even supporting robotic loading systems (e.g., at Schiphol Airport).

Flight Build and Aircraft Loading

Bags that have been processed and sorted ultimately are delivered to outputs where they are loaded either into ULDs or trailers. ULDs are containers into which bags and cargo can be loaded.

The number of makeup positions allocated per flight will depend on the expected volume of baggage, the flight build time, and the number of segregations into which bags have to be sorted. This can vary from one or two positions for small aircraft to 10 or more for larger aircraft with complex terminating and transfer products.

Smaller aircraft (e.g., B737s, B757s, and A319s) are not containerized, and bags for these types will be loaded into trailers. These trailers then are towed to the aircraft side, and the bags are loose loaded into the aircraft hold using a belt loader. Since this type of operation is relatively slow and labor-intensive, it becomes unsuitable for dealing with the number of bags carried by larger aircraft.

FIGURE 7.8   AKE and AKH ULDs.

Larger aircraft (e.g., A330s, A340s, B777s, B747s, and A380s) are equipped to carry ULDs. A ULD might be able to contain 30 to 50 bags depending on bag size and ULD type. There are many varieties of ULDs, but two are very commonly used: AKH and AKE (Figure 7.8). A single AKH can fit across the width of the hold of an A320, whereas a pair of AKEs can fit across the width of a hold of a B777, a B747, and an A380.

The flight build process can be very simple, particularly for small, non-containerized aircraft where there are not many bags to be loaded. However, with larger, containerized aircraft and for airlines with more complex products, the flight build involves ensuring that bags are sorted and loaded by segregation. Segregations might include some or all of the following:

- Premium terminating
- Economy terminating
- Crew bags
- Short-connect transfers
- Long-connect transfers
- Inter-terminal transfers (by departure terminal)
- Onward transfers (by transfer destination)

Loading bags according to these types of segregation assists the speed and ease of handling at downstream stations, but at a price. The flight build operation becomes larger and more complex, and the filling efficiency of ULDs generally will be poorer because some ULDs will be only partially filled. Thus, build segregation policies depend on airline priorities and products, handling operations, and facilities at originating, terminating, and transfer airports.

Irrespective of how bags are loaded into trailers or ULDs, most control authorities require airlines to ensure that all hold-loaded baggage is accounted for. This means recording which bags have been loaded and ensuring that the required security processes have been complied with for each and every bag.

At its simplest, this can be managed by the *bingo-card method*. This just means removing one of the self-adhesive bag tags from the loaded bag and sticking it onto a record sheet and reconciling the resulting list of bags against passengers. For small flights, especially without any inbound transfer connections, this is often sufficient.

However, for larger fights and those with inbound connecting passengers, this becomes increasingly impractical. A typical reconciliation system will consist of a number of hand scanners for use by handlers that are connected to a database and message-handling system. The handler scans the bar code on the bag tag and waits for confirmation that the bag may be loaded. The scanned tag number is matched against records in the database, and if the security status is satisfactory, the reconciliation system will indicate, usually via the hand scanner, that the bag can be loaded.

The reconciliation system generally will record other data about the bag, such as the registration number of the ULD into which it will be loaded and the sequence number of the bag within the ULD. This additional information is useful for identifying where to locate a bag if the bag has to be offloaded because, for example, the owner of the bag fails to board the aircraft. The reconciliation system usually will exchange messages with the baggage-handling system and an airline's departure control system (DCS) in order to maintain an up-to-date status of both bag and passenger. Filled ULDs then are taken to the departure stand and loaded onto the aircraft (Figure 7.9).

Arrivals Reclaim

The function of reclaim is to reunite passengers and their baggage. Since the arrival processes for passengers and baggage are very different, the reclaim hall functions as a buffer space—for passengers to wait for bags and for bags to wait for passengers.

Ideally, the appearance profiles of passengers and bags at reclaim should be similar. This ensures that neither the reclaim device nor the reclaim hall becomes too busy with bags and passengers, respectively. This can be assisted by inbound segregation of premium baggage.

Figure 7.9   Loading ULDs onto an aircraft.

Such baggage is unloaded from the aircraft first and delivered promptly to reclaim so that premium passengers, who usually leave the aircraft first and may take advantage of fast-track routes, have little or no wait.

However, there are times when the appearance profiles of passengers and bags are not so well matched. There are two extreme cases: All passengers arrive before any bags are delivered, and all bags arrive before any passengers arrive. In the first case, all terminating passengers have to wait for their bags (and queuing space has to be provided for all these passengers adjacent to a reclaim). If there is not sufficient space for passengers in the hall, then operational measures have to be taken to limit access to the hall to prevent overcrowding. A side effect is that reclaims may be filled with bags whose passengers cannot enter the hall, leading to gridlock.

The second case is that passengers are delayed (perhaps at passport control and immigration checks) and cannot reach the reclaim hall. Initially, bags can be delivered and accumulate on the reclaim device, but because a reclaim typically may be large enough to hold only around 25 percent of the all bags from a flight, baggage handlers then will be unable to deliver further bags. The operational response to this type of situation is to have staff in the reclaim hall remove bags from the reclaim device and stack them in an orderly fashion adjacent to the reclaims ready for passenger collection. This allows the handlers to complete the delivery operation and to be redeployed for subsequent tasks. If this does not happen, handlers cannot be redeployed, and as a result, subsequent arrival and departure activities may be delayed, leading, in extremis, to another form of gridlock.

Transfer Input

Transfer bags need to be processed and, if on a minimum connection time, processed rapidly. To enable this, bags should be loaded into segregated ULDs on the inbound aircraft at the outstation. These short-connect ULDs then can be unloaded as a priority from the aircraft and taken to transfer input locations. Bags then are removed from the ULDs and input into the baggage-handling system. Once the bags have been accepted by the system (oversize and/or overweight bags will be rejected and need to be processed manually), the baggage system will transport and process them (including screening) so that they are delivered to flight build locations, much like locally checked-in baggage. In some cases, special provision is made for the most urgent bags. This may result in the bag being delivered to an alternative output from which it can be expedited, by vehicle, to a departing flight.

In some jurisdictions, certain categories of transfer bag can be unloaded from an inbound flight and taken directly to the connecting flight without the need for screening. This operation is known as a *tail-to-tail transfer* and can support a very short minimum connection time. This operation is permitted, for example, on connections between domestic flights within the United States and, in Europe, for bags that have been screened by a European airport (although some national authorities within Europe impose additional measures that mean that tail-to-tail transfers are not permitted). By their very nature, tail-to-tail transfers are not processed through an automated baggage system.

Interterminal Transfers

At multiterminal airports, transfers can occur between two different terminals. In this case, baggage typically is put into the automated baggage system of the inbound terminal, where it will be sorted to a vehicle loading dock for transport to the terminal of departure, where the bag will be processed and, ultimately, delivered to the connecting flight.

A vehicle link between terminals is a simple and effective option, but it does have the disadvantage that bags generally will have to wait for a vehicle to arrive and for loading onto the vehicle and unloading at the outbound terminal. Such an operation is not well suited for relatively short minimum connection times. To overcome this waiting, batching, and unbatching, some airports (e.g., Heathrow London, Changi Singapore) have installed automated baggage links between terminals.

7.4   Equipment, Systems, and Technologies

This section describes the equipment, systems, and technologies that are used to implement and support the processes outlined earlier.

Baggage-Handling-System Configurations

The design of the passenger terminal complex itself can radically affect the configuration of the outbound-baggage system. A number of design considerations are covered in IATA (2004).

Conventional centralized-pier finger airports, such as Chicago O'Hare, Schiphol Amsterdam, and Manchester International, operate on one or more central bag rooms in the main terminal area. These require elaborate sorting systems, but can be efficient in the use of personnel who are released when not needed in off-peak periods. Decentralized facilities, such as Frankfurt (Germany) and Dallas–Fort Worth, have a number of decentralized bag rooms that are closely associated with a few gates. The sorting requirements of these makeup areas are minimal, but it is more difficult to use staff efficiently in the decentralized situation, where there are substantial variations in workload between peak and off-peak periods. A third concept of baggage makeup area is the remote bag room. In an airport such as Atlanta, where three-quarters of the traffic is transfer, there is considerable cross-apron activity. Remote bag rooms provide for the complex sorting necessary without transporting all baggage back to the main terminal. In Terminal 5 at Heathrow, the baggage system actually consists of two elements: (1) a bulk, centralized system for dealing with all but the most time-critical of bags (which brings the benefit of economies of scale for staffing and other resources) and (2) a distributed delivery system to most stands that is used to deliver just the time-critical bags (which brings the benefit of swift delivery right to the aircraft, giving handlers the best chance of loading last-minute bags).

Irrespective of the arrangement of the baggage system, most baggage systems consist of some or all of the following components.

Check-in and Bag Drop

Traditional check-in and bag-drop desks can be arranged in a number of ways:

- Linear
- Island
- Flow-through

Schematics of these three configurations are shown in Figure 7.10. Both linear and island check-in have the disadvantage that the flow of

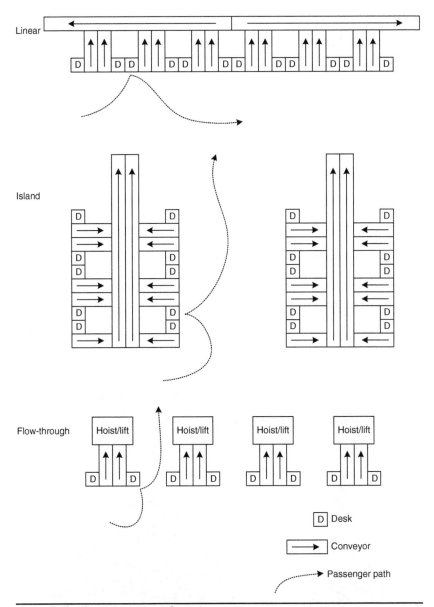

FIGURE 7.10   Check-in desk configurations.

passengers leaving the desks can conflict with queues of passengers waiting to reach the desks. Flow-through arrangements, however, avoid this difficulty but are feasible only where the terminal has the space to accommodate vertical movement of bags within the check-in floor plate.

Sorting

Once baggage has entered a system (other than the simplest), it has to be sorted. Destinations include screening equipment, manual encoding stations, and bag storage or flight makeup locations. There are several methods of sorting bags, the choice of which is governed by a combination of factors, including

- Space
- Cost
- Required capacity

For low-capacity applications, conveyor-based merges and diverts may be chosen. For somewhat higher capacities, vertical sorting and merge units may be employed because these can switch sufficiently quickly to allow adjacent bags to be sorted to two different locations with a throughput of over 1,000 bags per hour. By their nature, vertical sorting units require greater vertical space than horizontal merges and diverts, so they may not always be a feasible solution in some restricted locations. For higher capacities still, tilt-tray sorters can be used (Figure 7.11). These operate at around 400 ft/min (2 m/s) and typically have a tray size of about 4 feet (1.2 m), giving a tray rate of 6,000 per hour.

FIGURE 7.11   Tilt-tray sorter.

In cases where loose baggage is handled, every merge, divert, incline, and sorter in-feed or output has the potential for a bag to become snagged or trapped with the risk of damage to the system and/or bag. Careful design and tuning of the system become necessary to minimize this risk; otherwise, there will be frequent system stoppages and the associated cost of staff being needed to free jams.

An alternative approach that reduces the risk of bag jams is to use a toted system. In such a system, bags are not carried directly on conveyors but are first placed in a carrier or tray (Figure 7.12). With the provision of a secure container, each bag is less likely to catch on equipment, and by providing a standard base, the transport system can be optimized to deal with a single type of tote. Baggage tracking and storage are also made easier with totes. A bag can be identified once and then is linked in the baggage system with a given tote. The tote (rather than the bag) then is tracked using RFID tags, and this is more effective than trying repeatedly to read a bar code attached to a bag. However, tote-based systems require return routes to bring empty totes back to the baggage inputs, so they tend to require more space and, as a result, are initially more expensive to buy and install than untoted systems.

Hold-Baggage Screening

As screening technology develops, new and better machines become available. The control authorities build this into their regulations to ensure the best-possible chance of detection of known

Figure 7.12   Tote-based system.

and potential threats. To date in Europe, three standards of x-ray screening equipment have been identified:

1. *Standard 1*—a single-view technology
2. *Standard 2*—a multiview technology
3. *Standard 3*—a computed tomographic technology

During 2012 in Europe, standard 1 machines will no longer be acceptable, and there have been major programs of work at airports to replace standard 1 equipment. While the precise dates are subject to change (somewhere around 2018–2020), standard 2 machines will themselves become unacceptable and will have to be replaced by standard 3 machines. The changeover program will not be trivial because standard 3 machines weigh 6 to 8 tons and are over 17 feet (5 m) in length. An example of a computed tomographic machine is shown in Figure 7.13.

Bag Storage

Bag storage can take one of several forms. At its simplest is a manual store in which bags are grouped, by hand, by flight or departure time. This involves little more than space on the ground or racks to accommodate the bags. Automated stores vary in functionality. At one extreme,

FIGURE 7.13   Hold-baggage screening equipment.

they simply automate the manual process—accumulating groups of bags in conveyor lanes by flight or build open time. Such a store does not readily lend itself to the retrieval of a single, particular bag—a whole lane of bags would have to be released to access just one specific bag.

More sophisticated stores allow random access to any particular bag. These stores usually depend on bags being carried in totes, which enable them to be transported and tracked effectively. One type of store involves setting up long conveyor loops on which the toted bags circulate slowly. As the bags pass outputs, they can be diverted so that they leave the store. Another type of store makes use of a warehouse crane and racking approach (Figure 7.14). Toted bags entering the

Figure 7.14 Crane-served bag store.

store are taken by crane and placed in a slot in a lane of racking. This, too, allows single bags to be retrieved and thereby offers the most flexible of storage systems.

Flight Build

The type and configuration of manual makeup devices are varied, including

- Chutes
- Carousels
- Laterals

Each offers a combination of advantages and drawbacks. Chutes can be arranged space efficiently, thereby ensuring a one-to-one mapping between chute and ULD and/or trailer. However, they suffer from poorer handling ergonomics than laterals. Carousels offer a flexible means of distributing bags to several makeup positions, but there can be concerns over the ergonomics of picking bags from a moving device. Laterals (Figure 7.15) can be set at an

FIGURE 7.15   Build lateral.

optimal height for operators and are compatible with modern manual handling aids.

New ways of handling flight build are being implemented, and these require different makeup devices. Of particular note are fully automated, robot-based build cells and semiautomated batch build devices.

A build cell employs a robotic arm fitted with a specialized handling tool to receive a bag from the baggage-handling system and, using a machine vision system, then will place the bag into a trailer or ULD. The work rate achieved by such systems is typically three to four bags per minute—not necessarily faster than a human operator, but it is sustainable indefinitely and relieves handlers of the physical load. A build cell cannot operate unsupervised. In the course of filling a ULD with a capacity of, say, 40 bags, the supervisor may have to intervene a couple of times to reseat a bag that has slipped or fallen. The robotic system can fill ULDs to around 80 percent capacity. A baggage handler usually can fill the remaining space by hand. Practical build-cell designs recognize this and integrate both the automated element and the manual topping-up element, combining the cell supervisor's role with that of the baggage handler.

A semiautomated batch build arrangement employs a steerable, extendable conveyor controlled by an operator. This device is used to deliver bags into a trailer or ULD. The speed of placement can be much greater than that of a robot-based system, given that bags are delivered to the device sufficiently quickly—10 bags per minute can be achieved. Increasing the build rate allows build open times to be reduced. A conventionally built long-haul flight might be open for three hours, during which time 12 ULDs might be filled. Assuming about 40 bags per ULD, the average work rate is two to three bags per minute. The batch build arrangement, in theory, could be completed in less than an hour. Practical considerations mean that such a reduction in build time actually will not be possible, but halving the build time is conceivable given appropriate controls and logistic support (e.g., delivery and removal of ULDs from the makeup area). This can translate into reductions in both staffing and infrastructure, although this will depend on the specific pattern of flights and staff shifts.

To be used efficiently, both robot-based and semiautomated approaches require the baggage-handling system to be able to store, batch, and deliver bags for a single segregation (i.e., ULD or trailer). Figure 7.16 shows an example of a batch and compressed build process. The cost-benefit assessment of these concepts greatly depends on the cost of labor and the impact of health and safety regulations. For this reason, early adopters have been European airports.

ULDs that are filled with bags in a baggage makeup facility will be transported to the departure stand on dollies (Figure 7.17).

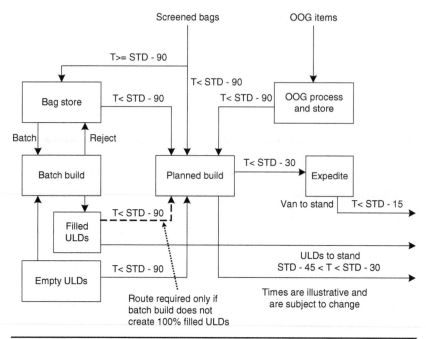

FIGURE 7.16   Batch and compressed build process.

FIGURE 7.17   Tug and dolly train.

Reclaim

The most common baggage reclaim device is a carousel, of which there are several variants. The two principal choices are

- Flatbed or inclined
- Direct or indirect infeed(s)

Flatbed carousels (Figure 7.18a) are preferred, if space permits, because bags are more easily picked off by passengers. An inclined carousel (Figure 7.18b) accommodates more bags per unit length—0.75 bag/foot (2.5 bags/meter) rather than 0.5 bag/foot (1.5 bags/meter) for a flatbed—but at the expense of bags being piled one upon another. This can make it difficult for passengers to retrieve their bags, particularly if theirs is trapped by a heavy bag that has fallen on top of it.

Bags can be loaded directly onto the device, or they can be fed indirectly via one or more conveyor routes. Direct loading has the advantage that with careful placement, a higher linear density of bags can be achieved than is possible with indirect feeds. However, by using indirect feeds, the adjacency between the reclaim carousel and the vehicle docks (where the bags are actually unloaded) can be relaxed. This may be desirable or even necessary to fit with a terminal building design.

(a)

FIGURE 7.18   (a) Flatbed reclaim. (b) Inclined reclaim.

(b)

Figure **7.18**   (Continued)

7.5   Process and System Design Drivers

Appearance Profiles

The appearance profile of bags at an airport is an important factor that influences the need for facilities to be open and available (e.g., check-in and transfer inputs), as well as the need for bag storage. The appearance profiles shown in Figure 7.19 are taken from a European hub airport for the major types of destinations. At first glance, the results suggest that the longer the journey, the earlier the bags will appear. In practice, the appearance profiles are also influenced by the consequences of missing a flight—if there are frequent flights to a destination, then passengers may be prepared to run the risk of missing one. If there is only a single flight a day, then passengers are more likely to play safe and arrive early.

The appearance of transfer bags is also shown. The profile shows the twin characteristics associated with transfer bags: early bags (which require storage) and late bags (which have very little time to reach the departing flight).

Bags per Passenger

Bag-per-passenger ratios are a key component in the design basis for baggage facilities, and they vary considerably by type of passenger

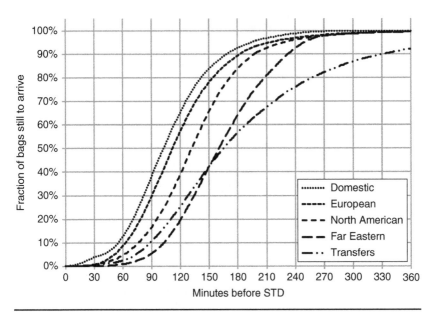

FIGURE 7.19   Appearance profiles.

and route. Direct, short-haul traffic generally is characterized by a relatively low bag-per-passenger ratio (0.5 to 0.9 bags per passenger), whereas long-haul and transfer traffic tends to have a higher bag-per-passenger ratio (1 to 2 bags per passenger). But such generalizations can be deceptive because baggage loads can be affected by time of year (e.g., ski season) or holidays (e.g., passengers traveling to stay at a destination for a week or more who have more baggage than, say, someone traveling on a shorter business trip). Some routes to and from developing countries also attract disproportionately large baggage loads because hold baggage is used to transport trade goods that otherwise might be handled as cargo.

Transfer Ratios

The *transfer ratio* is calculated by dividing the number of transfer passengers by the total number of passengers on a flight. Since direct and transfer passengers (and their bags) usually make use of different facilities, an understanding of how baggage demands are split between these two processes is important in sizing facilities. Overall transfer ratios are often quoted for a terminal or airport. While such single figures provide an indication of the nature of demands, they conceal a large variability that is vital to understand when designing systems and processes.

Parameter	Value
Check-in process	1–2 minutes per person
Bag drop process	0.5–2 minutes per bag (prelabeled to full-service)
ULD build rate	3–4 bags per minute
ULD break rate	8–12 bags per minute
In-line baggage-screening rate	15–20 bags per minute per machine (standards 1 and 2)
Aircraft ULD un/loading process	3 minutes per pair of AKE ULDs (one hold)
Reclaim input rate	20 bags per minute

TABLE 7.1 Processing Times

Processing Times

The number of facilities required to service a given demand depends on the processing times associated with that particular facility. Table 7.1 lists a number of important parameters.

7.6 Organization

Growth in the volume of baggage handled, coupled with the constant search for economies by airports and airlines, has led to gradual changes in the organization for this task. There has been a growing tendency for airlines and airports that have previously carried out the task of baggage handling to transfer it to handling agents, whether to another airline's handling company (e.g., Emirates' Dnata) or to an independent company (e.g., Menzies at London Heathrow). The tendency for airports in Europe to enjoy monopoly handling rights was challenged by the European Commission (EC). There is increasing pressure for the establishment of competing companies to carry out ground handling, including baggage handling, based on the argument that such competition will result in lower costs to airlines together with improved efficiency. Where an airline is a major operator at a particular airport, however, it is more usual for it to use its own personnel for baggage handling (e.g., British Airways at London Heathrow).

Staffing

As with all other aspects of air transport operation, the peaks and troughs of traffic so typical of the industry present problems to management when attempting to determine the level of staffing needed for any operation. There are obvious constraints in terms of costs, and as a result, there can be only limited response to the possibility of diversions or bunching of arriving flights. Where premium service is demanded and paid for, then special effort can be made, and a high level of staffing is assigned. Normally, however, there will be a compromise and a tacit acknowledgment that there probably will be a few occasions when staffing levels will be inadequate in the face of abnormal demand.

The largest group of personnel engaged in handling baggage consists of those who deal with it on the ramp, transporting baggage to and from the aircraft and loading and unloading the hold. Ramp personnel must be allocated by some system to individual flights, and this necessitates an oversight of ramp activity.

The basic method of allocating staff to flights is tackled in a variety of ways. At low-activity stations, this is not a complicated procedure and merely requires the lead hand (head loader) personally to allocate staff based on personal experience. At higher-activity stations, where handling staff might number several hundred, it is usual to find specialist staff employed as allocators. Their task is not only to ensure the necessary number of staff for a particular flight but also to ensure a reasonably fair distribution of the workload. In order to satisfy these requirements, it is essential for staff allocators to have available up-to-the-minute details of flight arrivals and departures, as well as prior notice of the load on board an arrival or the load planned for a departure. There is less of a problem in this respect if an airline is doing its own handling, but information easily can be delayed or forgotten when it has to be passed to another organization. All too often this is manifested by the unannounced flight. The establishment of a direct link between staff allocators and air traffic control (ATC), where possible, should ensure that accurate, up-to-the-minute times are available.

Increasingly, computerized resource-management systems are being used to manage handler task allocations. These involve a centralized management system linked to mobile data terminals in handlers' vehicles. The handlers respond to tasks that are presented to them in the vehicle cab—acknowledging the task, confirming that they are undertaking the task, and indicating when the task is finished so that a new task can be allocated.

Of course, the availability of mobile radios and telephones also has greatly assisted with on-the-spot last-minute changes or problems being encountered at the planeside. The whole system of allocating staff for baggage handling plays a vital part in achieving the effectiveness of the overall operation of aircraft turnaround.

7.7   Management and Performance Metrics

Well-defined performance metrics are an important part of the management of baggage-handling processes and systems. There are measures of the overall end-to-end performance of the baggage process, as well as subsidiary measures that focus on particular elements within the end-to-end process.

Overall

The industry-standard measure of success is the *short-landed rate*. This is the number of bags reported missing at the destination per 1,000 passengers flown—the lower this ratio, the better is the performance of the end-to-end baggage process. This ratio varies by airline, but is typically on the order of 1/1,000 for direct bags.

The short-landed rate for transfer bags is higher than for direct bags. This varies greatly by airline, route, and other factors, but typically is on the order of 5 to 50/1,000. This reflects the fact that a transfer bag is at greater risk of missing its connecting flight than one that is checked in directly. The reason for this is that the inbound leg of a transfer bag's journey is more variable. Factors include

- Late inbound aircraft, leading to little or no time to make the connection

- Poor segregation and loading of time-critical bags on inbound aircraft

- Poor handler performance in unloading and delivering bags to the baggage system

- Poor bag tag quality, leading to the need to manually code the bag

- Lack of data from the inbound airline, leading to the inability to sort the bag to the correct makeup position

At a hub airport, the overall short-landed rate is dominated by the transfer short-landed rate. For example, if the *transfer ratio* (the fraction of transfer passengers to the total of all passengers) at an airport is 50 percent, the direct short-landed rate is 1/1,000, the transfer short-landed rate is 40/1,000, then the overall rate is 20.5/1,000. Even if the direct rate were reduced to 0/1,000, the overall rate still would be 20/1,000. This explains why baggage performance-improvement programs at hub airports have to focus on the transfer-baggage product. It also shows that when comparing the baggage performance of different airports, it is vital to understand each airport's transfer ratio.

When comparing transfer short-landed rates between airports, it is important to bear in mind that different airports can and do offer different minimum connection times. Thus, a performance of 20/1,000 with a minimum connection time of 45 minutes will involve much better processes, systems, and operations than the "same" performance of 20/1,000 with a minimum connection time of 75 minutes.

This illustrates that there is a tradeoff between shorter minimum connection times and lower short-landed rates. Indeed, at one point,

Emirates, for a while, chose to increase its minimum connection time through its hub in Dubai in order to improve its short-landed rate. In other markets, though, there is perceived to be a competitive advantage to offering lower minimum connection times, resulting in a challenge to manage short-landed rates within tolerable levels.

Baggage System

Under normal circumstances, baggage-handling systems contribute only a very small fraction to the overall short-landed rate. The system-related measure is the system-attributable mishandled-bag rate. This is the number of bags that are mishandled by the system (e.g., delivered late or to the wrong output) per 1,000 bags handled by the system. Values depend on the complexity and extent of the system but typically are on the order of 0.1/1,000—in other words, an order of magnitude smaller than the direct short-landed rate.

The time it takes a bag to be processed through a baggage system can be important. For a small, simple direct system, the time from check-in to output may be only a few minutes and so is only a minor element of the end-to-end process. In contrast, a large baggage system with distributed inputs and outputs across several terminal and concourse buildings typically will have an in-system time of 10 to 20 minutes depending on its scale and the processing required. Such times become a significant part of a minimum check-in time of, say, 30 minutes before departure or a minimum connection time of, say, 45 minutes and therefore need to be monitored.

For systems that have no integrated bag storage, a simple measure of in-system time is likely to be sufficient to monitor system performance, although this has to be coupled with a measure of availability of sufficient input capacity, whether at check-in or at transfer inputs.

For systems that do have storage and buffering (and especially systems that make use of some form of the batch-building concept), an in-system time is of little relevance for the majority of bags that enter the system with plenty of time to go. They are simply held within the system until such time as they are ready to be delivered and made up. Nevertheless, in-system times remain vital for time-critical bags and should be monitored.

Arrivals Delivery Performance

The speed of delivery of bags from an inbound aircraft to either a reclaim device (for terminating bags) or the input of the baggage-handling system (for transfer bags) is the key measure of handler performance. Historically, this has been measured by first and last bag delivery times—for example, first bag on reclaim within 15 minutes and last bag on reclaim within 25 minutes of aircraft arrival on chocks. Such measures have the benefit of simplicity and can be used to encourage good handler performance, but three trends mean that more refined targets are becoming necessary at some airports:

- An increase in the number of very large aircraft
- A desire to reduce minimum connection times
- An increase in the size of airports and hence distances between facilities

The implications of these trends are described in turn. First, a performance standard based on delivering, say, 250 bags from a medium-sized aircraft becomes challenging to achieve for a very large aircraft with 500 or more bags. Second, the need to achieve reliable, short-transfer connection times (especially from very large aircraft with many transfer bags) means that a tighter performance standard needs to be applied to the time-critical transfer bags while allowing more time for non-time-critical bags. Third, large airports (without distributed arrival baggage systems) inevitably lead to longer driving times from some stands to reclaims than from others, making a "one size fits all" standard inappropriate.

In order to deal with the growth in size and scale, different priorities can be assigned to the four main categories of inbound bags:

- Premium terminating (e.g., first class, business class, frequent-flyer cardholders)
- Economy terminating
- Short-connect transfers (with scheduled connection times of less than about 2 hours)
- Long-connect transfers (with scheduled connection times of more than about 2 hours)

Logic dictates that premium bags should be delivered before economy bags and that short-connect bags should be delivered before long-connect bags. The only remaining choice is whether to prioritize premium bags over short-connect bags or vice versa. Long-connect bags should be given the lowest priority in any case. Of course, the ability to fine-tune the delivery of these different categories depends on the appropriate segregation and loading of the inbound aircraft.

For reclaim, it is desirable to set targets for the delivery of bags *relative to the arrival of passengers* in the reclaim hall. For example, the aim might be to deliver all premium bags before the first passengers reach the reclaim hall so that no premium passengers have to wait for their bags. A maximum-waiting-time target might be set for economy passengers. In practice, this can be hard to measure and control. While processes and systems can be put in place to log when a bag is delivered to the reclaim device, it is much harder to monitor the arrival times at reclaim of specific passengers. Another difficulty is the spread in passenger processes from disembarkation to arrival in the reclaim hall. A small aircraft, parked at the main terminal building with domestic passengers who do not need to clear immigration, can mean passengers reaching the reclaim hall within a few minutes of arrival on stand. In contrast, a large aircraft, parked remotely, with many international passengers requiring complex immigration and/or customs checks, can mean passengers taking an hour or more to reach the reclaim hall.

This illustrates rather clearly that passengers' perceptions of the performance of the baggage-reclaim function are influenced not so much by the absolute time it takes for bags to be delivered but by whether or not their bags are waiting for them—a long immigration process can make a mediocre baggage-delivery performance appear to be very good.

References

International Air Transport Association (IATA). 2004. *Airport Development Reference Manual*, 9th ed. Geneva: IATA.

Transportation Security Administration (TSA). 2011. *Planning Guidelines and Design Standards for Checked Baggage Inspection Systems*. Washington, DC: TSA.

CHAPTER 8

Passenger Terminal Operations

8.1 Functions of the Passenger Terminal

Analysis of the operation of an airport passenger terminal leads to the conclusion that three principal transportation functions are carried out within the terminal area (Ashford et al. 2011):

1. *The processing of passengers and baggage.* This includes ticketing, check-in and baggage drop, baggage retrieval, governmental checks, and security arrangements.

2. *Provision for the requirement of a change of movement type.* Facilities are necessarily designed to accept departing passengers, who have random arrival patterns from various modes of transportation and from various points within the airport's catchment area at varying times, and aggregate them into planeloads. On the aircraft arrivals side, the process is reversed. This function necessitates a holding function, which is much more significant than for all other transport modes.

3. *Facilitating a change of mode.* This basic function of the terminal requires the adequate design and smooth operation of terminal facilities of two mode types. On the airside, the aircraft must be accommodated, and the interface must be operated in a manner that relates to the requirements of the air vehicle. Equally important is the need to accommodate the passenger requirements for the landside mode, which is used to access the airport.

An intimation of the complexity of the problem can be grasped from an examination of Figure 8.1, which is admittedly a simplification of the flow process for passengers and baggage through a typical domestic-international airport passenger terminal. When examining a chart of this nature, it must be remembered that the representation can only be in generalized terms and that the complexities of operation are introduced by the fact that flows on the

213

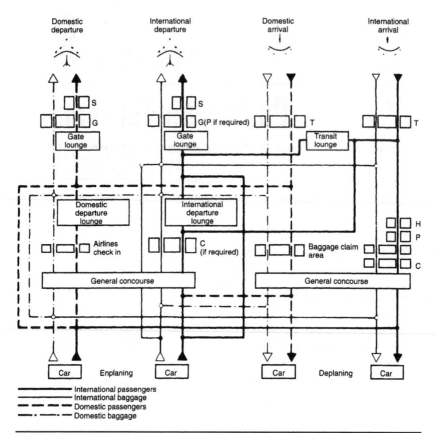

Figure 8.1 Schematic of the passenger baggage flow system (G = gate control and airline check-in, if required; P = passport control; C = customs control; H = health control, if required; T = transfer check-in; S = security control). (*Source:* Ashford et al. 2011.)

airside are discrete and those on the landside are continuous. The substantial growth rate of air transportation since World War II has meant that many airports around the world are now large operations. Unlike the pre-1940 period, when air transportation was a fringe activity on the economy, the air mode is now a well-established economic entity. The result on passenger terminals has been dramatic (Hart 1985). More than a score of large international airport terminals are handling more than 30 million passengers per year, and the number continues to grow. Operations of this scale are necessarily complex.

The relatively recent development of large air passenger volumes has required the provision of increasingly large facilities to accommodate the large peak flows that are observed routinely (see Section 2.2). Single terminals designed for capacities in the region of 10 million passengers per year often have internal walking distances of 3,500 feet

(1,100 m) between extreme gates. Where capacities in excess of 30 million annual passengers are involved, largely single-terminal complexes, such as Chicago O'Hare and Schiphol Amsterdam, are likely to have internal gate-to-gate distances in the region of 5,000 feet (1,500 m). To overcome problems such as this, and to meet International Air Transport Association (IATA) recommendations on passenger walking distances, several "decentralized" designs were evolved, such as those now in operation at Kansas City, Dallas–Fort Worth, and Paris Charles de Gaulle II. Decentralization is achieved by

1. Breaking the total passenger terminal operation into a number of unit terminals that have different functional roles (differentiation can be by international-domestic split, by airline unit terminals, by long-haul–short-haul divisions, via airline alliance terminals, etc.)

2. Devolving to the gates themselves a number of handling operations that previously were centralized in the departure ticket lobby (e.g., ticketing, passenger and baggage check-in, seat allocation, etc.)

Coupling a decentralized operational strategy with a suitable physical design of the terminal can result in very low passenger walking distances, especially for routine domestic passengers. Where considerable interlining takes place, or where the passenger's outbound and inbound airlines are likely to differ, decentralization is likely to be less convenient to travelers. For example, one of the earlier decentralized designs, Dallas–Fort Worth (Figure 8.2a), can be less convenient for an interlining passenger who has to change terminals than the newer Atlanta design (Figure 8.2b). International operations significantly affect the design of terminal facilities and the procedures used. From this viewpoint, the airport planner and operator must be extremely careful in extrapolating U.S. experience, which, although well documented, is likely to be based overwhelmingly on domestic operations (FAA 1976, 1980, 1988). The infusion of governmental requirements necessarily associated with international operations (i.e., customs, immigration, health and agricultural controls, and especially security) can add considerable complications to the layout and operation of a terminal. Separation is required in some European Union countries for operations that are within the Schengen group and those which are not.

The Eurohub terminal at Birmingham, the United Kingdom has a most complex arrangement of the interlocking doors to allow for flows among international, domestic, and "common travel" passengers,[1] who

---

[1]*Common travel* is travel between the Channel Islands and Britain. These journeys are subject to customs inspection, but not to immigration controls.

Figure 8.2 (a) Decentralized terminals of Dallas–Fort Worth International Airport. (*Dallas–Fort Worth International Airport*). (b) Aerial view of Atlanta Hartsfield International Airport.

must be segregated (Blow 1996, 2005). The complicated door system is centrally operated by a computerized control room with extensive closed-circuit-television monitoring. Figure 8.3 shows conceptualized processing outbound flow patterns for centralized and decentralized

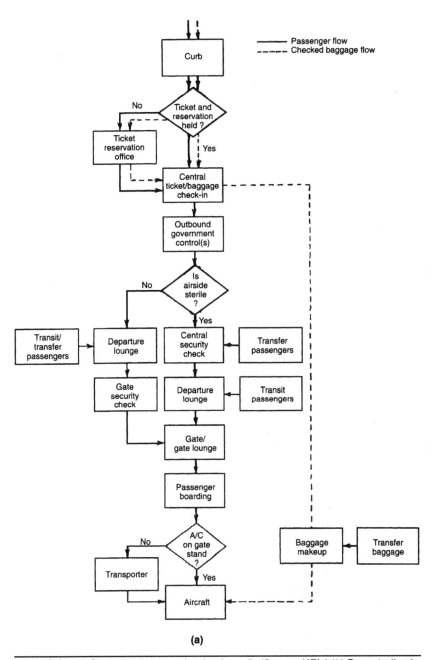

(a)

FIGURE 8.3 (a) Centralized processing (outbound). (*Source:* IATA.) (b) Decentralized processing (outbound). (*Source:* IATA.)

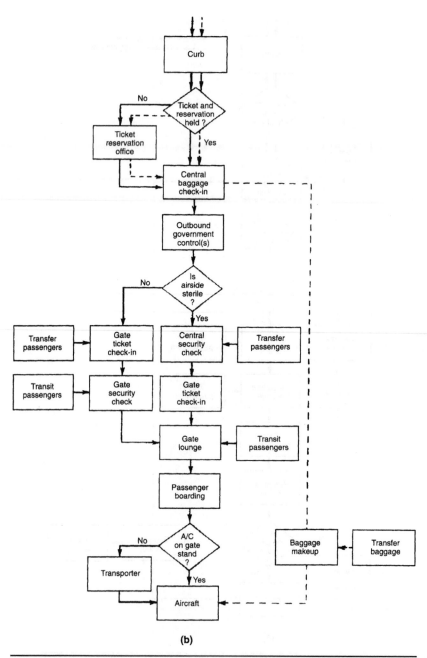

(b)

FIGURE 8.3   (Continued)

facilities. In almost all countries, it is not possible for outbound passengers to pass back through the governmental controls, and universally, airport visitors are now precluded from both domestic and international departure lounges. As a result, many passenger-related facilities must be duplicated, as will be discussed in Sections 8.4 and 8.5. In many countries, there is also a governmental requirement for security purposes to separate international arriving and departing passengers. In terms of space, this has been found to be necessary but very expensive, leading to considerable duplication of facilities and staff. Mixed arrival-departure areas are no longer accepted at most European airports. Where they are accepted, such as at Schiphol Amsterdam, which has a large number of international transit passengers, such passengers must submit to a security search at the gate. As a general rule, the inclusion of international operations must be seen as a complication of terminal processing activities that cuts down on the use of multiple-purpose space, requires duplication of facilities, necessitates additional processing space, and inevitably increases the number of languages involved in the operation.

8.2   Terminal Functions

Transportation planners use the term *high-activity centers* to describe facilities such as airport terminals that have a high throughput of users. In the peak hour, the largest passenger airports process well in excess of 10,000 passengers. With the increased security measures since 2001, departing international passengers are likely to spend 1½ to 2 hours in the terminal facility, and arriving international passengers spend at least 30 minutes. During the period that they spend in the terminal, passengers are necessarily engaged in a number of processing activities and are likely to use a number of subsidiary facilities put in the airport for their comfort and convenience, as well as for the airport's profit. Before discussing in some detail these individual activities, it is worth classifying the terminal activities into five principal component groups:

- Direct passenger services
- Airline-related passenger services
- Governmental activities
- Non-passenger-related airport authority functions
- Airline functions

Either directly or indirectly, these functions, where conducted in the passenger terminal area, will involve some responsibility on the part of the terminal manager. Figure 8.4 shows the typical organization of these responsibilities for the terminal operation at a major airport.

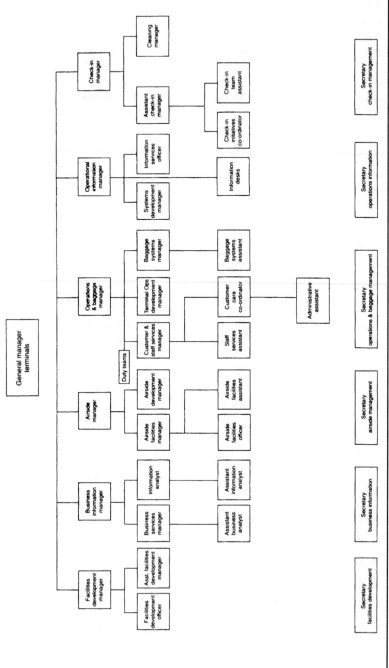

Figure 8.4 Organizational structure of terminal management.

The individual terminal functions are discussed in more detail in Sections 8.4 to 8.12.

8.3   Philosophies of Terminal Management

Although the basic operational procedures of airports as they relate to safety are generally similar throughout the world, the manner in which those procedures are operated and the organization used to effect them can differ quite radically. Perhaps nowhere in the airport do the operational philosophies differ as much as in the terminal area. The two extreme positions may be designated as

- Airport-dominant
- Airline-dominant

Where terminal operations are airport-dominant, the airport authority itself provides the staff to run terminal services. Apron, baggage, and passenger handling are either entirely or largely carried out by airport-authority staff. Services and concessions within the terminal are also mainly authority operated. Airport-dominant operations are sometimes called the *European model*, although similar arrangements are found throughout the world. Frankfurt is perhaps the best example of this form of operation, which involves high airport-authority staffing levels and high authority equipment costs with concomitant savings to airlines.

Most major airports around the world work on a mixed model, where the airport authority takes care of some terminal operations, and airlines and concessionaires operate other facilities. In some airports, competitive facility operation is encouraged to maintain the high service standards usually generated by competition. In the European Union, European Commission (EC) directives are forcing airports to introduce competition at airports where operation previously has been by a single organization. This trend away from single-authority operation has been aided by the increasing trend toward total airport privatization, either by outright transfer of ownership outside the public sector or by the granting of long-term concessions for the operation of entire airports.

Competitive handling operations are also less vulnerable to a complete shutdown by industrial action. The final choice of operational procedure will depend on a number of factors, including

- Philosophy of the airport authority and its governing body
- Local industrial relations
- International and national regulations
- Financial constraints
- Availability of local labor and skills

8.4   Direct Passenger Services

Terminal operations that are provided for the convenience of air travelers and are not directly related to the operations of the airlines are normally designated as direct passenger services. It is convenient to further divide this category into commercial and noncommercial services. There is no hard-and-fast division between these two subcategories, but noncommercial activities are usually seen as being entirely necessary services that are provided either free of charge or at some nominal cost. Commercial activities, on the other hand, are potentially profitable operations that are either peripheral to the transportation function of the airport (e.g., duty-free shops) or avoidable and subject to the traveler's choice (e.g., car parking and car rental).

Typically, at a large passenger terminal, the following noncommercial activities will be provided, usually by the airport authority:

- Porterage[2]
- Flight and general airport information
- Baggage trolleys
- Left-luggage lockers and left-luggage rooms[2]
- Directional signs
- Seating
- Toilets, nurseries, and changing rooms
- Rest rooms
- Post office and telephone areas
- Services for people with restricted mobility and special passengers[3]

Depending on the operating philosophy of the airport, commercial facilities will be operated either directly by the authority itself or leased on a concessionary basis to specialist operators. Typically, at a large airport, the following commercial activities can be expected to play and important part in the operation of the passenger terminal:

- Car parking
- Restaurants, cafés, and food bars
- Duty-free and tax-free shops
- Other shops (e.g., book shops, tourist shops, boutiques, etc.)
- Car rental
- Internet service

---

[2]Airports usually make commercial charges for these facilities.
[3]Some airports make charges for some of these facilities.

- Insurance
- Banks and exchange services
- Hairdressers, dry cleaners, and valet services
- Hotel reservations
- Amusement machines, lotteries
- Advertising
- Business-center facilities

Figures 8.5 and 8.6 show examples of commercial duty-free shops and advertising at airports having an aggressive commercial policy. The degree of commercialization of airports varies substantially. Airports that have adopted policies promoting such activities, such as Frankfurt, Singapore, Amsterdam, London, and Orlando, have commercial revenues that account for up to 60 percent of total revenues. Other airports that have no strong commercial development, either as a policy decision or owing to a lack of opportunity, typically would expect up to only 10 percent of their income to come from commercial sources.

Early arguments in the aviation world against the commercialization of airport terminals are now clearly lost. It is generally accepted that there is a demand for such facilities generated by the high volumes of passengers who can spend on average two hours in a terminal, and

Figure 8.5 Duty-free shops.

FIGURE 8.6   Advertizing display case.

of this time, perhaps only 30 percent is required for processing. The high volumes of passengers, meeters, senders, and visitors constitute a strong potential sales market that invariably can be developed, if desired. Furthermore, the revenues generated by commercial operations can cross-subsidize airside operations, which often are only marginally profitable. Passenger terminals are recognized as part the generation of the airport's revenue stream that can make the facility self-sustaining or even profitable. Large passenger terminals are generators of large commercial profits. If commercial exploitation of the airport is decided on, a number of operational policy decisions must be made. First, a decision must be made on the mode of operation. Five different modes are common; these are operation by

- A department of the airport authority directly
- A specially formed, fully owned commercial subsidiary of the airport authority
- A commercial subsidiary formed by the airport authority and the airlines
- A commercial subsidiary formed by the airport authority and a specialist commercial company
- An independent commercial enterprise

Some publicly owned airports choose to retain direct control of commercial operations. This option, however, is unusual. Most airports that run highly successful commercial operations, such as Dubai, Heathrow, Atlanta, and Frankfurt, generally prefer to use an approach of granting controlled concessions to independent enterprises with commercial experience in the particular area. However, Aer Rianta, the Irish Airports Authority, operates many of its own concessions through its highly successful commercial division, which also acts as a concessionary management organization to other airports. The contractual arrangements between the concessionaires and the authority ensure certain standards of service to the consumer and guaranteed profit levels to the authority: Beyond these guarantees, the concessionaire is free to use his or her enterprise to maximize commercial opportunities and therefore profit. Hybrid arrangements in which the authority collaborates either with the commercial departments of an airline or directly with specialized enterprises have been equally successful. Table 8.1 shows how various concessionary arrangements have been handled according to a survey of approximately 70 international airports.

It is also interesting to compare the ways in which concessionaires are selected. Some governmental airport authorities are required by law to accept the highest bid for a concession. Schiphol Airport in Amsterdam developed a successful commercial policy based rather on maximizing the level of airport control on operating standards and pricing. In this way, the airport authority feels that it is more able to attain its own commercial ends while still using the expertise of the individual concessionary enterprises. Concessions at airports may be leased in a number of ways:

- Open tender
- Closed tender
- Private treaty

Of these three, it is most likely that the second option, closed tender, will meet a publicly owned airport's requirements. Private treaty is likely to be seen to be a too restrictive manner of handling public funds, leading to charges of preferential treatment. Open tender, on the other hand, while giving a free hand to competition, may well lead to bidding by organizations that will prove to be incompetent in reaching necessary performance standards. In some countries, however, open tenders are legally required where public funds are involved. Under these conditions, it is sometimes permissible to have a prequalification arrangement to ensure that only competent and financially stable enterprises enter the bidding process. At privatized airports, the airport can use any legal means of granting the concessions it chooses.

Facility	Concessionary Operation	Operation Directly by Airport Management	Operation by Subsidiary Company of the Airport	Operation by Airline	Not Available at this Airport
Duty-free liquor and tobacco	69%	8%	4%	0%	18%
Duty-free shops (other goods)	72%	10%	4%	0%	14%
Specialized shops and facilities	79%	6%	3%	1%	11%
Local souvenir shop	80%	7%	3%	1%	8%
Gift shop	84%	7%	1%	1%	6%
Jewelry and gems	75%	6%	0%	1%	17%
Clothing	72%	3%	0%	1%	24%
Confectionery	73%	6%	0%	0%	21%
Pharmacy	48%	1%	0%	0%	51%
Perfumes	72%	7%	3%	1%	17%
Medical center	26%	21%	2%	0%	52%
Catering	60%	9%	6%	10%	16%
Specialist food	61%	6%	0%	2%	32%
Flowers	38%	4%	0%	0%	58%

Photographic equipment	41%	9%	0%	0%	51%
Electrical goods	54%	7%	0%	0%	38%
Hairdressing	18%	1%	0%	0%	81%
Nail bar	13%	3%	0%	0%	84%
Shoeshine	25%	1%	0%	1%	72%
Games room	10%	6%	0%	0%	84%
Crèche	3%	12%	0%	0%	85%
Casino	3%	0%	0%	0%	97%
Internet access	36%	49%	6%	0%	9%
Car park	39%	60%	1%	0%	0%
Car rental	83%	7%	0%	0%	10%
Hotel reservations	38%	12%	0%	3%	48%
Television	11%	47%	3%	0%	39%
Cinema	3%	0%	0%	0%	97%

TABLE 8.1 Operational Mode of Concessions at Surveyed Airports

Other methods of control that have been used successfully include

1. *Length of lease.* Medium-term leases of 5 to 10 years have several advantages. They permit the concessionaire to run an established operation with medium-term profits. Successful operators usually are able to renegotiate for renewed concessionary rights. Unsuccessful operators can be removed before long-term financial damage accrues to the airport.

2. *Exclusive rights.* In return for exclusive rights on the airport, the authority can demand contractual arrangements that protect the airport's financial and performance interests. There is a significant recent move away from granting exclusive rights in shopping concessions in order to encourage competitive pricing.

3. *Quality of service.* Many airports require contracts that restrict the concessionaire's methods of operation. These constraints include authority control over the range of goods to be stocked, profit margins and prices, and staffing levels, as well as detailed operational controls on such items as advertising, decor, and display methods.

Where the airport is privately owned, there are no limits on how the concessionary contracts can be drawn up. If the operator of the airport is itself a concession, the government may impose limits on how subconcessions are to be arranged.

Advertising is an area of financial return that has not been fully explored by many airports. The advertising panel shown in Figure 8.6 is an example of a very satisfactory modern display that adds to the decor of the terminal without clutter while paying a handsome return to the authority from a little financial outlay. Care must be taken in selecting advertising so that the displays do not interfere with passenger flow or obstruct necessary informational signs. Significantly, there are airports that ban internal advertising on aesthetic grounds, but these are growing fewer in number.

8.5 Airline-Related Passenger Services

Within the airport passenger terminal, many operations are usually handled entirely by airlines or their handling agents,[4] including

- Airline information services
- Reservations and ticket purchases
- Check-in, baggage check-in, handling of bag drop and storage

---

[4]At many airports, although visually the passenger is led to consider that the passenger handling is being carried out by the airline, often these operations have been consigned to the airline's handling agent.

- Loading and unloading of baggage at the aircraft
- Baggage delivery and reclaim (reclaim is often under authority control)
- Airline passenger "club" areas, sometimes called *commercially important persons (CIP) facilities*

These areas are part of the service offered to the traveler by the airline and, as such, the airline has an interest in retaining a strong measure of control over the service given. Such control is obtained most easily by carrying out this particular area of the operation. It is important to remember that the basic contract to travel is between the airline and the passenger. The airport is a third party to this contract and, as such, should not intrude into the relationship more than is necessary. Where airports remove the general handling responsibility from the airlines, there might be an adverse impact on passenger service because there is no overt contract between the passenger and the airport. Service levels are more likely to be maintained where the direct customer relationship has some influence on services performed.

The relationship becomes complicated when the airport is privatized or has extensive terminal commercial operations. The passenger in this case also becomes the airport's client in a very real sense. Figures 8.7 and 8.8 show check-in and baggage-delivery areas where the design of the facility emphasizes that the passenger is under the care of the airline. A more common arrangement for baggage-claim areas outside the United States is that the claim

Figure 8.7 Check-in showing area under lease to airline.

FIGURE **8.8**   Designated baggage-delivery system.

area is operated by the airport authority; the airlines have the responsibility of delivering bags to the claim area. This more common arrangement often results in authority staff receiving passenger abuse for delayed, lost, or damaged baggage when, in fact, the receiving airport has had no involvement in its handling prior to delivery to the baggage area and, thus, bears no blame for the default.

In recent years, airports have attempted to obtain better usage of the check-in desks by the adoption of common-user terminal equipment (CUTE) in the check-in area.[5] Use of CUTE technology permits the switching of desks among airlines according to their real demand for desks, which is likely to vary both seasonally and over the day. Many airlines resisted the introduction of CUTE because it prevented the airline from having a permanent presence in the terminal, whether or not it had operations at a particular time. Most new terminals are being designed with CUTE or CUSS systems, where there are shared facilities (IATA 2004).

8.6   Airline-Related Operational Functions

Flight Dispatch

A major preoccupation for airline management in relation to airport terminal operations is the achievement of on-time departures. Many of the activities associated with this, such as the refueling and cleaning

---

[5]See also Section 6.2 for explanations of CUTE and CUSS.

of aircraft, together with the loading of food supplies, are carried out on the ramp and are familiar to most airport staff. There is, however, a less familiar procedure that covers all the necessary technical planning without which a flight could not depart. The main activities associated with this procedure of flight dispatch are

- Flight planning
- Aircraft weight and balance
- Flight-crew briefing
- Flight watch

In the United States, this is a long-established procedure, and the work is carried out by aircraft dispatchers who work in close cooperation with the aircraft captain. However, in the case of large airlines, the flight-planning process is carried out more often as a central function at the airline's home base (for American Airlines, this is Dallas). Although aircraft dispatchers are used by many international airlines, there is also the designation of flight operations officer for staff members who carry out this work.

The airline departments at airports concerned with flight dispatch will need access to airport operations departments, air traffic services, meteorological services, and communications facilities, including email, internet, teleprinters, telephones, and radios. Depending on the extent of their activities, many airline operations offices also will use a variety of computer facilities, although these latter may not necessarily be in-house systems.

Flight Planning

The primary purpose of flight planning is to determine how long an individual flight will take and how much fuel will be required. For long-range flights, there will be a variety of options in terms of altitudes, tracks, and aircraft power settings and speeds. Variations in weather, wind, and temperature also will have to be taken into account. Of course, computerized flight-planning tools are used by major airlines to perform these optimizations. Such tools examine feasible options so that a decision can be taken as to the most appropriate of the several alternatives. The evaluation might include an indication of comparative costs: A slower flight might prove desirable from a cost point of view. The analysis would include several altitude options. This often proves useful if, owing to the density of traffic, air traffic control (ATC) has to impose a last-minute altitude change.

For short-range flights, there are generally very few options, and in areas of very dense traffic, routings for all practical purposes are predetermined by the structure of the airways. In such cases, such as, for example, in Europe, the flight plans usually will be standardized to the extent that relevant extracts can be placed on permanent file with ATC. These are referred to in Great Britain as *stored flight plans* and are automatically printed out from ATC computer files in advance of flight departures. The airline flight plans, the operational or company flight plans, give a great deal of information, including the en-route consumption of fuel. Such details are not the concern of ATC, which requires altitudes and times in relation to the ATC system checkpoints, together with certain safety details (e.g., number of persons on board the aircraft and the detail of the instrument-flying aids and safety equipment carried by the aircraft). The international format for the ATC flight plan is shown in Figure 8.9.

Aircraft Weight and Balance

The *dry operating weight* of an aircraft is taken as the starting point for weight calculations. To this is added the anticipated payload, which consists of

- Cargo load
- Passengers
- Baggage

This provides the *zero-fuel weight*. The total fuel load is added, less an allowance for fuel used in taxiing before takeoff, to calculate the *takeoff weight*. The fuel that is expected to be consumed during the flight is deducted from the *takeoff weight* to calculate the *landing weight*.

It should be noted that these calculations may be in either pounds, which is the case in the United States, or in kilograms. However, before any actual load calculations can be carried out, account must be taken of the physical weight limitations, the design limits, of the aircraft structure in the various operation phases.

Takeoff

There is a *maximum takeoff weight* (i.e., at brake release) that the available power can lift off the runway and sustain in a safe climb. The value is established by the manufacturer in terms of ideal conditions of temperature, pressure, runway height, and surface conditions. Along with these values, the manufacturer will provide performance details for variations in any of these conditions.

In Flight

There are limits on the flexibility of the wings of each aircraft design. These are imposed by the upward-bending loads that the wing roots can sustain without breaking. The greatest load would be imposed if there were no fuel remaining in the wings (fuel cells), which is why the *zero-fuel weight* is taken as a limitation on fuselage load.

FLIGHT PLAN

PRIORITY: <<≡ FF →
ADDRESSEE(S): <<≡

FILING TIME → ORIGINATOR <<≡

SPECIFIC IDENTIFICATION OF ADDRESSEE(S) AND/OR ORIGINATOR

3 MESSAGE TYPE: <<≡ (FPL
7 AIRCRAFT IDENTIFICATION: -
8 FLIGHT RULES: -
TYPE OF FLIGHT: <<≡

9 NUMBER: -
TYPE OF AIRCRAFT
WAKE TURBULENCE CAT: /
10 EQUIPMENT: - / <<≡

13 DEPARTURE AERODROME: -
TIME <<≡

15 CRUISING SPEED: -
LEVEL →
ROUTE

<<≡

16 DESTINATION AERODROME: -
TOTAL EET HR. MIN
ALTN AERODROME →
2ND ALTN AERODROME → <<≡

18 OTHER INFORMATION

) <<≡

SUPPLEMENTARY INFORMATION (NOT TO BE TRANSMITTED IN FPL MESSAGES)

19 ENDURANCE HR MIN: -E /
PERSONS ON BOARD: → P /
EMERGENCY RADIO: UHF → R / U; VHF V; ELT E

SURVIVAL EQUIPMENT: → S; POLAR / P; DESERT D; MARITIME M; JUNGLE J
JACKETS: → J; LIGHT / L; FLUORES F; UHF U; VHF V

DINGHIES
NUMBER → D /; CAPACITY →; COVER → C; COLOUR → <<≡

AIRCRAFT COLOUR AND MARKINGS
A /

REMARKS
→ N / <<≡

PILOT IN COMMAND
C /) <<≡

FILED BY

SPACE RESERVED FOR ADDITIONAL REQUIREMENTS

Please provide a telephone number so our operators can contact you if needed

CA48/RAF2919 VER 1.5.3

Figure 8.9 International flight plan.

Landing

Depending on the shock-absorbing capabilities of the aircraft under-carriage, there is a *maximum landing weight* that it can support on landing without collapsing. Thus the three design-limiting weights are maximum takeoff weight, maximum zero-fuel weight, and maximum landing weight. Typical examples of these values for a Boeing 747-300 are

- Maximum takeoff weight 883,000 pounds (377,850 kg)
- Maximum zero-fuel weight 535,000 pounds (242,630 kg)
- Maximum landing weight 574,000 pounds (260,320 kg)

The completed flight plan will provide two fuel figures:

Takeoff fuel. This is the total amount of fuel on board for a particular flight. This does not include taxiing fuel but will include required fuel reserves for flight to an alternative destination or for holding or delay before landing.

Trip fuel. This is the fuel required for the trip itself, that is, between the takeoff and the point of first intended landing (it is also some-times referred to as *burnoff*).

In order to arrive at the maximum permissible takeoff weight, we compare three possible takeoff weights:

- Takeoff weight' = maximum takeoff weight
- Takeoff weight" = zero-fuel weight + takeoff fuel
- Takeoff weight'" = landing weight + trip fuel

The lowest of these three values is the maximum allowed takeoff weight, and this value minus the operating weight will give the allowed traffic load. These and other values are used in relation to aircraft weight calculations and load, and they also appear on the load sheet, for which there is a format agreed on by the IATA. Together with the values for takeoff fuel and trip fuel, the following operational figures are included in a load-sheet calculation:

- *Dry operating weight.* The weight of the basic aircraft, fully equipped, together with crew and their baggage, pantry/commissary supplies, and flight spares, but not including fuel and payload
- *Operating weight.* The sum of dry operating weight and take-off fuel
- *Takeoff weight.* The operating weight plus payload (traffic load)
- *Total traffic load.* The sum of the weights of the various types of load, that is, passengers, baggage, cargo, and mail, as well as the weight of any unit-load devices (ULDs, containers) not included in the dry operating weight

All these various weights appear on the load sheet together with a breakdown of the weight distribution.

Balance/Trim

Having ensured that the aircraft load is within the permitted weight limitations, it is then necessary to distribute the load in such a way that the center of gravity is within the prescribed limits. This is calculated by means of a trim sheet, which might be a separate form or part of a combined load and trim sheet (Figure 8.10). On the trim diagram, each of the aircraft's compartments is given a scale graduated either in units of weight, for example, 1,100 pounds (500 kg), or blocks of passengers (e.g., five passengers). Starting from the dry-operating-index scale, the effect of weight in each compartment then is indicated by moving the required number of units along the scale in the direction of the arrow and dropping a line down from that point to the next scale, where the process is repeated, ending up with a line projecting down into the center-of-gravity (CG) envelope, where its value is noted as a percentage of the wing mean aerodynamic chord (MAC). The outer limits of the envelope are clearly indicated by the shaded areas. Certain sections of the load-sheet side of the form are also shaded to indicate data that should be included in a load message to be transmitted to the aircraft destination(s). These functions are now almost universally computerized.

Loading

The distribution of the load into various compartments must be detailed for the information of ramp loading staff, and this is achieved by the issue of loading instructions, usually in the form of computer-drawn diagrams. In Figure 8.11, the details are given of the various container positions. Where containers are not used, it will be necessary at this stage to take into account limitations in respect to dimensions, vis-à-vis the measurements of the hatch openings and also maximum floor loadings, and the loading instructions will be drawn up accordingly.

All matters relating to the load carried on an aircraft and the position of the CG have such a direct influence on flight safety that the documents used are of considerable legal significance, reflecting as they do the regulations of each country. For this reason, they have to be signed by the airline staff responsible for these various aspects.

Flight-Crew Briefing

The purpose is to present to flight crew appropriate advice and information to assist them in the safe conduct of a flight. The information will include a flight plan and load details together with information regarding en-route and destination weather and notices regarding any unserviceabilities of navigation or landing aids. This latter information is contained in *Notices to Airmen* (NOTAMs), an internationally

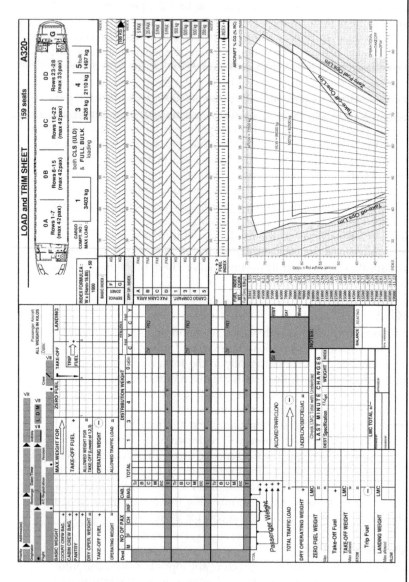

Figure 8.10 Load and trim sheet.

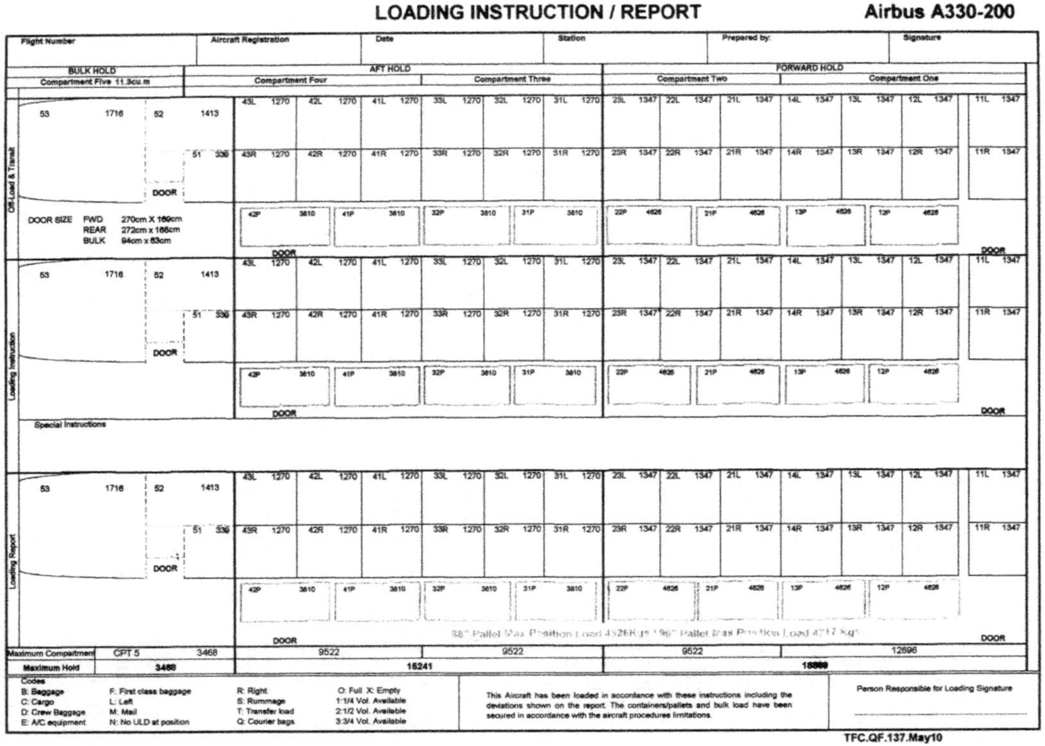

FIGURE 8.11 Loading instructions.

```
LAXFO
..LAXFOUA 231828 2970/ROS
WBM 108-23        LAX--ORD        RT: 19          ALTNT MKE
MAP FEATURES WESTERN U.S. 231652Z-240600Z
SFC TROF XTNDS NWWD FROM THE GULF OF CALIF THRU CALIF TO WRN
OREG.HIGH CNTRD OVER XTRM NW MONT DRFTG SE.FOG AND ST OVT THE
PAC NW LIFTING.MARINE ST CONTRL CALIF COAST LIFTING AND
RETURNING TOWARD 06Z.


LAX 1748 150 SCT E280 OVC 4H 110/90/51/1704/986/ 802 1087 65
LAX FT23 231616  180 SCT 250 SCT. 20Z 180 SCT 250 SCT 2512. 02Z
    CLR. 10Z VRF..
LAX NO 9/4 7L/25R OPEN 1600
LAX NO 9/5 EFF 1600 THR 25R DSPLCD 962
LAX NO 9/10 VORTAC OTS 20-2200
LAX NO 13/1 GATE 80 TOW-IN GATE. GATE 72 D737 ONLY NO
    ACCU-PARK, WILL HAVE SIGNALMAN. GATE 74 IS TOW-IN GATE FOR
    DC10 WHEN 83A OR 83 OCCUPIED.
LAX NO 13/2 RWY 25R/07L OPEN FOR ALL UAL ACFT, NO STRUCTUAL.
    WT RSTN, WIDEBODY'S DO NOT APPLY T/O THRUST UNTIL TAXIWAY OJ


ORD 1750 E100 DKN 250 DKN 15 164/64/44/2016/001/ 717 1031 40
ORD FT23 231515 80 SCT 250 --DKN 1810 SCT V BKN. 18Z C80 DKN 250
    DKN 2112 LWR DKN V SCT CHC C30 DKN 3TRW AFT 21Z. 097 VFR..
ORD NO 9/98 14R--32L CLSD 02--1400
ORD NO 9/106 14L--32R CLSD 16--1800
ORD NO 9/108 9R--27L CLSD 241100--1600
ORD NO 9/109 14R--32L CLSD 02--1400 NIGHTLY THRU 11/24 EXCP
    SUN


UA571 /OV HCT 1708 F390/TA MS56/WV 300 TO 305/WV AT 105 TO 110
    KNOTS/TD SMTH/SK NO CLDS BLO CLR
UA235 /OV GCK 1711 F390/WX OVER OCK WIND 045125 TS PLUS3... OVER
    CIM 1743Z WIND 035100 CLEAR WEST OF GCK
UA709 /OV SLN 1643 F350/TA MS26/WV 30090/SK CLD TOPS FL360/TD
    LT TURBC IN CLDS
UA235 /OV SLN 1649 F390/TD FL 390 SMOOTH ACFT BLG REPORTING MOD
    CHOP.TTSM  ACTVTY 40 DME N.IRK TOPS 890 EASY DETOURABLE
```

FIGURE 8.12   United Airlines flight-crew briefing sheet.

agreed-on system whereby the civil aviation authorities of each country exchange information on the unserviceability of any of the facilities in their country (e.g., navaids and airports). Airline flight dispatch staff will obtain NOTAMs from the appropriate governmental agency, edit them, and where necessary, add details relating to any company facilities. Weather information also will be obtained from the meteorological department at the airport and might be augmented by in-flight reports received from other flight crews. An example of the presentation of briefing information is given in Figure 8.12 for a flight from Los Angeles (LAX) to Chicago (ORD). Further details of the NOTAM system and of the various kinds of weather information available are given in Chapter 11.

Flight Watch (Flight Control)

This is a procedure by which flight dispatch/flight operations personnel monitor the progress of individual flights. For this reason, it is also sometimes described as flight following [not to be confused with the flight following by ATC in the United States for visual flight rules (VFR) aircraft]. Owing to the worldwide nature of air transport, it is carried out using Greenwich Mean Time (GMT), sometimes written as Z *time*. Flight watch is not intended to be entirely passive; however, information about any unexpected changes in weather or serviceability or facilities is transmitted to aircraft in flight. Depending on the extent of an airline route network, the responsibility for flight watch may be divided into areas. In addition, most larger airlines have one centralized coordinating operations center equipped with comprehensive communications facilities providing the latest information on the progress of all their aircraft. The center for United Airlines is located at Chicago O'Hare; for Air Canada, at Toronto International Airport; and for British Airways, at London Heathrow Airport. It is useful for airport operations management to know the locations and telephone/telex addresses of such centers for airlines using their airports, as well as the organization of flight-watch responsibility.

8.7   Governmental Requirements

Most airports handling passenger movements of any reasonable scale will be required to provide office and other working space in the vicinity of the passenger terminal for the civil aviation authority and the ATC authority, if this is constituted separately. At major airports where international passengers are handled, it is also possible that up to four governmental controls must be accommodated:

- Customs
- Immigration
- Health
- Agricultural produce

In most countries, the facilities necessary for health and agricultural inspection are not particularly demanding. Conversely, customs and immigration procedures can be lengthy, and the requirements in terms of operational space for the examining process can be very great. Figure 8.13 shows the layout of an immigration hall at a major international airport. Because of the filtering effect of immigration and the relatively speedy processing at most customs examination halls, customs facilities are not usually extensive. The use of red/green customs procedures, especially in Europe, has materially improved customs processing time without any apparent deterioration in enforcement. Some countries, however, still have very time-consuming and involved

Figure 8.13   Arrivals, immigration-desk area.

customs examination procedures that require the provision of many desks and extensive waiting areas. In addition to their processing areas, most governmental agencies require office and other support space, such as rest, changing, and toilet areas.

8.8   Non-Passenger-Related Airport Authority Functions

It is often convenient at smaller airports to locate within the terminal building for ease of intercommunication all the airport authority's non-passenger-related functions. These include

- Management
- Purchasing
- Finance
- Engineering
- Legal
- Personnel
- Public relations
- Aeronautical services
- Aviation public services (e.g., noise monitoring)
- Plant and structure maintenance

At larger airports, it is customary to separate these authority functions into distinct buildings or buildings away from the terminal building to ease the level of congestion associated with busier terminals. At multiple airport authorities, such as Aeroports de Paris, the Port Authority of New York and New Jersey, and the privatized BAA in the United Kingdom, many of the management and staff functions are carried out entirely off-airport, only the line-operating functions being staffed by airport-based personnel. The detailed design of a terminal should take great account of the way in which the authority intends to operate its facility because space requirements revolve around operational procedures.

8.9 Processing Very Important Persons

Air travel is still a premium method of travel, attracting important, famous, and very rich individuals. Some of the busier airports process a large number of very important persons (VIPs). For example, more than 1,000 groups of VIPs pass through London Heathrow every month. This requires special facilities and staff to ensure that the arriving and departing party can pass through the terminal with all necessary courtesies, sheltered from the conditions of the average traveler. Consequently, VIP facilities have separate landside access, a fully equipped and comfortable lounge in which the party can wait for either landside or airside transport, and a separate access to the apron. The facility must be capable of holding fairly large parties; often traveling heads of state have VIP parties in excess of 25 persons. In addition to the need for sufficiently large and adequately equipped accommodation, the facilities must be safe from the security viewpoint because they may become the target of unlawful acts. Figure 8.14 shows the VIP lounge at a large airport. At multiterminal airports, there is the choice of either several VIP lounge facilities or one central facility to minimize congestion and inconvenience. The choice will depend on the ease of accessing aircraft across the apron for the range of flights involved.

8.10 Passenger Information Systems

Passengers move through airport terminals under their own power. They are not physically transported in a passive manner, as is freight, although in larger terminals mechanical means are used to aid in movement through the facility (see Section 8.12). This, of course, does not refer to people with restricted mobility, who need special ramps and other necessities, which are beyond the scope of this book. Equally important, a large number of passengers reach airports in their own personal vehicles. There is therefore a need to ensure that the passenger has sufficient information both in the access phase of the journey and in passing through the terminal to

FIGURE 8.14   VIP lounge facilities. (*Source:* Bahrain International Airport.)

reach the correct aircraft gate at the right time with a minimum of difficulty and uncertainty. Additionally, the passenger requires information on the location of many facilities within the terminal, such as telephones, toilets, cafeterias, and duty-free shops. Information therefore is usually functionally classified into either directional-guidance or flight-information categories. Directional guidance commences some distance from the airport and normally involves cooperation with some local governmental authority to ensure that suitable road signage is incorporated into the road system on all appropriate airport access roads (Figure 8.15). Often such signs include an aircraft symbol to help the driver to identify directions rapidly. Nearer the airport, terminal-approach road signs will guide the passenger to the appropriate part of the terminal. It is essential that the driver be given large, clear signs in positions that permit safe vehicular maneuvering on the approach-road system. The driver must obtain information on the route to be taken with respect to such divisions as arrivals/departures and domestic/international flights and often to airline-specific locations (Figure 8.16). In multi-terminal airports, there will be signage to each individual terminal, either by terminal designation or by airline groups. Within the terminal, departing passenger flows are guided principally by directional-guidance signs, which indicate check-in, governmental controls, departure lounges, gate positions, and so on. Other terminal facilities that must be identified are concessionary areas and public service facilities such as telephones, toilets, and restaurants (Figure 8.17).

Figure 8.15   Road sign to airport in Arabic and English with pictogram.

Figure 8.16   Signage with directions to specific airline terminal areas.

FIGURE 8.17   Information signs in terminal.

It is essential that the signage is carefully designed. The International Civil Aviation Organization (ICAO) has a set of recommended pictograms for signage inside terminals. Many airports have adopted their own signage convention. In some cases, the signage used falls short of acceptable standards. Sufficient signage must be given to enable the passenger to find the facility or the direction being sought; equally, there cannot be such a proliferation of signs that there is confusion. It is essential that the signage configuration be designed to conform to available internal building heights, which itself must be set recognizing that overhead signage is essential. Once in the terminal, passengers receive information concerning the status and location of departing flights by the departure side of the flight information system. Historically, this information has been displayed on mechanical, electromechanical, or electronic departure flight information boards. However, these largely have been supplanted by cheaper visual display units (VDUs), which can be located economically at a number of points throughout the terminal. Figure 8.18 provides an example of a modern bank of VDUs.

The arriving passengers are given similar guidance information, which helps to convey them to the baggage-reclaim area and to the landside access area, stopping en route at immigration and customs in the case of international arrivals. It is necessary to have adequate

FIGURE 8.18 Bank of VDUs for flight information.

exit signing within the terminal for all passengers and on the internal circulation roadways for passengers using the car mode. An example of an airport road exit signing is shown in Figure 8.19. Meeters who have come to the airport to greet a particular flight are informed of flight status and location either by an electromechanical arrivals board or by VDUs (Figure 8.20). Arrival and departure VDUs have the advantage that they are readily compatible with computerized information systems and can be updated easily. The units themselves, which are relatively inexpensive, are easily removed, replaced, and repaired in the case of failure.

Most airport operators supply at least one airport information desk per terminal on the departures side and less frequently on arrivals. This worker-staffed desk, an example of which is shown in Figure 8.21, supplies information that goes beyond that supplied by the visual systems. Also, it is capable of assisting those unable to use the automatic system for one reason or another. In the case of failure of the automatic systems, the only means of providing flight status and location might be through a manned desk.

In an attempt to make information more available, airports are introducing self-service information kiosks. These have the advantage that they are relatively inexpensive, take up little space, and can be positioned flexibly to suit the needs of users.

FIGURE 8.19 Terminal roadway exit signs.

8.11 Space Components and Adjacencies

Earlier it was stated that the organization of a terminal must closely follow operational strategies and requirements if the terminal is to function adequately. Consequently, no hard-and-fast rules can be set down for the overall division of terminal space. However, Figure 8.22 provides a rough guide on the functional distribution of terminal

FIGURE 8.20   Arrivals board.

FIGURE 8.21   Staffed information desk.

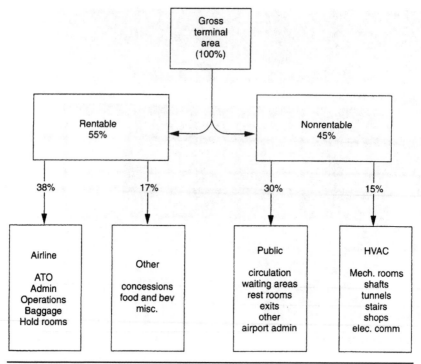

FIGURE 8.22   Terminal space distribution.

space in a typical U.S. airport. At privatized facilities, considerably more space is provided for commercial concessions. More than half the terminal area is likely to be rented out if baggage rooms are included in the figure of rented areas. For detailed estimates of terminal space requirements, it is suggested that the reader refer to design texts and guides (Ashford et al. 2011; Horonjeff et al. 2010; TRB 2011a, 2010b). However, there still remains the question of the interrelationships of the spaces provided, that is, adjacencies that are operationally desirable. In a typical terminal layout, there are several facilities that ideally should be grouped in close proximity, whereas juxtaposition of other facilities is nonessential. For example, grouping is desirable for concession areas but nonessential for airline administration areas. Clearly, it is essential that customs-check areas be in the immediate vicinity of the baggage claim. Figure 8.23 indicates a functional-adjacency chart published by the IATA to aid in the location of terminal facilities. These adjacencies are of as much interest to the authority that must operate a facility as they are to the designer (Hart 1985; Blow 1996, 2005).

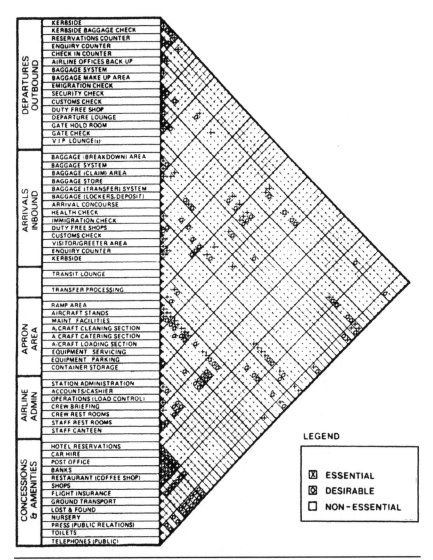

FIGURE 8.23 Functional adjacency chart. (*Source:* IATA.)

8.12 Aids to Circulation

Large airport terminals with multiple gate positions for large transport aircraft necessarily involve large internal circulation distances. At some largely single-terminal airports, such as Chicago O'Hare, the distance between extreme gate positions is close to 1 mile (1.6 km), and the distance from the center of the car parking area to the extreme gate is about the same. To ease the burden of walking long distances, it is now becoming common for airports to install some form of mechanized circulation aid. In airports with multiterminal designs (e.g., Kansas City, Charles de Gaulle, Seattle, New York JFK, and Houston), remote piers (e.g., Atlanta, Madrid, and Pittsburgh), and remote satellites (e.g., London Gatwick, Terminal 5 at London Heathrow, Miami, Tampa, and Orlando), the distances can be very large, and mechanized movement becomes essential. For example, if the ultimate construction of the New Seoul International Airport (NSIA) is built to the master plan, there will be more than 5 miles (8.5 km) between the extreme terminals. In any case, it is now becoming common practice to provide mechanized assistance where practical when walking distances exceed 1,500 feet (450 m). Three main methods of movement assistance are used:

1. *Buses*—used to link unit terminals in multiterminal operations (e.g., Paris Charles de Gaulle, New York JFK, Los Angeles, and London Heathrow)

2. *Pedestrian walkways*—used within piers and to connect to remote satellites or railway stations (e.g., Amsterdam, London Heathrow, Los Angeles, Atlanta, and Barcelona)

3. *Automatic people movers*—used to make connections with remote piers, railway stations, or between terminals (e.g., Miami, London Gatwick, Frankfurt, Tampa, Atlanta, Houston, and Singapore)

Pedestrian walkways are an older and now fairly widely used technology in which there is a great deal of experience. Their great limitation is their speed, which must, for safety reasons of boarding and alighting, be kept to approximately 1.5 miles per hour (2.5 km/h). For very long distances, therefore, they are unsuitable. Another disadvantage is the fact that there are technical reasons that limit their length. There is also the likelihood that at least one in a chain of walkways will be inoperable owing to failure as the devices age. They also must be operated in one direction, which means that unlike a two-track people-mover system, one direction cannot operate in the shuttle mode should there be a failure in the other direction. Under conditions of equipment failure, walking might be the only other option.

A number of larger airports now use people movers, automatic vehicles acting essentially as "horizontal elevators" that are capable of moving passengers at top speeds of approximately 30 miles per hour (45 km/h). Figure 8.24 shows the subterranean tunnel connecting the airside remote piers to the main terminal area in Atlanta. Passengers can connect to the piers by walking, by using the moving

Figure 8.24   Walkway tunnel connecting piers to terminal at Atlanta Hartsfield-Jackson International Airport.

walkways, or by using the loop people-mover system, which can be entered at one of the pier stations. One of the first such connector vehicles was the kind used to connect the terminal to the satellites at Atlanta, as shown in Figure 8.25, and terminal to terminal, as shown in Figure 8.26. Such automatic systems reduce personnel but require extensive control systems (Figure 8.27). It is usual to provide maintenance areas such as that shown in Figure 8.28 either within the terminal area or close to one of the satellite areas it feeds. Where such systems are used, it is necessary to provide station areas, track, control room, maintenance areas, appropriate emergency evacuation areas, and escape points in addition to alternative methods of travel in case of failure.

System reliability is extremely important because without the people mover, the design of the terminal area is no longer coherent—passenger would be subjected to intolerable walking distances. Therefore, the airport authority sets high performance standards for such equipment. It is usual to require several months of break-in operation prior to carrying passengers. Authorities then specify system availability of 98 percent during the first few months of operation and subsequent performance at 99.5 percent availability. From operating systems, it is apparent that 99.9 percent availability is possible with current systems. A common arrangement is for the equipment manufacturer to operate and maintain the system for a period of the first two years and to perform subsequent maintenance on contract.

FIGURE 8.25 Station for people mover between terminal and satellites.

FIGURE 8.26 People mover on track between main terminal and satellite at Tampa International Airport.

FIGURE 8.27   Control room for people mover at Atlanta Hartsfield-Jackson International Airport. (*Source:* Bombardier.)

FIGURE 8.28   People mover in maintenance area. (Note the pit for maintenance under the rails.)

8.13   Hubbing Considerations

In the last 15 years, particularly since deregulation, airlines have tended to set up hub-and-spoke operations to improve service frequency, load factors, and available destinations. Consequently, a number of airports in the United States and elsewhere have become hub airports, where passenger transfers are common and may amount to more than two-thirds of the total traffic (e.g., Dallas–Fort Worth and Atlanta). In some cases, the hub operation is airline-driven (e.g., Pittsburgh with USAir). In other cases, the policy is airport-driven, where interlining as well as online transfers are encouraged (e.g., London Heathrow).

Hub terminals differ considerably from origin-destination terminals. They must accommodate large numbers of passengers moving between gates at the terminals rather than from the landside to the gate and vice versa. Similarly, a large proportion of passenger baggage must be handled for online or interline transfer rather than being originating or destined baggage.

A hub terminal must be designed and operated to handle waves of passengers fed by banks of arriving and departing aircraft. During a single day at a major hub, there might be as many as 12 such waves. Recognizing that the intergate transfers may require considerable distances to be covered in relatively short connection times, large hubs require mechanized aids to circulation that are speedy and reliable (e.g., Pittsburgh, Atlanta, Singapore, and Madrid). Where the facility has to act as a hub between international and domestic flights (e.g., Birmingham, the United Kingdom), particular attention must be paid to customs and immigration facilities to ensure that connections can be made. International hub terminals (e.g., Amsterdam, Singapore, Dubai, and Hong Kong) often develop extensive commercial facilities for tax-free and duty-free shopping with the knowledge that passengers are likely to have some free time for shopping during the connection. Even domestic hubs have developed extensive commercial facilities that are designed to attract impulse buyers with time to spare (e.g., Pittsburgh).

The requirements for baggage handling at hub terminals differ greatly from origin-destination airports (see Chapter 7, in which baggage handling is described more completely). It is essential that there is a rapid and accurate online and interlining baggage-transfer capability. The operational cost to airlines of mishandled baggage is unacceptably high where this cannot be guaranteed. The situation becomes even more complicated where domestic and international flights are concerned. ICAO regulations require that passengers and their baggage be reconciled to ensure that unaccompanied bags of no-show passengers are not permitted on international flights. Should a passenger not make the connection, loaded bags must be unloaded from the aircraft, a costly and time-consuming source of aircraft delay.

References

Ashford, N. J., S. Mumayiz, and Paul H. Wright. 2011. *Airport Engineering*, 4th ed. Hoboken, NJ: Wiley.

Blow, C. 1996. *Airport Terminals*, 2nd ed. London: Butterworth.

Blow C. 2005. *Transport Terminals and Modal Interchanges*. Amsterdam: Elsevier.

Federal Aviation Administration (FAA). 1976. *The Apron and Terminal Building Planning Report* (FAA-RD-75-191). Washington, DC: FAA, Department of Transportation.

Federal Aviation Administration (FAA). 1980. *Planning and Design of Airport Terminal Facilities at Non-Hub Locations* (AC 1 SO/5360-9). Washington, DC: FAA, Department of Transportation.

Federal Aviation Administration (FAA). 1988. *Planning and Design of Airport Terminal Facilities* (AC150/5360-13). Washington, DC: FAA, Department of Transportation.

Hart, W. 1985. *The Airport Passenger Terminal*. New York: Wiley-Interscience.

Horonjeff, R., F. X. McKelvey, W. J. Sproule, and S. B. Young. 2010. *Planning and Design of Airports*. New York: McGraw-Hill.

International Air Transport Association (IATA). 2004. *Airport Development Reference Manual*, 9th ed. Geneva: IATA.

International Air Transport Association (IATA). 2004. *Airport Development Reference Manual*, 9th ed. Montreal, Canada: IATA.

Transportation Research Board (TRB). 2010. *Airport Passenger Terminal Planning and Design*, Vol. 1: *Guidebook*. Washington, DC: ACRP, TRB.

Transportation Research Board (TRB). 2010. *Airport Passenger Terminal Planning and Design*, Vol. 2: *Spreadsheets and Users Guide*. Washington, DC: ACRP, TRB.

CHAPTER 9

Airport Security

9.1 Introduction

Airports, in common with other public facilities, have always been vulnerable to conventional crime such as vandalism, theft, breaking and entering, and even crimes against the person. Lately, as part of a worldwide transport system, they also have become the focus of terrorism. Terrorist acts have included exploding bombs aboard aircraft in flight, ground attacks on aircraft and ground facilities, the use of firearms and missiles, the hijacking of aircraft, and the use of hijacked aircraft to attack prominent buildings and facilities. Hijacking usually results in the taking of passengers and crew as hostages and the subsequent involvement of an airport in attempts to free the hostages and apprehend the hijacker(s). Starting in the 1980s, large aircraft were brought down in midflight, resulting in each case in hundreds of fatalities.[1] In the 9-11 incidents of 2001, hijacking moved into another dimension when four aircraft were attacked simultaneously in a coordinated terrorist atrocity, for use in suicide attacks on four different targets in the Northeast of the United States in New York and Washington, DC. The immense loss of life and the associated loss of property changed the world of airport and aviation security overnight.

Nationally and internationally, there is considerable concern to provide continuous protection against the possibility of attacks on civil aviation; airports stand in the last line of defense. The occurrence of a severe security incident is as unpredictable and as unlikely as the probability of an aircraft accident, but both have the serious potential for loss of life and injury or damage to property.

Airport management, in common with others involved in the operation of elements of the civil air transportation system, are required

---

[1]On June 23, 1985, a bomb exploded on board a B747 Air India 182 in Irish airspace killing 329 passengers and crew. Iran Air Flight 655, an Airbus A300, was shot down on July 3, 1988 into the Persian Gulf with a loss of 290 passengers and crew. On December 21, 1988, Pan Am 103, a Boeing 747, was brought down by a baggage bomb over Scotland, with the loss of 243 passengers and 16 crew members.

to take measures that will provide a high level of protection of buildings and equipment (including aircraft) in addition to ensuring the safety and personal security of passengers and staff using the system. This must be done in a manner that disturbs the normal operating patterns as little as possible while maintaining acceptable security standards throughout the whole of the airport system. The achievement of this basic goal of the modern security operation requires the commitment and cooperation of central and local government agencies, airport authorities and companies, airlines, other airport tenants, police and security staff, and the public itself. This chapter discusses how security procedures affect airport operation and describes in general terms airport security requirements. For obvious reasons, descriptions of detailed procedural arrangements will be avoided, as will the identification of the procedures and arrangements at particular airports.

9.2   International Civil Aviation Organization Framework of International Regulations

The basis of international regulation dates back to the International Civil Aviation Organization (ICAO) Convention on International Civil Aviation in 1944, which superseded the earlier Paris Convention on Aerial Navigation of October 1919 and the Havana Convention on Commercial Aviation of 1928. In the 1944 convention, little account is taken of the need for security of civil aviation, but the document does briefly indicate, in Article 64, the need to enter into arrangements for world security to preserve peace.

During the 1960s and 1970s, hijacking was seen as the most significant problem affecting aviation. Over time, the emphasis of those attacking civil aviation has moved to the destruction of aircraft in flight. A series of conventions has addressed the issue of aviation security:

Tokyo 1963. Convention on Offenses and Certain Other Acts Committed on Board Aircraft—concerned with the whole subject of crime on aircraft and in particular with the safety of the aircraft and its passengers.

The Hague 1970. Convention for the Suppression of Unlawful Seizure of Aircraft—dealing with hijacking, specifically recommending that it be made an extraditable offense.

Montreal 1971. Convention on the Suppression of Unlawful Acts Against Civil Aviation—enlarging the Hague Convention and adding the offense of sabotage.

Safeguarding International Civil Aviation Against Acts of Unlawful Interference, Annex 17 to the Chicago Convention of 1944, dated 1974 (ICAO 1974)—established 40 international aviation standards and 17 recommended practices.

Montreal Protocol for the Suppression of Unlawful Acts of Violence at Airports Serving International Civil Aviation, 1988—a supplement to the Montreal Convention intended to cover acts of violence against civil aviation that occur at airports and ticket offices, which were overlooked in 1971.

Mexico Convention 1991—produced regulations for the marking of plastic explosives for the purposes of detection of their sources.

ICAO Document 8973: Security: Safeguarding International Civil Aviation Against Acts of Unlawful Interference—first published in 1971 and frequently updated, this manual documents, in detail, procedures for preventing acts of violence against aviation. It is restricted, being made available only to those who are considered to need to know this type of information.

9.3 Annex 17 Standards

Annex 17 sets out a number of standards and recommended practices for securing civil aviation. These include standards involving the setting up of a national organization with overall responsibility for aviation security, the requirement that each airport should run an airport security program, that there is a responsible authority at each airport, that there is an airport security committee, and that the design requirements for airport security are adhered to. These standards are, respectively:

Standard 2.1.2: "Each Contracting State shall establish an organization and develop and implement regulations, practices and procedures to safeguard civil aviation against acts of unlawful interference taking into account the safety, regularity, and efficiency of flights."

Standard 3.2.1: "Each Contracting State shall require each airport serving civil aviation to establish, implement, and maintain a written airport security programme appropriate to meet the requirements of the national civil aviation security programme."

Standard 3.2.2: "Each Contracting State shall ensure that an authority at each airport serving civil aviation is responsible for coordinating the implementation of security control."

Standard 3.2.3: "Each Contracting State shall ensure that an airport security committee at each airport serving civil aviation is established to assist the authority mentioned under 3.2.2 in its role of coordinating the implementation of security controls and procedures as specified in the airport security programme."

Standard 3.2.4: "Each Contracting State shall ensure that airport design requirements, including architectural and infrastructure-related requirements necessary for the implementation of the security measures in the national civil aviation security programme, are integrated into the design and construction of new facilities and alterations to existing facilities at airports" (ICAO 1974).

These major security standards are met in different ways in individual countries among which there are significant differences.

9.4 The Structure of Planning for Security

Clearly, the subject of security has wide implications that reach far beyond the jurisdictional limits of the airport to central government itself. Planning to meet the needs of security emergencies and to ensure deterrence of unlawful acts against civil aviation requires the involvement of a number of organizations, such as

- The airport administration
- The operating airlines
- The national civil aviation organization
- The national security services
- Police
- Military
- Medical services
- Labor unions
- Customs
- Government departments

Internationally, ICAO requires that each Member State initiate a national aviation security program that can be developed by a national aviation security committee formed from representatives of these organizations. If this body and the airports themselves are to be effective in countering security threats, there must be a clearly established process that starts with the issue of a policy at the national level and is operationally apparent in the procedures adopted at the individual airports. National policy is translated into a national aviation security plan, a necessity if airports and government wish to do other than react post hoc to a security incident.[2] The national plan is implemented by the provision of staff, equipment, and training at airports and other sensitive aviation areas. System-wide and at individual facilities, security operations are tested, evaluated, and modified to ensure adequate performance standards.

---

[2]However, it is acknowledged that no security program guarantees that incidents will not occur, so contingency plans are prepared and exercised at both national and individual airport levels.

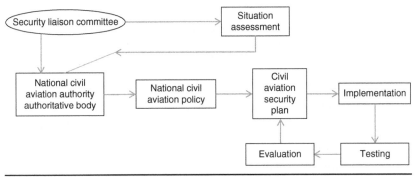

Figure 9.1 Security planning cycle. (*Adapted from ICAO.*)

Reviews of this nature must be carried out by qualified security officers and operations personnel, and assessments should include information on the severity of any deficiency and how it relates to airport security as a whole. In particular, efforts should be made to determine whether unsatisfactory conditions reflect individuals' carelessness or the existence of systematic problems. By applying an analytical approach, a security system's strengths and weaknesses can be evaluated. Alterations in major policy directions are made by a continuous situation assessment of the changing security climate. Factors that can radically alter the security threat in a particular country or at a particular airport are political agitation or unrest and widespread publicity of other security incidents. Figure 9.1 indicates the conceptual structure of the security planning cycle. Reassessment of threat should take into account not just the level of threat but also perceived trends, especially the types of weapons used and the techniques and tactics employed. Reassessment, if it is to be of any value in the preventative sense, should be based on accurate and timely intelligence concerning the intentions, capabilities, and actions of terrorists before they reach the airport. Here too, international cooperation has a part to play, a fact underlined when the United States signed into law the Foreign Airport Security Act as part of the International Security and Development Act of 1985. To support the intelligence requirements of this act, new interdepartmental offices were set up in the United States under the provisions of the Aviation Security Improvement Act of 1990.

9.5   Airport Security Program

Because national aviation security programs are set up as required by the conventions of ICAO, there is an overall similarity in the structure of these programs among the many signatories to the conventions. However, there are also striking differences in the manner in which security programs are structured. The main differences lie in the degree of involvement of central government in aviation security and the degree to which this is delegated to regional or state governments and the airport authorities. The overall structure can be described briefly as follows (Tan 2007):

- Legislation and sources of regulations
- Airport security committee
- Communications structure and physical description of airport
- Security measures and controls
- Security equipment
- Response to acts of unlawful interference
- Security training
- Quality control

It is in the detail of the assumption of responsibilities that the different program structures vary among jurisdictions.

9.6   U.S. Federal Involvement in Aviation Security

Within the United States, the overall responsibility for airport security now is lodged within the Department of Homeland Security (DHS), administered by the Transportation Security Administration (TSA). Prior to the major terrorist attacks of 2001 on the World Trade Center and the Pentagon, security screening of passengers was the responsibility of the airlines. The operational aspects of screening generally were carried out by private security firms. The Aviation and Transportation Security Act of November 2001 was in direct response to the earlier 9-11 attacks. As part of the terms of this act, the TSA was set up in the Department of Transportation but was rapidly transferred into the DHS, which was set up by the Homeland Security Act of 2002.

The regulations that relate to the establishment and operation of security at airports are published in "Title 49: Transportation" of the *Code of Federal Regulations*. They are as follows (US 2012):

49 Code of Federal Regulations, Part 1540: Security: General Rules
49 Code of Federal Regulations, Part 1542: Airport Security
49 Code of Federal Regulations, Part 1544: Aircraft Operator Security
49 Code of Federal Regulations, Part 1546: Foreign Air Carrier Security

9.7   Airport Security Program: U.S. Structure

To conform to the requirements of the *Code of Federal Regulations* as they apply to securing at U.S. airports, each airport is required to prepare an airport security program that must contain the following major elements:

- The name, duties, and training requirements of the airport security coordinator (ASC)

- A description and map of the secured areas detailing boundaries, identification media to be used, procedures to control personal and vehicular movement, and access control (see Figure 9.2)

- A description and map of the air operations area (AOA), including boundaries, activities that affect security, access control, movement control within the area, and signing requirements

- A description of the security identification display area (SIDA), including a description and map, pertinent features, and activities within or adjacent to the SIDA

- A description of the sterile areas, including a diagram with dimensions detailing boundaries and pertinent features and access controls

- Compliance with the requirements of criminal history record checks

- Description of the personnel identification system

- Escort procedures for nonemployees requiring escort

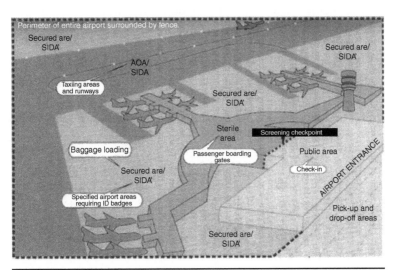

FIGURE 9.2   General depictions of the security areas of an airport. (*Adapted from TSA.*)

- Challenge procedures

- Training programs for personnel

- A description of use and operation of law-enforcement support

- Contingency plans for incidents such as terrorist attacks, bomb threats, civil disturbances, air piracy, and suspicious/unidentified items

- Procedures for the distribution, storage, and destruction of classified and unclassified security information

- Public advisories

- Incident-management procedures

- Alternate security procedures in the event of natural disaster or other non-security-related emergencies

- Agreements related to airport tenants' security programs and exclusive area agreements

9.8   Airport Security Planning Outside the United States

ICAO requires that each Member State establishes a national organization responsible for aviation security and that each airport have an airport security plan. It does not, however, specify the relationships involved or the procedures to be followed. The U.S. approach, based entirely on the responsibilities of the federal government's TSA and the procedures that it mandates, is a strong top-down structure. Other nations have chosen much less centralized arrangements. In the United Kingdom, the structure of security planning is a bottom-up arrangement based on a proper assessment of local risk at a particular airport. British security plans are based on existing arrangements that are seen to be operating satisfactorily and the Multi-Agency Threat and Risk Assessment (MATRA) methodology.

Figure 9.3 indicates how in the United Kingdom a risk advisory group (RAG), consisting of airport management and the local chief officer of the police, produces an airport security risk report for the security executive group (SEG).

SEG is constituted of

- The airport operator

- The local police force

- The local police authority

- The airlines

From the airport security risk report, SEG produces an amended airport security plan that is reviewed cyclically. Where a policing

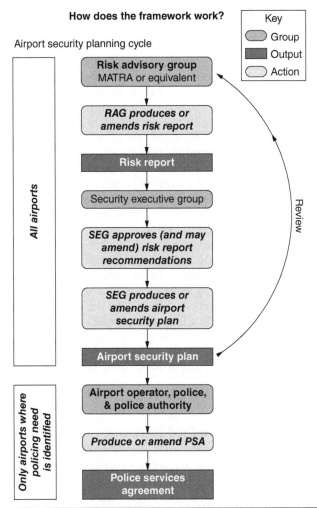

How does the framework work?

Airport security planning cycle

Key
- Group
- Output
- Action

All airports

Risk advisory group
MATRA or equivalent

RAG produces or amends risk report

Risk report

Security executive group

SEG approves (and may amend) risk report recommendations

SEG produces or amends airport security plan

Airport security plan

Review

Only airports where policing need is identified

Airport operator, police, & police authority

Produce or amend PSA

Police services agreement

Figure 9.3   Airport security planning in the United Kingdom. (*Source: Wheeler 2002.*)

need is identified, the airport operator, the police, and the local police authority produce a police services agreement that determines the level of policing and how it is funded (Wheeler 2002).

9.9   Passenger and Carry-On Baggage Search and Screening

Perhaps more than in any other part of the airport, security measures are perceived to be the most effective in preventing subsequent unlawful acts in the air. If the public is made aware in general terms that a security program is in operation, the incidence of attacks is lowered, indicating a deterrent effect. It is entirely possible to publicize the fact that security systems are operating in a terminal without disclosing their nature. A number of international airports make a point of placing public notices to this effect in the terminal. The less well understood the security measures are, the greater is the likelihood that the program will succeed in heading off all but the most determined attackers. Ideally, a security system operates throughout the whole passenger facilitation process of ticketing, passenger and baggage check-in, and boarding (IATA 2004; TSA 2006). Abnormal behavior at the ticketing and check-in stage should alert staff to potential problems. In the boarding process, security procedures must ensure that no would-be assailant is able to convey any weapon to the aircraft. The mere presence of visible security systems is likely to reduce the occurrence of incidents.

Successful security necessitates that the airside-landside (sterile area–public area) boundary be well defined and continuous through the passenger terminal, with a very clear definition of where the security-cleared sterile area is. The number of access points to the airside must be very severely limited; those which are available to passengers must be staffed with security personnel. Access to the sterile areas through staff entrances must not be direct and must be signed as closed to the public. Staff access also should incorporate the same level of security screening that is used for the passenger entry points to the sterile area.

Centralized and Decentralized Screening

The form of security screening carried out on passengers depends on the location of the security screening checkpoint in the passenger terminal. The two basic forms of screening are centralized and decentralized. *Centralized* security before entry into a sterile departures area serving multiple departure gates generally requires fewer security staff and less but more sophisticated equipment. Its main drawback is that unscreened individuals may be able to access the sterile area from the apron or through unattended staff routes. *Decentralized* screening or gate search is carried out directly before the airplane is boarded; after screening, the passengers are held in a sterile lounge. Some operators believe that gate screening achieves maximum security. However, gate search requires more staff and more screening equipment, tends to cause more boarding delays, and suffers the disadvantage that the challenge to any armed person and group is performed in the vicinity of the aircraft. However, at some international airports with large numbers of international transfer passengers, where there is a requirement that all boarding passengers are searched no matter from where they have come, decentralized screening at the gates is the most feasible solution (e.g., Schiphol Amsterdam and Changi Singapore). In the United States, where international transit without entry into the United States is not permitted, the matter does not arise. Some of the advantages and disadvantages of centralized and decentralized search are indicated in Table 9.1. Figure 9.4 shows in diagrammatic form the structure of the sterile and public areas for centralized, sterile pier, and decentralized security systems.

Advantages	Disadvantages
Centralized search	
Favored by passengers	Requires search of staff entering the sterile area
Minimum security staff and equipment needed to process a given number of passengers	Food , merchandise, and other materials must be scrutinized
Encourages passenger spending in restaurants and other commercial areas	Separation of arriving and departing passengers is difficult to achieve
Easier to concentrate security personnel in one location	Only one standard of search is possible, whereas
	High-risk flights may require a more thorough search
	Surveillance of passengers is difficult at busy airports
Gate search	
The separation and surveillance problems are eliminated	
	Requires earlier call forward of passengers
	Results in loss of spending time and revenue from restaurants, bars, shops, etc.
Risk of collusion of staff and potential attackers is minimized	
Allows special measures to be taken on high-risk flights	
	Involves long waiting times in crowded gate lounges with no amenities
	Requires more personnel and more equipment to process a given number of passengers
	Creates problems of search team availability if flight schedules go awry

TABLE **9.1**   Advantages and Disadvantages of Centralized and Decentralized Search Areas

Advantages	Disadvantages
	Makes an armed police presence difficult depending on the number of gates in use at one time
	Allows a potential terrorist to get close to aircraft before search and risk of access to the apron is much raised through emergency exits
	Enables terrorists to identify specific passenger groups and assembles them in attackable queues and groups
	Gate lounges must be enlarged to accommodate
Holding area/pier search	
Combines the advantages and disadvantages of the other configurations Could be the best option if space is available to set up required search points in appropriate locations	

TABLE 9.1   Advantages and Disadvantages of Centralized and Decentralized Search Areas (*Continued*)

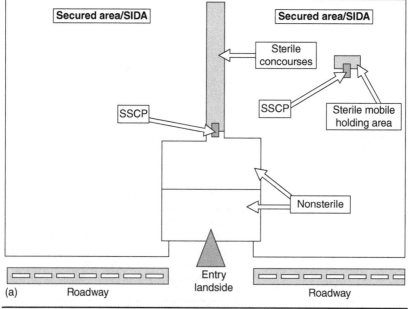

FIGURE 9.4   Sterile and public areas in (*a*) sterile pier and (*b*) centralized and (*c*) decentralized terminals. (*Adapted from TSA.*)

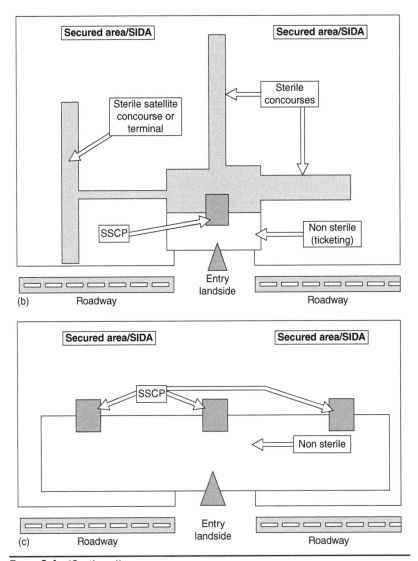

FIGURE 9.4   (Continued)

Security Screening Checkpoint

The layout of a typical security screening checkpoint (SSCP) is shown in Figure 9.5 (TSA 2006). The figure shown shows a layout that conforms to the standards set out by the TSA in the United States and should not be taken as satisfying the requirements of all jurisdictions or being necessary in all jurisdictions. The TSA requires that all checkpoints are formed of modules of either single- or double-lane team modules. Table 9.2 indicates how individual screening lanes are

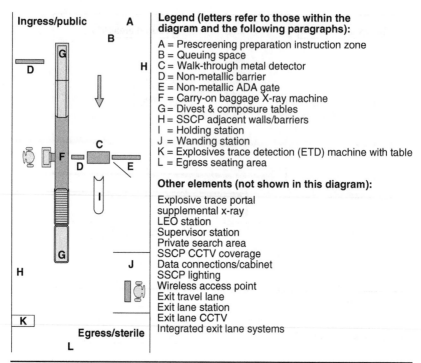

Legend (letters refer to those within the diagram and the following paragraphs):

A = Prescreening preparation instruction zone
B = Queuing space
C = Walk-through metal detector
D = Non-metallic barrier
E = Non-metallic ADA gate
F = Carry-on baggage X-ray machine
G = Divest & composure tables
H = SSCP adjacent walls/barriers
I = Holding station
J = Wanding station
K = Explosives trace detection (ETD) machine with table
L = Egress seating area

Other elements (not shown in this diagram):

Explosive trace portal
supplemental x-ray
LEO station
Supervisor station
Private search area
SSCP CCTV coverage
Data connections/cabinet
SSCP lighting
Wireless access point
Exit travel lane
Exit lane station
Exit lane CCTV
Integrated exit lane systems

FIGURE 9.5   Typical security screening checkpoint (SSCP) layout. (*Source: TSA.*)

Checkpoint Area	Equipment Elements
Per single lane	Enhanced walk-through metal detector (WTMD)
	Carry-on baggage x-ray with roller extensions
Per module (one or two lanes)	Explosives trace detection (ETD) machine
	(1 or 2) Bag-search tables
	Wanding and holding stations (one or two sides)
Per checkpoint	Law enforcement officer station
	Supervisor station (at larger airports only)
	Private search area
	(0 to 2) Supplemental x-ray machines
	1 or 2 lanes: None
	3 to 5 lanes: One
	6 or more lanes: Two
	Data connections/cabinet

TABLE 9.2   Elements of a Standard TSA Checkpoint

combined to form checkpoints. At a very small airport, a single channel will be required. At a larger airport requiring eight search lanes, this can be achieved with four modules each formed of a double-lane layout. The checkpoints are comprised of

Enhanced walk-through metal detector (WTMD). This device investigates the entire body as the passenger walks through the detector arch (Figure 9.6). Any suspect metal material triggers an alarm, which requires a more detailed personal search.

Handheld metal detector (HHMD). Passengers triggering the alarm are routinely examined with a handheld metal detector, sometimes called a *wand*. The devices, which are light and easily manipulated, detect both ferrous and nonferrous metals. Figure 9.6 also shows a typical wand.

Holding stations. Where these are provided, passengers triggering an alarm at the WTMD archway can be diverted to a holding station until they are passed through secondary inspection by an HHMD. These stations ideally have transparent walls so that passengers are able to keep a watch on their carry-on cabin baggage, which is being scanned simultaneously by the carry-on baggage x-ray machine.

Figure 9.6 Walk-through metal detector (WTMD) and wand for personal search. (*Garrett.*)

FIGURE 9.7   Explosive trace detector. (*GE Security.*)

Explosive trace detector (ETD). These devices are often available to support two lanes of search. They will determine the presence of traces of explosive material in the material being carried on by passengers. Figure 9.7 shows a typical ETD, which can detect the traces captured on a swab of the suspect object.

Explosive trace portal (ETP). These are used to provide the capability of detecting explosive traces on the person of a potential passenger. These devices, which are constructed in the form of a walk-through portal, have the ability to determine the presence of explosive that could evade detection at the WTMD stage of search. Unlike in the WTMD, the passenger is required to pause briefly in the ETP portal. ETP technology is capable of detecting a range of explosives, liquids, and narcotics. In size and aspect, the ETP is very similar to the ordinary WTMD.

Supplemental x-ray equipment is required at larger checkpoints to examine shoes and other items scanned in the secondary screening process. A private search area should be provided for passengers requesting a discreet search.

Around 2010, larger airports both in the United States and elsewhere, particularly in Britain and the Netherlands, introduced *whole-body scanners*. These come in two types that depend on very different technologies:

Backscatter x-ray scanners

Terahertz scanners

Figure 9.8   Whole-body scanner: x-ray backscatter. (*Courtesy: Smiths Detection*)

Backscatter x-ray scanners (also called *soft x-ray scanners*) operate on the use of a very low ionizing radiation dose. This dosage is stated to be equivalent to one additional hour of background radiation and is much lower than the exposure received when flying for one hour at 35,000 feet. A modern scanner is shown in Figure 9.8.

Terahertz scanners use extremely high-frequency radio waves that are capable of penetrating clothing. Because they do not use x-rays, there is no radiation dose to which it can be considered to be equivalent.

When introduced in the United States, the personal intrusiveness of body scanners, which produce images of unclad passengers for examination by security personnel, caused considerable public unease and hostility. Travelers refusing to use full-body scanners are required to undergo an extensive hand search, known as a *pat down*. The introduction of this regulation caused even more concern because of the intimate areas of the body that had to be examined physically by security staff.

The *carry-on baggage x-ray machine*, which is continuously monitored by a trained operator, scans the items placed on a belt as it

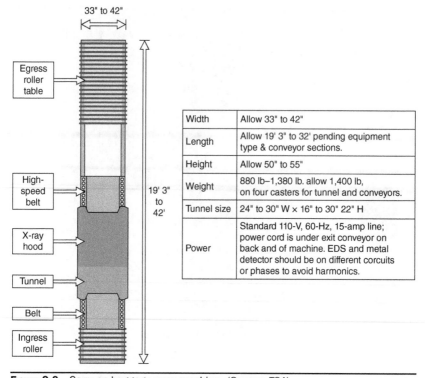

33" to 42"

| Egress roller table |
| High-speed belt |
| X-ray hood |
| Tunnel |
| Belt |
| Ingress roller |

19' 3" to 42'

Width	Allow 33" to 42"
Length	Allow 19' 3" to 32' pending equipment type & conveyor sections.
Height	Allow 50" to 55"
Weight	880 lb–1,380 lb. allow 1,400 lb, on four casters for tunnel and conveyors.
Tunnel size	24" to 30" W × 16" to 30" 22" H
Power	Standard 110-V, 60-Hz, 15-amp line; power cord is under exit conveyor on back and of machine. EDS and metal detector should be on different corcuits or phases to avoid harmonics.

FIGURE 9.9   Carry-on baggage x-ray machine. (*Source: TSA*)

moves through the machine. The device, such as the example shown in Figure 9.9, has tables before and after the rollers at both its ends to allow passengers to assemble the luggage and other items to be examined and to retrieve them after examination.

SSCPs are similar in size and layout throughout the world. The United States has published an extensive and detailed manual covering the planning, design, and operation of these secured areas (TSA 2006). Guidelines are also available from International Air Transport Association (IATA) publications (IATA 2004). The availability of such guidance material will tend to make security facilities more uniform in standards and efficiency. Table 9.2 shows the elements of a standard TSA checkpoint.

Many of the attacks on aviation have come through penetration of the system through access via passenger terminals. At the larger airports, especially those with international operations, very strict security measures have been introduced to effect safe screening of passengers. A homogeneous level of aviation security is still not universal; it is tempting to believe that at smaller airport, especially small airports with only domestic operations, security measures, with respect to passenger search, can be relaxed below standards observed in the large airports. This is a specious argument. The air transport system is interconnected, and lax search at an airport on one continent can introduce an armed and dangerous passenger to flights departing on another. Unless security of passenger search is maintained at a level that ensures acceptable minimum standards throughout the aviation system, the air transport world is divided into airports that are secure and those which must be classified as insecure. This requires security measures against all aircraft arriving from insecure origins.

9.10   Baggage Search and Screening

Following the bombing of Air India Flight 182 over Irish waters in June 1985, the aviation industry introduced rules with respect to checked baggage search and the carriage of checked baggage when a passenger fails to board. The bombing of Pan Am Flight 103 over Lockerbie, Scotland, in December 1988 alerted the industry to the fact that the rules were not being enforced adequately. Subsequently, all checked baggage has to be subject to x-ray scanning, and baggage belonging to a passenger who fails to join a flight must be offloaded if it has already been loaded when the aircraft doors are closed. Satisfactory enforcement of these rules requires the airline to have an exact knowledge of the location of all loaded baggage in the containers in the hold to permit rapid offloading.

To conform to the checked-baggage search requirements meant some initial difficulties for the airlines because the baggage-handling systems often had inadequate space or capability to accommodate the x-raying of bags after they had been accepted into the baggage-loading system at check-in. At many airports, passengers were required to check in and then carry their bags to an x-ray machine installed in the vicinity of the check-in area, where they were subject to x-ray and dispatched into the baggage-loading system. Most airports now operate an in-line baggage-examination system. In Europe, bags are subject to a three-level examination. Level 1 is an x-ray scan, which, if passed, allows the bag to go to loading. Bags that are rejected at Level 1 are subject to more detailed scrutiny of the x-ray image at Level 2. Those which pass are allowed to go on to loading. Bags that fail Level 2 are passed to a detailed search at Level 3. Only when the Level 3 search is satisfactory is the bag allowed to progress. In the United States, in-line screening involves a single high-level (CTX) machine integrated into the baggage system. European standards are to be brought in line with those of the United States by 2018. Figure 9.10 provides examples of checked-baggage x-ray layouts for in-line systems for multiple baggage streams. In some countries and regions, baggage scans still are carried out within the ticket counters themselves (TSA 2006).

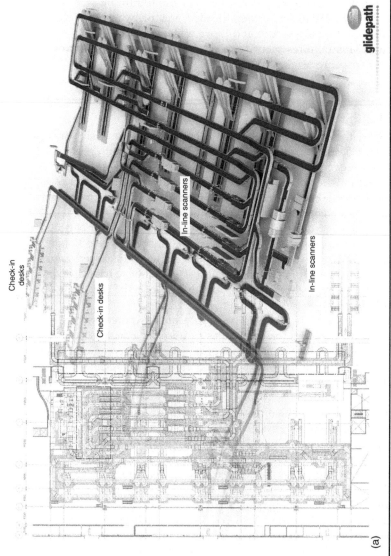

Figure 9.10 (a) Inline hold baggage x-ray scanning system. (*Courtesy: Glidepath, Ltd.*)

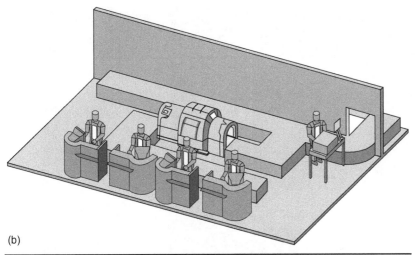

(b)

FIGURE 9.10   *(Continued)* (*b*) Near check-in inline hold baggage x-ray scanner. (*Source: TSA.*)

9.11   Freight and Cargo Search and Screening

The screening of freight and cargo also has been routine since the introduction of hold baggage screening soon after the Air India bombing in 1985. To screen hold baggage without also screening the freight that is normally carried on passenger flights would be pointless. The x-ray screening equipment must accommodate very large packages in unit loading devices (ULDs), and an example of such a machine is shown in Figure 9.11. Such machines have color displays of container contents and are capable of detecting contraband such as concealed weapons, explosives, narcotics, and currency.

9.12   Access Control Within and Throughout Airport Buildings

Recognizing that the airside is potentially an area vulnerable to terrorist attack, access, for other than passengers, should be restricted to identified personnel. All airport and airline ground staff, all other airport workers needing access to perform their jobs, and all necessary airline staff and crews should have passes authorizing entry to the airport security identification area (SIDA). Identification cards, usually known as *SIDA passes*, should have a photograph and details in a tamper-proof badge that should be issued by the unit responsible for airport security. Badges should be displayed at all times. In the United States, prior to obtaining a SIDA badge, an individual must undergo a fingerprint-based criminal record check. Visitors needing to visit the SIDA areas should have special passes and must be accompanied

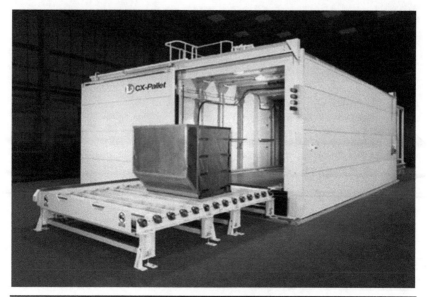

Figure 9.11   Freight x-ray machine. (*L-3 Security & Detection Systems.*)

at all times. Most airports have detailed rules on the escorting of visitors specifying the maximum number of visitors per escort.

In any jurisdiction, the control of the issue of SIDA badges is necessarily strict if the system is to be secure. Badges should be recovered from any employee who ceases to be employed or no longer has need to work in the SIDA area. Failure to do this resulted in the destruction of Pacific Southwest Airlines Flight 1771 from Los Angeles to San Francisco in December 1987. All 43 persons on board were killed when a recently dismissed employee shot the pilot and crew, having reportedly brought the weapon aboard after bypassing preboarding screening by showing his elapsed company ID.

In any system, the SIDA badge is made more secure if it is necessary to key into the computer-controlled pass system a *personal identification number* (PIN) when passing through the access point (Figure 9.12). Even if the pass is not handed in on cessation of employment or reassignment to a nonrestricted area, the PIN can be immediately deleted from the approved list.

Although properly designed, operated, and maintained pass and PIN systems have been universally successful, newer systems relying on biometric identification are under trial and operation. These include iris and retina scanning, fingerprint recognition, and voice and other biometric measurements. Biometric identification involves a digital biometric scan of the individual wanting access, comparison with stored digital security records, and the decision either to permit or to deny entry. Because most security systems also require the display of a SIDA pass, biometric systems are used in conjunction with SIDA badges.

Figure 9.12   Access gate for staff. (*Courtesy: Kaba, Ltd.*)

9.13   Vehicle Access and Vehicular Identification

Similarly, airside access should be granted only to vehicles that must be airside to perform their functions. Access can be restricted by the issue of individual vehicular passes, the control of which is maintained by the central airport security authority. These passes should be fixed term and subject to immediate cancellation when no longer required. Individuals within the vehicle must have individual passes in addition to the vehicle pass. The register of pass holders and the vehicles must be checked periodically and maintained up-to-date at all times.

9.14   Perimeter Control for Operational Areas

It is essential in a properly secured airport that the secured areas are positively separated from those which are unsecured. The two basic elements of perimeter control are *fencing* and *controlled-access gates*.

Fencing

The airside must have an adequate security fence that serves the multiple functions of clearly defining the protected area, providing a deterrent to an intruder, delaying and possibly inhibiting unlawful entry, and providing controlled access points at gates. The fence must be a real deterrent to unlawful entry. It must be high, solidly built of nonscalable metal construction, and normally topped with barbed or razor wire. Care must be taken to secure all conduits, sewers, and other ducts and pipes that pass under the fence to ensure that entry to the airside is not possible. Examples of suitable designs for the United States are available in the literature (IATA 2004; TSA 2006). However, local conditions will apply elsewhere.

Access Gates

Controlled access gates must be provided to the movement area and other parts of the airside. These should be kept to a minimum, and where access is other than by key or automatic control, the gates must be manned, illuminated, and provided with alarms. The access gates are normally equipped with barrier systems with retractable devices that will disable and severely damage a vehicle attempting to force entry, as shown in Figure 9.13.

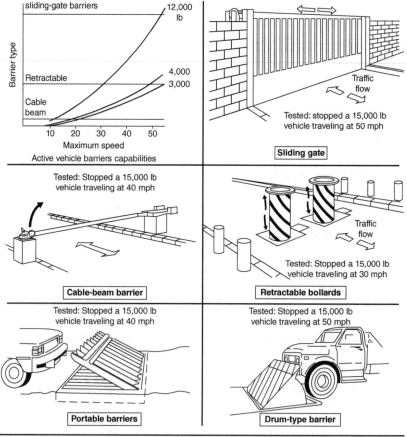

Figure 9.13 Five types of vehicle barriers for airport controlled access points. (*Source: TSA.*)

Depending on the threat assessment relating to the size, importance, and location of the airport, other perimeter security measures will include some or all of the following:

- Security lighting
- Patrols
- Closed-circuit-television (CCTV) monitoring
- Electronic intruder-detection systems—electronic sensors, motion detectors, infrared microwave sensors

Even with the strongest of measures taken to maintain a secure perimeter, the airport and aircraft using the facility still can come under attack from outside the perimeter. London Heathrow was subject to a terror attack from mortar fire over the security fencing from a firing position on open land outside the perimeter. In Mombasa, a missile was fired unsuccessfully at an Israeli aircraft from a launch site outside the perimeter.

By 2013, there was strong evidence that as security measures to prevent terrorists gaining access to the movement area were strengthened, terrorist attacks were refocused on the nonsecured landside of the terminal. The failed passenger terminal bomb attack at Glasgow Airport in June 2007 and the successful suicide bombing of the Moscow Domodedovo Airport passenger terminal in January 2011 indicate that as security is tightened in one area of the airport, terrorists move to the more vulnerable areas open to the public.

9.15   Aircraft Isolated Parking Position and Parking Area

An airport should designate an isolated aircraft parking position that can be used for parking an aircraft when sabotage is suspected or when an aircraft appears to have been seized unlawfully. This position should be at least 325 feet (100 m) from any other aircraft parking position, building public area, or utility (IACO 2010). Furthermore, a disposal area should be designated on the airport for disposal or exploding of any device found in the course of sabotage or unlawful seizure. The disposal area also should be clear of all other used areas, including the isolated parking position, by at least 325 feet (100 m). An airport might need several designated isolated positions to be used for different kinds of incidents. Some positions should be amenable to surreptitious approach.

9.16   Example of a Security Program for a Typical Airport

By way of general guidance for airport management in drawing up an airport security program, a suggested outline is offered here. It is emphasized, however, that such an outline is not intended to be followed rigidly; individual airports will need to modify it to suit their own particular requirements and, of course, the requirements of the jurisdiction to which they are subject.

Security Program for (Official Name of Airport Goes Here)

1. General
    a. Objective. This security program has been established in compliance with Standard 3.1.1 of Annex 17 to the Chicago Convention and in accordance with national legislation and regulations, namely, laws, decrees, etc. The main purpose of the provisions and procedures contained in this program is to protect the safety, regularity, and efficiency of civil aviation by providing, through regulations, practices, and procedures, safeguards against acts of unlawful interference

2. Organization of security
    a. Name and title of the official(s) responsible for airport security
    b. Organizational details of services responsible for the implementation of security measures, including
        (1) Airport security officers
        (2) Central government security officials (if appropriate)
        (3) Police
        (4) Government inspection agencies
        (5) Airline operators
        (6) Tenants
        (7) Municipal authorities

3. Airport security committee
    a. An airport security committee must be established to comply with Standard 3.1.1 of Annex 17 to the Chicago Convention. This committee is responsible for providing advice on the development and implementation of security measures and procedures at the airport. It must meet regularly
        (1) To ensure that the security program is kept up-to-date and effective
        (2) To ensure that the provisions it contains are being satisfactorily applied
        (3) To coordinate the activities of all the bodies concerned with security measures (e.g., police, gendarmerie, operators, airport management, central government, etc.)
        (4) To maintain liaison with the various security services outside the airport (e.g., responsible government departments, bomb-disposal service, etc.)
        (5) To give advice to the airport management on any reorganization or extension of the facilities

       (6) Minutes must be kept for every meeting of the airport security committee, which, after approval of the members, are circulated to the main authorities concerned.

    b. Composition of the committee. The airport security committee should be made up of representatives of all the public and private bodies concerned with operation of the airport. In addition, the airport manager normally will act as the chair, with the chief of airport security as stand-in for those occasions when the chairperson is unable to attend. The following would be appropriate members:

       (1) Airport manager
       (2) Airport security chief
       (3) Representative of the National Aviation Security Agency (if appropriate)
       (4) Police
       (5) Military
       (6) Customs
       (7) Immigration
       (8) Air traffic services
       (9) Fire services
      (10) Communications representatives
      (11) Health service
      (12) Postal service
      (13) Operators
      (14) Cargo companies and forwarders
      (15) Tenants

    The names, titles, and other useful details of all members should be included.

4. Airport activities
    a. Name, location, official address, telephone/fax/e-mail number, or address of the airport and identification code
    b. Hours of operation of the airport
    c. Description of the airport's location with respect to the nearest town or province
    d. Attachments including a location map and plan of the airport with particular emphasis on the airside indicating the various security-restricted areas (e.g., air operations area, secured area/SIDA, sterile area, public area)
    e. Name of the airport owner
    f. Name of the airport manager
    g. Airport operating services
    h. Administration
    i. Air traffic services
    j. Maintenance
    k. Others
    l. Airline operators and route/traffic details

5. Security measures at the airport
   a. Definition and description of the air operations area and secured area/SIDA and the measures designed to safeguard these areas (e.g., boundary fencing, guarded access points, lighting, alarm systems, closed-circuit television, walk-through units, patrols, etc.)
   b. Restricted areas
      (1) Restricted areas within the SIDA (e.g., air traffic services facilities, runways, taxiways, maneuvering areas, parking and ramp access, cargo areas, other operational areas)
      (2) Restricted areas within the sterile area in the terminal (e.g., departure lounges, transit lounges, commercial areas, immigration area, customs area, etc.)
   c. Public areas (nonsterile) (e.g., check-in, departures hall, arrivals hall, car parking areas, access and egress curbs, etc.)
   d. Access movement and control
      (1) Identification procedures for persons. Attach the text(s) that regulate the movement of persons at the airport.
         (a) Specify the access points where access passes are required.
         (b) Specify the criteria for granting access passes.
         (c) Describe in detail the format and contents of the various badges, cards, devices, and signs used for identification.
         (d) Specify the procedures for checking the access pass and the penalties for not complying with the regulations.
         (e) Specify the procedures to be adopted for the cancellation of an access pass.
      (2) Identification procedures for vehicles. Vehicles authorized to enter the SIDA and airport operations area shall be equipped with a pass. The pass will specify in which particular areas the vehicle is authorized to circulate and the applicable hours when this may occur. As in the case of personal access, the program specifies the allocation procedures, the form of the pass, and the procedure for cancellation.
   e. Security control for passengers and baggage
      (1) Passengers
         (a) Custody and control of flight documents (i.e., tickets, etc.)
         (b) Identification of passengers at check-in or other identified locations (e.g., passport check at boarding gate)
         (c) Agency implementing security controls
         (d) Equipment and procedures for passenger screening

(2) Control of hold baggage. Procedures: Search using security equipment, percentage of hand searches required on a random basis, identification and disposition of removed articles, procedures for off-airport checked baggage, procedures for "short shipped" and mishandled baggage

f. Security control of cargo, mail, and small parcels
   (1) Assignment of responsibility for security control
   (2) Screening of cargo, courier and express parcels, and mail
   (3) Nature of control procedures: hand searches, searches using security equipment
   (4) Measures for the treatment of suspect cargo, mail, and small parcels
   (5) Airline responsibilities in relation to the control of flight catering and other stores

g. Security control of VIPs and diplomats
   (1) National guidelines for special procedures
   (2) Procedures for VIPs and diplomats
   (3) Private or semiprivate arrangements for special passengers
   (4) Measures to limit arrangements to strict minimum
   (5) Procedures for dealing with diplomatic bags and diplomatic mail

h. Security control for certain categories of passengers
   (1) Staff members, including crew members in uniform
   (2) Facilities and procedures for disabled passengers
   (3) Procedures for inadmissible persons, deportees, escorted prisoners (these require notification to the operator and relevant captain)

i. Security control of firearms and weapons
   (1) National laws and regulations
   (2) Carriage of firearms on domestic and foreign aircraft
   (3) Authorized weapons carriage in the aircraft cabin (e.g., prisoner escort, VIP escort, sky marshals)

j. Protection of the aircraft on the ground
   (1) Responsibilities and procedures
   (2) Security measures for aircraft not in service
   (3) Positioning of aircraft
   (4) Use of intruder detection devices
   (5) Preflight security checks
   (6) Special measures available to operators on request

k. Security equipment
   (1) Responsibilities for the operation and maintenance of equipment

(2) Detailed description
 (*a*) X-ray equipment
 (*b*) Walk-through metal detectors
 (*c*) Handheld metal detectors
 (*d*) Whole-body imaging devices
 (*e*) Explosives detectors
 (*f*) Simulation chambers—location, type, and construction
 (*g*) Security gates, turnstiles, keys
 (*h*) Biometric devices for personal identification

6. Contingency plans to respond to acts of unlawful interference
 a. Categories
  (1) Reception of unlawfully seized aircraft
  (2) Bomb threat to an aircraft in flight or on the ground
  (3) Bomb threat to a facility on the airport
  (4) Ground attacks—ground to air and ground to ground
 b. Responsible organizations
  (1) Operational command and control
  (2) Air traffic services procedures
  (3) National aviation security agency (if constituted)
  (4) Special services (location day/night)
  (5) Explosive ordinance disposal units
  (6) Armed intervention teams
  (7) Interpreters
  (8) Hostage negotiators
  (9) Police authority
  (10) Fire brigade
  (11) Ambulances

7. Security training program. All personnel with direct responsibility for security and all staff members at the airport should attend either a training course or a security-awareness presentation adapted to the particular needs of the various levels.
 a. Training policy
 b. Training objectives
 c. Curriculum outline
 d. Course syllabuses
 e. Procedures for evaluating training

8. Appendices to the security program.
 a. Organizational diagram showing the structure of the airport administration and its relationship to the agency or department responsible for airport security.
 b. Security management map of the airport and peripheral area: Detailed maps: airside and landside areas, terminal layout, layout of all categories of areas.
 c. Agreements/instructions to tenants.

d. Instructions to air traffic services.

e. Legislative and regulatory texts relating to aviation security, including those in the national context, or any other documentation/references that would be of use to the program.

9.17   Conclusion

On the morning of September 11, 2001, nineteen hijackers attacked four different flights from Newark, Boston, and Washington, DC, all of which were to go to San Francisco and Los Angeles. Some of the hijackers were stopped at the airport terminal security points, but all were allowed to proceed to board their aircraft. No weapons were discovered, and nothing was removed from the hijackers' persons. There is no evidence that other teams should have boarded other aircraft. Apparently, the hijackers had a 100 percent success rate in avoiding the airport security measures in place. They must have been allowed to carry on the knives, mace, and box cutters that they used to overpower two of the crews and to interfere with the safe flight of the third aircraft. The subsequent commission of enquiry found major lapses in search and cockpit security procedures (US 2004). The conclusions that must come to the reader of the commission's report is that procedures that should have been used were applied with great laxity and that devices installed to prevent the subsequent hijacking were not activated. Unless security precautions are taken as seriously as they obviously merit, attacks on aircraft certainly will recur in the future, and there will be a repeated tragic and unnecessary loss of life.

References

International Air Transport Association (IATA). 2004. *Airport Development Reference Manual,* 9th ed. Montreal, Canada: IATA.

International Civil Aviation Organization (ICAO). 1974. "Safeguarding International Civil Aviation Against Acts of Unlawful Interference," Annex 17 to the *Chicago Convention of 1944.* Montreal, Canada: ICAO.

International Civil Aviation Organization (ICAO). 2010. "Aerodromes," Annex 14 to the *Chicago Convention of 1944,* Vol. 1, 5th ed. Montreal, Canada: ICAO.

Tan, S. H. 2007. "Airport Security." Presentation to CAAS Strategic Airport Management Program, April 2007, Singapore, Civil Aviation Agency of Singapore.

Transportation Security Administration (TSA). 2006. *Recommended Security Guidelines for Airport Planning, Design and Construction.* Washington, DC: TSA.

United States (US). 2004. *Final Report on Terrorist Attacks on the United States.* Washington, DC: U.S. Government.

United States (US). 2012a. *49 Code of Federal Regulations,* Part 1540: "Security: General Rules." Washington, DC: U.S. Government.

United States (US). 2012b. *49 Code of Federal Regulations,* Part 1542: "Airport Security." Washington, DC: U.S. Government.

United States (US). 2012c. *49 Code of Federal Regulations*, Part 1544: "Aircraft Operator Security." Washington, DC: U.S. Government.

United States (US). 2012d. *49 Code of Federal Regulations*, Part 1546: Foreign Air Carrier Security, Washington, DC: U.S. Government.

Wheeler, Sir John. 2002. "Independent Report on Airport Security for the Department of Transport and the Home Office." London: HMSO.

CHAPTER 10

Cargo Operations

10.1 The Cargo Market

For almost 60 years, air cargo[1] has been a steadily growing sector of the air transport market. During the late 1960s, the total world ton-kilometers doubled every four years, a growth rate of 17 percent (ICAO 2011). At that time, and ever since, the aviation world has been replete with extremely optimistic forecasts of a burgeoning air cargo market. In 1970, McDonnell Douglas projected that growth rates would increase and that the total air cargo market would grow from 10 billion revenue ton-kilometers in 1970 to 100 billion revenue ton-kilometers in 1980. In fact, this figure was not reached until 1995 owing to recurrent recessions and steep fuel cost rises in the 1970s and 1980s. In 1995, the Boeing Airplane Company forecast annual average growth rates of 6.5 percent over the coming 10 years. The actual growth rate between 1999 and 2009 was a mere 1.9 percent. In 2010, industry forecasts were still very sanguine, predicting growth rates of between 6.6 and 5.2 percent over the period 2009–2029 (Boeing 2011). Such growth may be difficult to achieve. A closer examination of the development of the demand for air cargo indicates that a number of factors are involved.

Gross Domestic Product

There is a very strong positive correlation between air cargo demand and gross domestic product (GDP). In times of economic buoyancy, air cargo grows rapidly, but cyclic recessions of the last 40 years have retarded air cargo growth in the Western industrialized nations. During times of high oil prices, cargo to the oil-producing countries of the

---

[1]For the purposes of clarity, *freight* is defined in this text as being composed of cargo, airline stores, unaccompanied baggage, and mail. The reader should be aware that not all references in other publications differentiate between the meanings of *freight* and *cargo*.

Middle East grows rapidly, falling as oil prices fade. The area of the world generating the highest growth rates of air cargo traffic are the developing Western Pacific Rim nations of the Far East (e.g., China, India, and Indonesia). It is in these nations that the GDP is growing the fastest.

Cost

In real terms, there has been a secular decline of the cost of air freight, but that rate of decline has been very irregular. Aided by the decreasing real cost of fuel and technological improvements, the trend of decreasing cargo costs was reasonably regular until 1974. Declining real costs ceased abruptly with several large oil price increases; the subsequent growth rates were significantly lower. For nearly 20 years following 1985, the real cost of oil again declined, and the demand for air cargo rose in a healthy manner. From 2002 to 2008, yield increased as fuel and security surcharges were imposed by the carriers, but the yield plunged again in 2008 with the rapid decline in traffic caused by the massive recession. The economic slump was accompanied by a rapid decline in fuel prices and fuel surcharges. By 2012, oil prices were again on the rise. For the 20 years following 1989, freight yield had fallen by an average 4.9 percent per year when adjusted for inflation (Boeing 2011).

Technological Improvements

Technological improvements are usually manifest in terms of lower freight costs through improved efficiency. Improvements to technology have taken place in three principle areas:

- In the *air vehicle*, with the introduction of wide-bodied large-capacity aircraft
- In the development of a *wide range of unit load devices (ULDs)* and the necessary subsidiary handling and loading devices on the aircraft, on the apron and in the terminal
- Finally, in *facilitation* with the maturing of freight-forwarding organizations, the rapid growth of integrated carriers, and the development of computerized control and documentation

Miniaturization

Various other secular trends have contributed to the increasing demand in air cargo. For example, *miniaturization* of industrial and consumer products has made items much more suitable for carriage by air. The expected continued growth of the silicon-chip market will continue this trend with the growth of personal telecommunications and other electronic consumer items.

Just-in-Time Logistics

Another factor is the increasing trend for industry to move away from regional warehousing and the high associated labor, construction, and land costs. Manufacturers find that centralized warehouses backed by sophisticated electronic ordering systems and air cargo delivery are as efficient as and less costly to operate than decentralized regional warehouses. Since the mid-1970s, the concept of *just-in-time* (JIT) delivery has revolutionized many industrial production processes. This concept requires reliable delivery procedures that are capable of adjustment on short notice. The consequent reductions in industrial inventories produce savings that can more than pay for air cargo charges.

Rising Consumer Wealth

Real incomes are rising in the industrial countries, and more wealth falls into the bracket that can be designated as discretionary. Such income is less sensitive to transport costs for goods purchased. Consequently, as the real standard of living rises, the higher cost of air transport is more easily absorbed into the prices of goods in either an explicit or implicit tradeoff between cost and convenience. By 2012, a dichotomy had arisen between the eastern and western economies; in the former, real incomes were rising, and in the latter, they were at best stagnant and in many countries were falling.

Globalization of Trade and Asian Development

With the liberalization and globalization of trade, some 40 percent by value of world trade is now carried by air. The emergence of the Asian economies, particularly China, India, Malaysia, and South Korea, has brought strong growth to intercontinental and intra-Asian trade. China, Japan, Taiwan, South Korea, and India have deliberately aimed their manufacturing and research and development (R&D) industries toward high-value products and components that are transported by air.

Loosening of Regulation

Liberalization of the policies regulating international air services and "open-skies policies" as advocated by the United States and supported by the International Civil Aviation Organization (ICAO), the International Air Transport Association (IATA), the Organization of Petroleum Exporting Countries (OPEC), and the Asia-Pacific Economic Cooperation (APEC) forum have drastically changed the provision of Asian and Pacific air services in the past 20 years. Carrier entry has been made easier, and deregulation has permitted the development of hub and spoke operations by conventional airlines and integrators.

Cargo Types

Air cargo is heterogeneous in character. It is often convenient to categorize it according to the manner in which it is handled in the terminal.[2]

1. *Planned.* For this type of commodity, the air mode has been selected as the most appropriate after analysis of distribution costs. It is either cheaper to move by air freight, or the added cost is negligible when weighed against the improved security and reliability. Speed of delivery is not of vital importance to this type of cargo.

2. *Regular.* Commodities in this category have a very limited commercial life, and delivery must be rapid and reliable. Newspapers and flowers are examples of regular commodities, as are fashion items under some conditions.

3. *Emergency.* Speed is vital and lives may depend on rapid delivery of emergency cargo such as serums and blood plasma.

4. *High Value.* Very high value cargo such as gemstones and bullion require special security precautions in terms of staffing and facilities.

5. *Dangerous* (ICAO 2001). The carriage by air of dangerous goods is a topic of much concern with airlines because of on-board hazard. Even on the ground, dangerous chemicals and radioactive materials, for example, require special storage and handling in the terminal. Especially when containerized loads are used, it is important that personnel are adequately trained in the handling of dangerous shipments. IATA includes within its definition of hazardous goods the following: combustible liquids, compressed gases, etiologic agents, explosives, flammable liquids and solids, magnetized materials, noxious and irritating substances, organic peroxides, oxidizing materials, poisons, and polymerizable and radioactive materials.

6. *Restricted articles.* In most countries, arms and explosives can be imported only under the severest restrictions, which include very strict security conditions.

7. *Livestock.* Where livestock are transported, arrangements must be made for animals to receive the necessary food and water and to be kept in a suitable environment. In a large terminal with considerable livestock movements, such as London Heathrow, the care of animals occupies a number of full-time staff.

---

[2]The first three categories include shipments from the following four categories, which might be carried by air on a planned, regular or emergency basis, owing to their particular nature and the special attention that becomes necessary.

Patterns of Flow

The cargo terminal, like the passenger facility, experiences significant temporal variations in throughput. Unlike the passenger terminal, freight facilities often demonstrate very large differences between inbound and outbound flows on an annual basis.

Cargo flow variations occur across the year, across the days of the week, and within the working day. The pattern of variation differs quite noticeably among airports and even may vary remarkably among airlines at the same airport. Figure 10.1a shows the monthly variations of flows observed in the same year at four major airports: JFK, London Heathrow, Schiphol Amsterdam, and Paris Charles de Gaulle. The patterns differ substantially even though they are all in the northern hemisphere and have some summer/winter seasonal similarity. Figures 10.1b and 10.1c indicate observed daily and hourly variations for a particular airline's cargo terminal. Careful analysis of the underlying data of the individual shipments can give an understanding of the reasons for the temporal patterns of the observed flows. This enables the operator to provide adequate and economic facilities and, furthermore, allows the planning of suitable staffing levels. Although terminal facilities are designed around peak rather than average conditions, they are not necessarily sized to cope with the immediate processing the highest peak flows. Freight, unlike passengers, can be held over from the peak hour. When unloading aircraft in an arrival peak, it is not unusual for containers to be offloaded

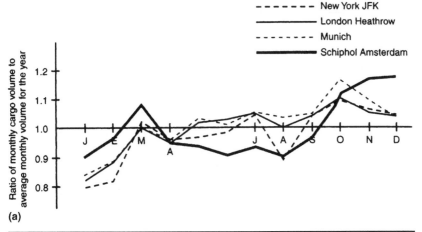

Figure 10.1   (a) Variations in cargo volume throughout the year. (*PANYNJ, BAA plc, Flughafen Muenchen GmbH and Schiphol Amsterdam.*) (b) Daily variations in cargo throughput at British Airways freight terminal, London Heathrow. (*British Airways.*) (c) Hourly variations in cargo throughput at British Airways freight terminal, London Heathrow. (*British Airways.*)

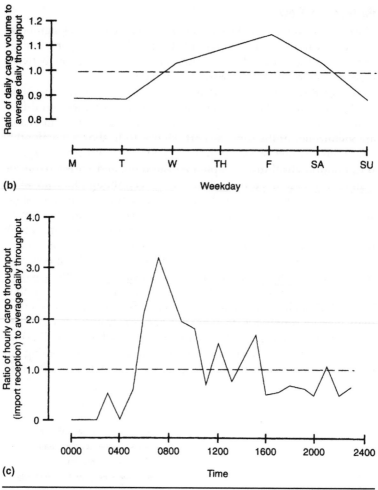

Figure 10.1 (*Continued*)

to a bed of lazy rollers outside the terminal building. They remain there until they can be processed after the peak.

Before the introduction of operational simulation models, freight terminals working close to capacity could be seriously inconvenienced and slowed when unexpected peaks occurred in the flows. These were unplanned and would lead to problems with understaffing and lack of storage space at the terminal. Modern logistics systems are able to predict the occurrence of overcapacity flows and are able to alleviate some of the effect of attempting to work with demand-capacity ratios of greater than 1.0. By taking active control of the flows and interfering with the arrivals patterns at the terminal, the effect of random uncontrolled flows can be smoothed into longer peaks, the demand-capacity ratio is reduced, and the effect of over-crowding can be partially alleviated. These are operational measures that work only in the short term. As demand reaches capacity, the only long-term solution is to increase capacity.

10.2   Expediting the Movement

Figure 10.2 shows diagrammatically the parties and organizations involved in the movement of freight. Freight is moved from the shipper to the consignee, usually through the agency of a freight forwarder, by one or more airlines using premises and infrastructure provided to some degree by the airports through which it passes. In many cases, on-airport facilities are provided not only for the airline but also for the freight forwarders. Even large firms frequently use the facilities of a freight-forwarding agency because air cargo requires rather specialized knowledge, and air cargo might be only a small part of a firm's total shipping operation. In order to provide an air shipping

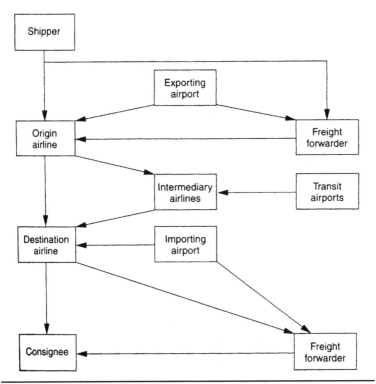

FIGURE **10.2**   Relationships among actors in the movement of air cargo.

service, the freight forwarder performs functions that are likely to be beyond the expertise or capability of the shipper. These include

- Determining and obtaining the optimal freight rate and selecting the best mixture of modes and routes
- Arranging and overseeing export and import customs clearance, including preparing all necessary documentation and obtaining requisite licenses (these are procedures with which the specialist freight forwarder is familiar)
- Arranging for the secure packing of individual consignments
- Consolidating small consignments into larger shipments to take advantage of lower shipping rates (the financial savings obtained by consolidation are shared between the forwarder and the shipper)
- Providing timely pickup and delivery services at both ends of the shipment

Most airlines see freight forwarders as providing a necessary and welcome intermediary service between themselves and the shipper and consignee. The freight forwarder, being familiar with the necessary procedures, permits the airline to concentrate on the provision of air transport and to avoid the time-consuming details of the facilitation and landside distribution systems. Shippers with large air cargo operations frequently use their own in-house expertise within a specialized shipping department.

To encourage shipments that are more economical to handle, airlines have a complex rate structure, of which the main components include

- *General cargo rates.* These apply to general cargo between specific airport pairs.
- *Specific commodity rates.* Often over particular routes, there are high movement volumes of a particular commodity. The IATA approves specific commodity rates between specific airports. For *general cargo* and *specific commodity rates*, there are quantity discounts.
- *Classified rates.* Certain commodities, because of their nature or value, attract either a percentage discount or surcharge on the general commodity rate. Classified rates frequently apply to the shipment of gold, bullion, newspapers, flowers, live animals, and human remains.
- *ULD rates.* This is the cost of shipping a ULD container or pallet of specified design containing up to a specified weight of cargo. ULDs are part of the airline's equipment and are loaned to the shipper or forwarder free of charge, provided that they are loaded and relodged with the airline within a specified period, normally 48 hours.

- *Consolidation rates.* Space is sold in bulk, normally to forwarders at reduced rates, because the forwarder can take advantage quantity and ULD discounts. The individual consignee receives the shipment through a break bulk agent at destination.

- *Container rates.* Containers in this context are normally owned by the shipper rather than the airline. They are usually nonstructural, of fiberboard construction, and suitable for packing into the aircraft ULDs. If a shipment is delivered to them in approved containers, airlines provide a reduction in air-freight rates.

The very rapid movement of air freight requires accurate documentation. This is provided in terms of the airline's *air waybill*, sometimes called the *air consignment note*, an example of which is shown in Figure 10.3. The air waybill is a document with multiple uses. It provides the following:

- Evidence of the airline's receipt of goods
- A dispatch note showing accompanying documentation and special instructions
- A form of invoice indicating transportation charges
- An insurance certificate, if insurance is effected by the airlines
- Documentary evidence of contents for export, transit, and import requirements of customs
- Contents information for constructing the loading sheet and flight manifest
- A delivery receipt

10.3   Flow Through the Terminal

Figure 10.4 is a system diagram that represents the various stages involved in cargo flow through a terminal (IATA 2004). On the import side, over a very short period of time, a large "batch" of cargo (i.e., the aircraft payload) is brought into the terminal operation. This cargo is then sorted, and that which is inbound and not direct transfer is checked in, stored, processed, and stored again prior to delivery in relatively small shipments (i.e., up to container size). The export operation is the reverse process. Small shipments are received, processed, stored, and assembled into the payload for a particular flight that is then loaded by a procedure that keeps the aircraft turnaround time to an acceptable minimum. Figure 10.5 shows for import and export cargo, respectively, how facilitation proceeds concurrently with the physical flow of cargo through the terminal. For very large terminals, the flow of documentation proved in the past to be a potential bottleneck. With electronic documentation, the old problems of large paper flows now has largely been eliminated in the most developed countries, resulting in faster, more efficient cargo processing.

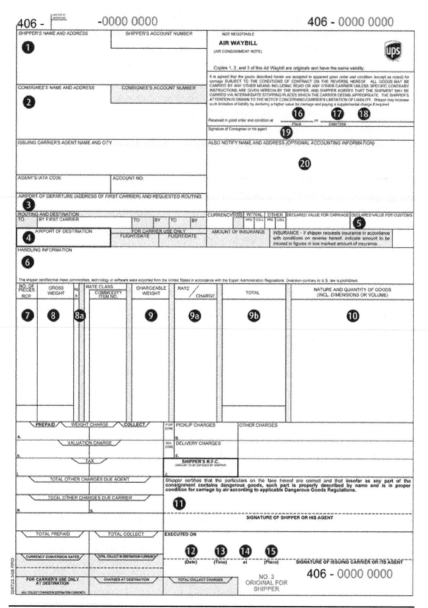

Figure 10.3 Air waybill.

Although most cargo terminals are similar in overall function, the nature of the traffic and therefore the actual mode of operation can differ depending on a number of factors:

1. *Type of cargo.* Handling depends on how much cargo arrives already unitized, how much in small shipments, and how much requires special handling.

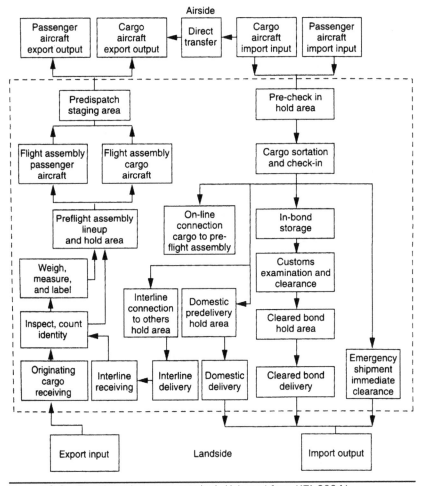

FIGURE 10.4   Flow through a cargo terminal. (*Adapted from IATA 2004.*)

2. *Type of shipper.* Terminal operations are simplified when receiving shipments through freight forwarders rather than private shippers. The forwarder will already have partially consolidated the handling and facilitation.

3. *Domestic-international split.* Domestic cargo requires less documentation and handling than international cargo. At some airports in the United States and other countries, suitably containerized domestic shipments physically bypass the terminal, the only handling being the facilitation. In some countries, international cargo also may be permitted, by special arrangements with customs.

4. *Transfer.* Terminals such as Hong Kong, London, Dubai, and Chicago have very high levels of transfer cargo that moves

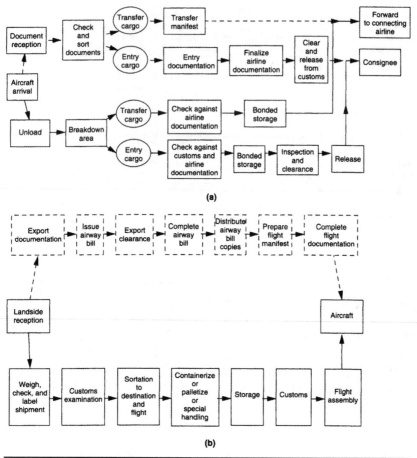

Figure 10.5   (a) Physical flow and documentation of import cargo. (b) Physical flow and documentation of export cargo.

between flights *across the apron*. In 2010, transfer tonnage at the giant Hong Kong Air Cargo Terminal (HACTL) terminal in Hong Kong amounted to 24 percent of the combined import and export cargo. The special requirement of transfer traffic should be reflected in terminal design and operation because it requires an apron handling capacity that is not needed in routine terminal processing. Some modern designs effect the so-called across-the-apron transfer within a special area, sometimes even within the cargo terminal building, which can provide shelter from extreme weather conditions.

5. *Surface shipments.* In many countries, the major air cargo terminals receive a large amount of "air" cargo by road. For example, cargo taken into the air freight terminal at Naples,

eventually bound for Singapore, can be assigned to a "flight" from Naples in Italy to Munich in Germany that is in fact a load on a road vehicle. After the first international leg by surface transport, the second leg from Munich to Singapore is achieved by air. London Heathrow provides a similar road-to-air transfer hub for many of the cargo services of British regional airports.

6. *Interlining.* Interlining traffic to other airlines is unlikely to require the same processing as other import cargo.

7. *All-freight operations.* Where freight is carried by all-freight aircraft, the airside operation is characterized by dramatic peaks in flow because flights may be concentrated in the night hours. Severe overloading of the freight apron is much more likely with this type of operation. Combined freight/passenger operations lessen the peak flows in the cargo terminal and transfer operations to the peak passenger hours. These are not likely to coincide with the major freighter movements. However, the use of passenger aircraft requires that a portion of the total freight traffic is remotely loaded at the passenger apron.

10.4   Unit Load Devices (IATA 1992, 2010)

Freight, which is cargo, mail, airline stores, and unaccompanied baggage, was originally carried loosely loaded in the cargo holds of passenger aircraft and in small all-cargo aircraft. Until the mid-1960s, all air freight was carried in this loosely loaded or "bulk" freight form. The introduction of large, all-freight aircraft, such as the DC8 and the B707, meant very long ground turnaround times owing to the lengthy unloading and loading times involved with bulk cargo. The ground handling process was substantially speeded up by combining loads into larger loading units on pallets. Pallets, in combination with suitably designed aircraft floors and loading equipment, permitted rapid loading and unloading times. However, pallets must be carefully contoured to prevent damage to aircraft interiors, where aircraft are also used for passengers. Goods on the pallets are themselves vulnerable to damage and even pilferage in carriage as well as being possibly inadequately protected from the weather while on the apron. Igloos, which are nonstructural shells or covers to the pallets, were introduced to overcome these drawbacks. Even with a cover, a pallet is a relatively unstable device that might shift during the handling process. A more substantial structural unit, the unit load device (ULD) air freight container, gives considerably more support to the cargo during the handling and carriage stages, but the device still must be lifted from below, unlike the intermodal container. With the introduction of wide-bodied aircraft, belly containers also were available in

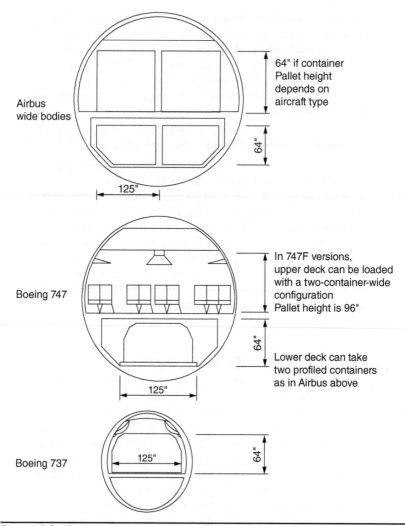

FIGURE **10.6**   Container arrangements in wide- and narrow-bodied aircraft.

shapes that cannot be built from pallets. Figure 10.6 shows how ULD containers can be used both on the main deck and in the belly loads of wide-body aircraft.

IATA recognizes a set of standard ULDs in the form of dimensioned pallets, igloos, and ULD containers (IATA 2010). These ULDs are each compatible with a number of different aircraft types and generally are compatible with the terminal, the apron, and the loading equipment. However as Table 10.1 indicates, there is serious incompatibility among aircraft types. This can cause considerable rehandling at air freight transfer airports.

ULD	LD1	LD2	LD3	LD3 INS	LD4	LD6	LD7	LD8	LD9	LD11	A2	88-Inch Pallet	96-Inch Pallet	Half Pallet	16-Foot Pallet	20-Foot Pallet	AO6	AO7	M-6	LD 26	LD 29
Aircraft																					
747	X		X	X			X		X	X	X	X	X		X	X				X	X
747F							X				X	X			X	X	X	X	X		X
747 Combi		X					X										X	X	X		
767	X		X	X	X		X	X	X			X	X								
777	X		X	X	X		X			X		X	X							X	
787			X			X				X											
707F														X							
727F																					
737F														X							
DC10			X	X			X		X	X	X		X							X	
MD11	X		X	X			X		X	X			X								
L1011			X	X			X		X	X			X								
A300			X	X		X	X		X	X			X								
A300F											X										
A310			X	X		X	X		X	X			X								
A320			X*																		
A330			X	X		X	X		X	X			X								
A340			X	X		X	X		X	X			X								
A380			X			X				X											

*With reduced height.

TABLE 10.1 Compatibility of ULD Standard Contours with Aircraft Loading Envelopes

303

Because all ULDs do not have an optimal fit with all aircraft, it is obvious that without some degree of standardization, aircraft can suffer from poor utilization of space, which must result in unnecessarily high cargo rates. Maximum space efficiency is achieved by use of the aircraft's optimal ULDs on each flight. For a container moving on a journey composed of several different flight legs, this could mean an additional break-bulk operation at any transfer terminal, which is extremely expensive. A tradeoff must be made between the use of a less than optimal container fit and additional break-bulk operations.

For some years there has been discussion of the use of intermodal containers that can be used for both surface and air modes. Air cargo containers, designed for the purpose of keeping tare weight low, have little structural strength. They must be lifted under the base. Surface containers are robust, have considerable structural strength, and can be lifted from the top. However, they are very heavy. There is little evidence of any significant growth in the use of intermodal containers.

10.5   Handling Within the Terminal

Unlike passengers, who merely require information and directions to flow through all but the largest terminals, freight is passive and must be physically moved from landside to airside and vice versa. The system used to achieve this physical movement will depend on the degree of mechanization to be used to offset personnel costs. The terminal types are threefold, and any particular terminal may well be made up of a combination of types.

Low Mechanization/High Manpower

Typically, in this design, all freight within the terminal is manhandled by workers over roller systems, which are either unpowered or partially powered. Forklift trucks are used only for building and breaking down ULDs. On the landside, freight is brought to the general level of operation in the terminal by a dock-leveling device. This operational level, which is maintained throughout the terminal, is the same as the level of the transporting dollies on the airside. Even heavy containers are fairly easily handled by the workers over the unpowered rollers. This system is very effective for low volume flows in developing countries where unskilled labor is cheap, where mechanization is expensive, and where there might be a lack of skilled labor to service mechanized equipment. Gradually, low-mechanization/high-manpower systems are being upgraded to fixed mechanized systems. The old systems are space extensive and are unsuitable for large flows simply from the number of unskilled workers in the terminal. Figure 10.7*a* shows a terminal in Brazil using extensive roller beds and rail transfer tables.

Open Mechanized

The open mechanized system has been used for some time in developed countries at medium-flow terminals. All cargo movement within the terminal is achieved using forklift vehicles of various designs that are capable of moving fairly small loads or large aircraft container ULDs. Forklift vehicles are capable of stacking up to five levels of bin containers. Many older terminals operated successfully with this system, but the mode is space extensive, and forklift operations incur very high levels of ULD and other container damage. As pressure has come on cargo terminals to achieve less-costly, higher-volume throughputs in existing terminal space, many open mechanized terminals have been converted to fixed mechanized operation. A typical open mechanized system is shown in Figure 10.7b.

Fixed Mechanized

The very rapid growth in the use of ULDs in aircraft has led to cargo terminal operations in which extensive fixed mechanical systems are capable of moving and storing the devices with minimum use of labor and low levels of container damage in the handling process. Such fixed-rack systems are known as *transfer vehicles* (TVs) if they operate at one level and *elevating transfer vehicles* (ETVs) if they operate on several levels. Because they can have very large storage capacities, they can level out the very high apron throughput peaks

(a)

Figure 10.7   (a) Air cargo interior—low-technology freight terminal. (b) INFRAERO cargo terminal, São Paulo International Airport. (*Courtesy of INFRAERO, Brazil.*) (c) An ETV. (*Courtesy of INFRAERO, Brazil.*)

FIGURE 10.7 (Continued)

that can occur with all-freight wide-bodied aircraft. ETV rack storage can absorb incoming ULD freight for several hours and conversely can feed departure flows on the airside that are greatly in excess of average flows. New and renovated terminals in major freight operations at airports such as Hong Kong, JFK, Seoul, Frankfurt, and Heathrow all include ETV systems as a matter of course. A typical ETV system is shown in Figure 10.7c. In the case of Hong Kong, the development of multistory cargo terminals was driven by the scarcity of land.

In developing countries, it is not unusual to build freight terminals with low mechanization and later to install first forklift vehicles and later, if volumes continue to grow, TVs and ETVs. In developed countries, labor costs preclude manhandling, and even the simplest terminals are equipped with forklift technology, which is rapidly further mechanized with traffic growth. Automated storage and retrieval systems (ASRSs) for storing bulk cargo also have been installed to improve efficiency, enhance security, and facilitate track and tracing.

10.6   Cargo Apron Operation

During the 1960s, forecasters considered that the future for air cargo was the virtual separation of passenger and cargo traffic and the rapid development of all-cargo fleets. Two major factors combined to ensure that this did not occur. First, wide-bodied aircraft were introduced very rapidly in the 1970s to achieve crewing and fuel efficiencies. The new wide-bodied aircraft had substantial and underused belly space that was suitable for the carriage of containerized cargo. Second, as indicated in Section 10.1, while exhibiting healthy growth rates, air cargo did not achieve the explosive growth rate forecast at the time. By the early 1980s, the air cargo market and operation had changed so much that a number of major airlines sold their all-freight airplanes in favor of using lower-deck space on passenger aircraft. By 2012, several major carriers had reintroduced all-cargo aircraft, mainly to serve airports where there was already all-cargo service by all-cargo airlines. There are many small airlines that have all-cargo operations. If one excludes the integrated carriers such as FedEx, UPS, TNT, and DHL, only a few major airlines have all-cargo aircraft. The operations of the integrators are discussed in Section 10.9.

Even though much freight is carried by other than all-cargo aircraft, very large volumes are moved by such operation through the air cargo aprons. All cargo aircraft are capable of very high productivity, provided that there is a sufficient level of flow to support these productivity levels. The maximum payload of a B747-8F is over 295,000 pounds (133,900 kg) (Boeing 2008). Figure 10.8 shows that with containerized cargo, the operator estimates that it is possible to off-load and load 245 tons in under an hour using the nose and side doors or an hour and a half using only the nose door.

The times given by the manufacturer must be regarded as ideal times, where the load is immediately available and sequenced for loading. Real-world apron operations often mean that load control of the aircraft seriously increases actual turnaround time. For a Series 100 747F aircraft with only side-door loading, a turnaround time of 1½ hours would be considered very good; for a Series 200 aircraft with nose loading only, 2½ hours is more likely. The latter time can be reduced by simultaneously loading the nose and side doors of the main deck, but this ties up two high-lift loaders. Minimum total turnaround time might well be seriously affected by outside considerations such as the availability of off-loading equipment or the necessity to wait for customs or agricultural inspection before any off-loading can be started. Average turnaround times are much greater than minimum times because frequently aircraft are constrained by schedules that provide a total ground handling time that is much greater than these cited minima. The following turnaround times have been cited for the B747-F operators at Melbourne Airport in Australia:

Cargolux	2 hours
Cathay Pacific	2 hours
Lufthansa	1 hour, 45 minutes
Air New Zealand	2 hours
Singapore Airlines	2 hours

Using containers, the effective cargo payload is decreased, but considerable gains are made in operational efficiency. Typically, the total payload of a B747-F would be constituted in the following way:

Main deck	168,000 pounds (containerized)
Lower lobe aft	21,700 pounds (containerized)
Lower lobe forward	24,800 pounds (containerized)
Bulk compartment	7,500 pounds
Total payload	222,000 pounds

Figure 10.9 shows the location of ground-handling and servicing equipment required for simultaneous upper- and lower-deck unloading and loading assumed in the Gantt charts for a typical all-freight large (Code E) flight, shown in Figure 10.8. Such an operation places a very heavy load on apron equipment and apron space. In all, the region of 48 containers and 14 m³ could be offloaded and a similar amount loaded. Two lower-deck low-lift loaders and one upper-deck high-lift loader will be needed, each requiring at least one and possibly two container transporters or a train of container dollies. Figure 10.10

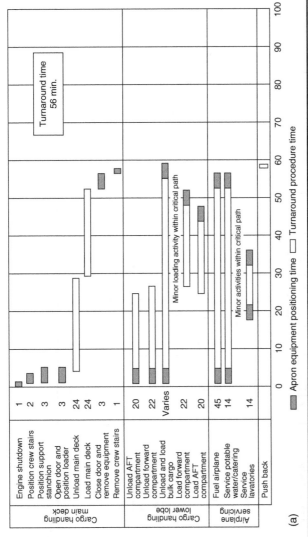

Figure 10.8 (a) Gantt chart for a typical turnaround for a large (Code E) all-freight flight, using nose- and side-door loading with 100 percent cargo exchange. (b) Gantt chart for a typical turnaround for a large (Code E) all-freight flight, using only nose cargo door with 100 percent cargo exchange.

309

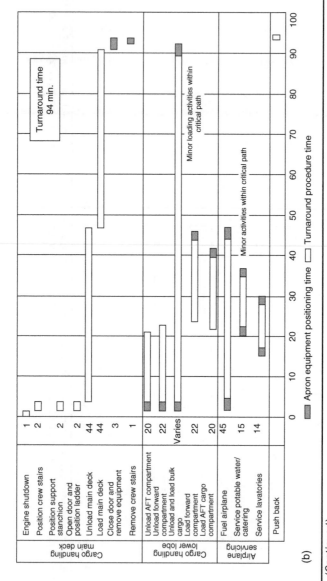

The chart contains the following activities, times (min), and legend:

	Activity	Time
Cargo handling main deck	Engine shutdown	1
	Position crew stairs	2
	Position support stanchion	2
	Open door and position ladder	2
	Unload main deck	44
	Load main deck	44
	Close door and remove equipment	3
	Remove crew stairs	1
Cargo handling lower lobe	Unload AFT compartment	20
	Unload forward compartment	22
	Unload and load bulk cargo	Varies
	Load forward compartment	22
	Load AFT cargo compartment	20
Airplane servicing	Fuel airplane	45
	Service potable water/catering	15
	Service lavatories	14
	Push back	

Turnaround time 94 min.

Minor loading activities within critical path

Minor activities within critical path

Turnaround procedure time

Legend: ■ Apron equipment positioning time   □ Turnaround procedure time

(b)

FIGURE 10.8 (Continued)

310

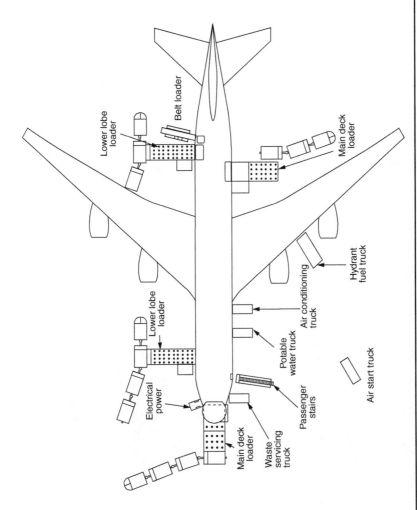

Figure 10.9 Airplane servicing arrangement for a typical turnaround for a large (Code E) all-freight aircraft.

Figure 10.10   (a) Upper-deck freight container loader. (b) Lower deck freight container loader (*Courtesy: Munich International Airport.*) (c) Nose loader for freighter in position. (*Courtesy: Chapman Freeborn Air Chartering.*) (d) Cargo container transporter.

Figure 10.10 *(Continued)*

shows the range of apron equipment required for the turnaround of a large air cargo flight. The bulk cargo hold is loaded using a bulk loader, fed by an apron trailer unit. Additionally for the general turnaround servicing, two fuel trucks are required if hydrant fueling is unavailable, a potable water truck, a truck to supply demineralized water for water injection, a sanitary truck for toilet servicing, a ground power unit, a compressed-air start unit, and a crew access stair. In the immediate vicinity of the cargo building, it is usually necessary to provide a bed of lazy rollers or slave pallets to accept containers that might arrive from the apron at peak times when the terminal is unable to accept all the flow. Peak apron capacity is often in excess of terminal throughput capacity. An ETV system is often also capable of absorbing apron peaks by storing received ULDs until they can be processed in the terminal.

10.7   Facilitation (ICAO 2005)

The secure and efficient shipment of a freight consignment can take place only when documentation keeps pace with the physical movement of cargo. This is a fairly straightforward matter with low volumes moving through a single air cargo building. At major airports with high cargo flows, numerous airlines, and multiple processing facilities, the control of facilitation becomes extremely complex and very necessary.

In the early days of air cargo, all necessary documentation was carried with the consignment. As flows became greater, this system became impractical, and documentation was necessarily moved by electronic means. When computer mainframe systems became more widespread in the 1970s, early freight tracking systems, such as London Airport Customs and Excise System (LACES) and Air Cargo Processing in the 80s (ACP80) at London Heathrow and Community System for Air Cargo (COSAC) at Hong Kong, became more common. These systems connected customs with the inventory-control systems of the airlines, and the airlines connected their own computers to a central bureau in order to transmit inventory-control data to a communal file. Information from the communal file was made available to the airlines, forwarding agents, and customs, provided that the information sought was within the authority of the operator requesting it. Information was transmitted by special air transport communication systems such as the Société Internationale de Télécommunications Aéronautiques (SITA) system.

Figure 10.11 indicates in a very simplified diagram the principal stages in exporting and importing freight. A computer record of the consignment is created initially either at the time of space reservation or at export reception. The initial file contains all details of the consignment from the air waybill, such as weight, contents, destination, carrier, shipper, and consignee. Additionally, files are set up that create

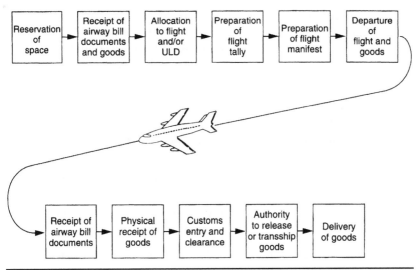

FIGURE 10.11   Principal stages in exporting and importing freight.

flight records up to seven days in advance of departure date, indicating maximum weights and volumes to be allocated to each particular destination. When the consignment is completely received (i.e., the number of packages agrees with the airway bill), the consignment is either allocated to a flight or to a ULD, which itself will be allocated to a flight. The computer then provides a flight tally file that indicates a shed storage location, ULD details, and any instructions on handling. Once all consignments have been tallied, a flight manifest is produced. The manifest is the working document reporting the movement of cargo. The manifest is used on the load control of the aircraft itself. The final input to the information is the statement that the flight and goods have departed. There is also the possibility of modifying initial input data to allow for consignment splitting, offloading, and coping with short shipments where a partial shipment is made when part of the consignment is delayed.

On the import side, a document is created on receipt of the waybill documentation. This record identifies the receiving shed, the airline, the airway bill number, and other necessary information contained on the airway bill. Status reports are made on the physical receipt of the goods—first that the goods have been received and second that the consignee is creditworthy. After customs entry and clearance, either a release note is produced after clearance, or removal authority is granted to permit transshipment to another airport for clearance or removal to another import shed. At this stage, a transfer manifest is produced for through-transit consignments. The final input in the life of the record is that the goods have been delivered. Several reports are automatically available that record discrepancies, such as the receipt of more or fewer packages than were expected. As a standard inventory-control measure, a number of daily reports are produced. These give the current status of problem areas such as

- Consignments not received within 24 hours of receipt of the air waybill
- Consignments not delivered within two days of customs clearance
- Through-transits not delivered to onward carriers within six hours
- Transshipments and interairport removals not achieved in seven days

At all stages, the status files may be interrogated by operators with necessary authority.

By the 1990s, the widespread introduction of personal computers (PCs) had caused the large cargo operators to set up or join cargo community systems (CCSs), which, using *electronic data interchange* (EDI), enabled computers to exchange data directly without human intervention. By the mid-1990s, for example the ICARUS system, used by British Airways, linked more than 50 major airlines and more than 1,500 forwarders. It also connected with other CCSs such as TRAXON centered in France, Cargo Community Network (CCN) in Singapore, Avex in the United States, Cargonet in Australia, and other systems in Austria, Italy, South Africa, and the United Kingdom. CCS was able to exchange information among computers without the need for paper-based documentation and could

- Access flight space availability
- Obtain electronic booking/reservation
- Transmit electronic documents (e.g., air waybills)
- Receive electronic documents
- Provide consignment tracking and status
- Create a community structure among forwarders
- Provide schedule information

Before the Internet was established, the airlines paid the CCSs or other message providers such as SITA to transmit the Cargo Interchange Message Procedures (Cargo-IMP). However, EDI was greatly facilitated by the development of the Internet in the 1990s. The practice of the independent CCSs communicating with each other by conventional telephone connections was replaced by the Logistics Management Systems (LMS), connecting by Internet and providing all the necessary links among forwarders and carriers. This development allowed free communication using the File Transfer Protocol (FTP).

In 2007, IATA launched the e-Freight Project, which aimed to replace all paper documentation for the carriage of freight with electronic messages. The aim was to reduce costs, improve transit times and accuracy, and improve the competitiveness of air freight. At the time, up to 30 paper documents could be required for a single consignment. Achieving paperless facilitation was to prove difficult for a number of reasons:

- As of 2011, many countries required advance submission of electronic cargo information for risk and assessment purposes.

- The legal and technical environment in many jurisdictions was averse to accepting paperless transfers.

- In many cases, customs in the country concerned had no electronic platform.

- Not all governments followed international standards relating to documentation and procedures.

- Different governments required the submission of documents in different formats (e.g., EDIFACT, XML, TXT, etc.)

- In some jurisdictions, shippers frequently had to submit the same data or documentation via multiple channels.

However, by 2010, IATA reported that 44 countries, 380 airports, and 32 airlines were using e-Freight. This amounted to 80 percent of all air cargo volume and a significant advance in the three years since launch of the project.

10.8 Examples of Modern Cargo Terminal Design and Operation

Lufthansa redesigned its Frankfurt air cargo terminal in 1995 to handle a capacity of approximately 1 million tons per year. Figure 10.12 shows a schematic layout of the facility, which, in common with most modern terminals, is extensively mechanized for the handling of ULDs. Frankfurt has a very high proportion of transfer freight, some of which requires reconsolidation within the cargo terminal itself. The handling system within the terminal saves both labor and space by the employment of extensive mechanization. The system is designed to handle the range of container types that the industry uses. Prior to consolidation, bulk cargo shipments were assigned to rollboxes in a continuous-handling section, where vertical stackers were able to assign the rollboxes to vertically racked rollbox slots. An ETV feeds a vertical container racking system that can hold both 10- and 20-foot units. The layout permits lateral, transverse, and round-the-corner movements by conveyor

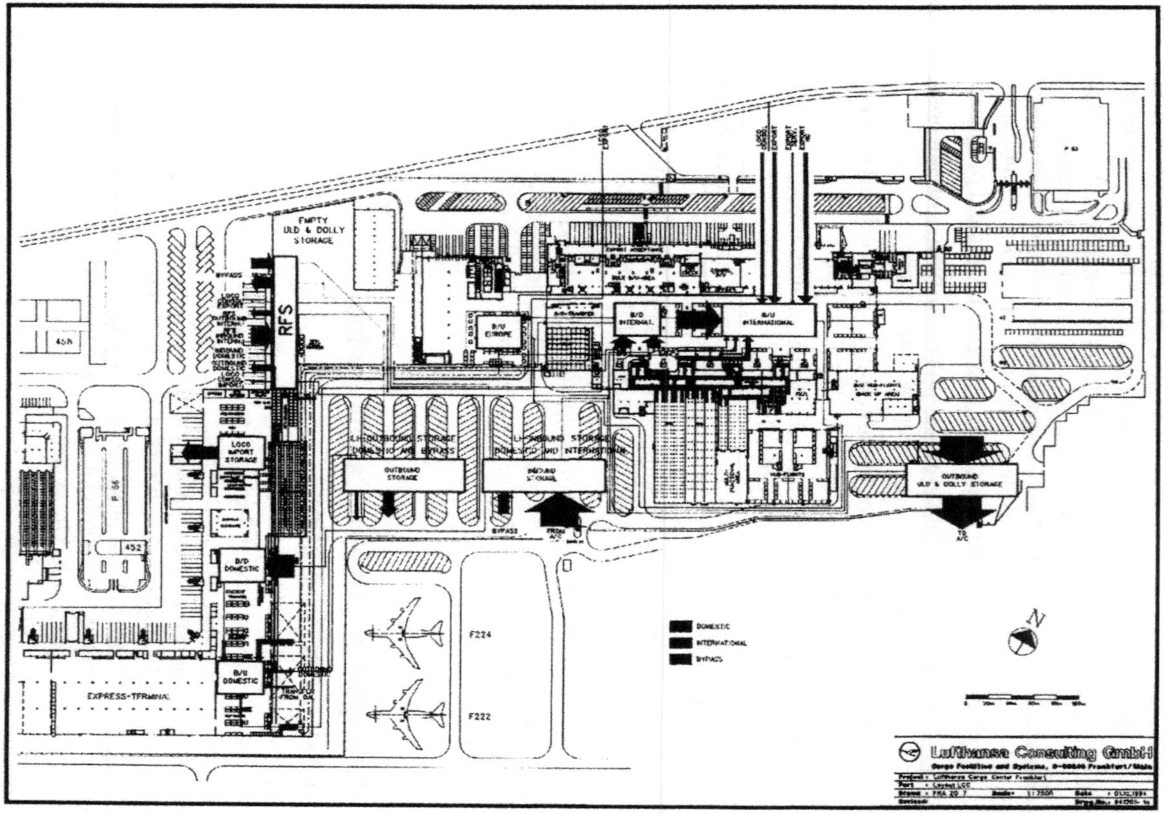

Figure 10.12 Schematic of Lufthansa terminal.

systems without the use of mobile wheeled equipment. The operation has three principal objectives:

- Minimization of accidents to personnel and cargo
- Minimization of damage to ULDs
- Maximal use of the cargo terminal space

In 1998, when Hong Kong relocated its airport from Kai Tak to the Chep Lap Kok site, it opened a new $1 billion cargo terminal, *HACTL SuperTerminal 1*. This is a six-story facility with a total floor area of 288,341 m^2 and a stated capacity of 2.6 million tons per year, giving a throughput of 9.0 tons/m^2 per year. This compares with the IATA recommendation of 17 tons/m^2 for a highly automated facility and 10 tons/m^2 for an average level of automation (IATA 2004; Ashford et al. 2011).The facility, shown in Figure 10.13, served 90 airlines and 1,000 freight forwarders in 2010, with 34 all-cargo stands for large aircraft. In the same year, Hong Kong for the first time overtook Memphis as the world's busiest airport cargo operation. The HACTL facility has 40 ETVs and TVs, 362 build-up/breakdown stations, a CSS with 3,500 container storage positions, and an automated box storage

Figure 10.13   HACTL Terminal, Hong Kong. (*Courtesy: Hong Kong International Airport.*)

system (BSS) of 10,000 storage positions. The CSS is capable of handling 10- and 20-foot containers.

In addition to SuperTerminal 1, the Chep Lap Kok facility has an adjacent express center with an area of 40,361 m² and a capacity of 200,000 tons per year. The express center and SuperTerminal 1 are built on a site of 170,000 m². This is an indication of the vigorous growth of the Asian cargo market that in 2006 the Asia Air Freight Terminal also developed a new, expanded terminal at Chep Lap Kok with a fully automated storage and retrieval system, and in 2013, a new cargo terminal for Cathay Pacific Services Limited is planned to open, also equipped with a state-of-the-art multistory cargo-handling system. The air cargo terminals in Hong Kong are one of a number in Asia that reflect the region's growing industrialization and increasing use of air cargo as a vital method of supplying goods to the rest of the world.

10.9   Cargo Operations by the Integrated Carriers

By 2012, twenty-one of the world's airports were handling more than 1 million tons of cargo. Of these, Memphis ranked as the world's second busiest freight airport and Louisville as the eleventh. The former is the base of FedEx and the latter of UPS, both so-called integrated carriers (ICs). The growth of the ICs over the last 40 years has been extraordinarily buoyant. Such carriers offer door-to-door service, usually within a stated time limit. FedEx, originally Federal Express, the largest of the integrated carriers, is also the world's largest all-cargo airline, serving 375 airports in 220 countries, owning in excess of 80,000 motorized vehicles, and having 290,000 employees in 2010. In the same year, the average number of daily shipments was 8.5 million.

Operation of the ICs' air cargo terminals is quite different from that of conventional terminals, and it is difficult to draw comparisons between the two operations even at the same airport. The terminals of the ICs have very high daily peaks and are characterized by lack of the required storage space because very little cargo dwells in the terminal for any significant time. The principal characteristics that differentiate this type of terminal from the conventional cargo terminal include

- Cargo is under the physical control of one organization throughout the whole of its journey.
- Delivery to the exporting airport and pickup at the importing airport are by the IC itself; the landside loading and unloading area is more easily controlled and organized.
- No freight forwarders or clearing agents are involved.
- ICs generally use their own fleets and but may use other commercial airlines in the international arena (DHL and TNT

frequently use available commercial aircraft to overcome the problems with international air services rights).

- All-cargo aircraft are used, and the characteristics of the fleet are well known to the terminal operator, the IC itself.
- The cargo is given a guaranteed delivery time; this implies rapid clearance through the airport facility.
- Documentation and facilitation are through one company's system.
- High security and limited access to the terminal are possible because no outside organizations are involved other than customs.

Typically, the large global ICs have set up intercontinental hubs such as Memphis, Newark, Hong Kong, and Paris Charles de Gaulle. Serving these intercontinental major hubs is a subsidiary network. For example, FedEx in Europe has a support network of 15 second-level airport hubs and 22 minor airport hub or spoke operations. Trucks from these facilities serve the remainder of Europe.

The location and operation of an IC's hub terminal will have little relationship with the operation of any passenger facilities at that airport. The IC terminal tends to operate in isolation, with peaking characteristics that are quite independent of the activities on the passenger apron. Often the IC terminal sees banks of arrivals around midnight and banks of departures in the early hours of the morning. As such, they have much in common with airmail terminals, although in scale they can be much larger.

The largest ICs have networks that can be classified as follows:

1. Hubs and subsidiary hub airports
2. Spoke airports

Hub and subsidiary hub airports are characterized as airports that have little destined or originating local traffic. There is a large amount of transfer air cargo traffic, much of which is across the apron. Flow through the terminal is small compared with total flow. Landside of the terminal, the parking, acceptance/dispatch truck docks, and access roads are also minimal. *Spoke airports* have no apron transfer. For these airports, the flow through the terminal is essentially the total flow. Landside parking, loading/unloading, and access road provision are directly proportional to the scale of the operation.

IATA recommends the following provision of terminal space (IATA 2004):

- Regional hubs/gateway hub terminals: 7 tons/m² per year
- Reliever hubs: 5 tons/m² per year

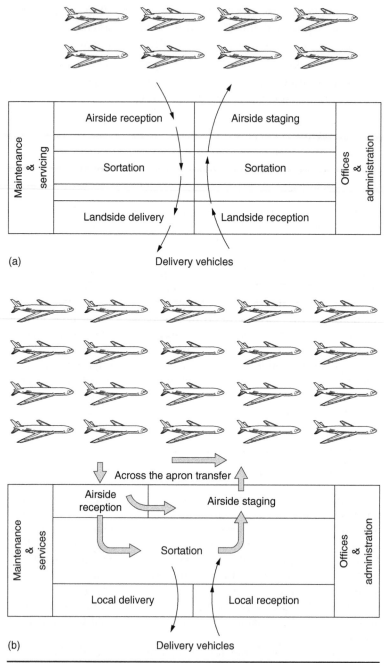

Figure 10.14 (a) Schematic of flows in a spoke terminal of an integrated carrier. (b) Schematic of flows in a hub terminal of an integrated carrier.

In a recent study carried out at the major hub of a European airport, it was determined that the expected capacity at an acceptable level of service with respect to space provision was 8 tons/m^2 per year. The average shipment weight was found to be 15 kilograms. Figure 10.14 shows schematics of the mode of operation of hub and spoke terminals of an integrated carrier.

References

Ashford, N. J., S. Mumayiz, and P. H. Wright. 2011. *Airport Engineering: Planning, Design, and Development of 21st Century Airports*, 4th ed. New York: McGraw-Hill.

Boeing. 2008. *747-8 Airplane Characteristics for Airport Planning*. Chicago: Boeing Commercial Airplanes.

Boeing. 2011. *World Air Cargo Traffic Forecast, 2010–11*. Seattle: Boeing Airplane Company.

International Air Transport Association (IATA). 1992. *Principles of Cargo Handling and Perishable Cargo Handling Guide*. Montreal, Canada: IATA.

International Air Transport Association (IATA). 2004. *Airport Development Reference Manual*, 9th ed. Montreal, Canada: International Air Transport Association.

International Air Transport Association (IATA). 2010. *ULD Technical Manual*. Montreal, Canada: IATA.

International Civil Aviation Organization (ICAO). 2001. *The Safe Transport of Dangerous Goods by Air*, 3rd ed. (Annex 18 to the Chicago Convention, Including Amendments 7 to 9). Montreal, Canada: ICAO.

International Civil Aviation Organization (ICAO). 2005. *Facilitation*, 12th ed. (Annex 9 to the Chicago Convention, including Amendment 21 of November 19, 2009). Montreal, Canada: ICAO.

International Civil Aviation Organization (ICAO). 2011. *Civil Aviation Statistics of the World*. Montreal: ICAO.

CHAPTER 11

Aerodrome Technical Services[1]

11.1 The Scope of Technical Services

Various operational services found at air carrier airports can be conveniently grouped together under this general heading. They are concerned with the safety of aircraft operations in terms of control, navigation and communications, and information. The attainment of these objectives is respectful, too, of the environmental and cost-related operational performance, and the service implementations will consider the attainment of optimal performance if limitations are unavoidable.

Clearly, these matters are of interest to all aviation nations, and they form the subjects of four of the technical annexes to the International Convention on Civil Aviation (Chicago Convention). The services are (in the order they are addressed in this chapter):

Annex 11: *Air Traffic Services*
Annex 10: *Aeronautical Telecommunications* (including navaids)
Annex 3: *Meteorology Services for International Air Navigation*
Annex 15: *Aeronautical Information Services*

In recent decades, a significant development has been adoption of the conclusions of the International Civil Aviation Organization's (ICAO's) Future Air Navigation Systems (FANS) Panel. The impact of that implementation of operations with respect to services conducted overall has been considerably affected already and is likely to be even more greatly affected in the next few decades.

In addition to the four preceding services, emergency service also has to be provided at all airports and general aviation airfields in

---

[1]This chapter was written by Mike Hirst.

order to provide firefighting and rescue capabilities in the event of aircraft accidents. This service is dealt with separately in Chapter 12.

In the past, it has been relevant to observe that a general aviation airfield may need only some of these services to fulfill its particular type of operations (e.g., flight training, executive aviation, and personal flying), but changes are affecting all aerodrome operators. While small airfields are lesser affected than large airports, the proportional impact will be similar overall, and the consequences could be much more significant than might be imagined.

11.2   Safety Management System

Regulatory demand that safety management systems (SMSs) must support all airport technical activity has increased considerably in recent years, and the shift of focus toward this topic is explained here with regard to the topics addressed in this chapter. There is a more detailed description of SMSs in Chapter 15.

New provisions in Annex 14, Volume I (applicable from November 2001) included the statement: "As of 27 November 2003, States shall certify aerodromes used for international operations in accordance with the specifications contained in this Annex as well as other relevant ICAO specifications through an appropriate regulatory framework." This led to an ICAO Universal Safety Oversight Audit Program conducted across international airports worldwide in the period 2003–2008. The audit was performed to determine where best practice was already achieved and to what extent it was necessary to provide support in those areas where safety practice was assessed to be below acceptable levels.

Airports Council International (ACI) agreed with the principle of certification in accordance with ICAO standards, but stated that "in the context of the ICAO programme of safety audits for airports, Recommended Practices, for airport design should not become 'de-facto' Standards for airport operation, since they do not have the same status as Standards, may be difficult to apply to existing airports (notably those concerned with airfield dimensions), and are not based on a defined and consistently-applied 'target level of safety.'"

Target level of safety (TLS) is fundamental to the provision of all technical services on an airport. The documentation pertaining to this is the product of applying the safety management system (SMS) process. The concept involves application of the criteria set out in the probability-of-likelihood classification (Figure 11.1), where acceptable failure is described according to the failure effect, and the transcribing of failure effects, using the risk-tolerance classification matrix (Figure 11.2), into severity categories.

The Civil Aviation Authority (CAA) Safety Regulation Group (SRG) in the United Kingdom, in a brochure providing an introduction to staff, states: "There is no recognized standard in aviation for defining

	Probability of occurrence definitions				
	Extremely improbable	**Extremely remote**	**Remote**	**Reasonably probable**	**Frequent**
Qualitative definition	Should virtually never occur	Very unlikely to occur	Unlikely to occur during the total operational life of the system	May occur once during total operational life of the system	May occur several times during operational life
Quantitative numerical definition	$< 10^{-9}$ per hour	10^{-7} to 10^{-9} per hour	10^{-5} to 10^{-7} per hour	10^{-3} to 10^{-5} per hour	1 to 10^{-3} per hour
Quantitative annual/daily equivalent (approximate)	Never	Once in 1,000 years to once in 100,000 years	Once in 10 years to once in 1000 years	Once per 40 days to once in 10 years	Once per hour to once in 40 days

Figure 11.1   Probability of likelihood classification. (*Source:* UK CAP 760 Guidance on the Conduct of Hazard Identification, Risk Assessment, and the Production of Safety Cases–Table 2.)

		Probability of occurrence (likelihood)				
		Extremely improbable	**Extremely remote**	**Remote**	**Reasonably probable**	**Frequent**
		$<10^{-9}$ per hour	10^{-7} to 10^{-9} per hour	10^{-5} to 10^{-7} per hour	10^{-3} to 10^{-5} per hour	1 to 10^{-3} per hour
ESARR 4 severity	Accidents	Review	Unacceptable	Unacceptable	Unacceptable	Unacceptable
	Serious incidents	Acceptable	Review	Unacceptable	Unacceptable	Unacceptable
	Major incidents	Acceptable	Acceptable	Review	Unacceptable	Unacceptable
	Significant incidents	Acceptable	Acceptable	Acceptable	Review	Unacceptable
	No effect immediately	Acceptable	Acceptable	Acceptable	Acceptable	Review

Figure 11.2   Tolerance classification matrix. (*Source:* UK CAP 760 Guidance on the Conduct of Hazard Identification, Risk Assessment, and the Production of Safety Cases–Table 3.)

a typical SMS. So it has been necessary to adapt best practice from other industries in order to provide guidelines for those parts of the aviation industry that wish to implement a formal SMS. SRG has drawn up a number of SMS Policy and Principles aimed at providing a simple SMS framework supported by clear definitions." This is an example of a procedure that many regulators have adopted and that supports the introduction of procedures as airports become subject to these newer stringencies.

Subsequently, the regulatory authorities must be expected to place on aerodrome license holders the full responsibility for using an SMS-based procedure to justify the adequacy of existing provisions and to show that their processes take into account changes that are occurring over time. For example, a safety notice issued by the U.K. CAA in March 2012 entitled, "Runway End Safety Areas (RESA) and Runway Excursion Guidance for Aerodromes," stated that "all Aerodrome Licence Holders are now required to assess the risk of a runway excursion on applicable runways where the RESA does not extend to the recommended distance for the runway code number." The note is very clear about responsibility, stating "The annual requirement for Aerodrome Licence Holders to review and determine the RESA distance, even if there were no actual changes to the operations at the aerodrome, is now withdrawn." This is indicative of the way that SMS is being forced upon operators, and in Europe it is expected that the European Aviation Safety Agency (EASA) will mandate a similar regime of SMS application in the next few years.

RESA compliance provides a useful example of the SMS process in the airport technical services regime. If a risk assessment that takes into account the traffic mix at the airport shows that the stipulated acceptable probability of an aircraft reaching the end of the RESA is no longer "acceptable" in the risk-tolerability matrix, the airport should seek either to extend the RESA or apply mitigation, such as installing an engineered material arresting system (EMAS).

An example of an analysis leading to a mitigation being required was an assessment carried out on runway 23 at Charleston Airport, West Virginia. The terrain beyond the runway threshold falls away gradually and then steeply (approximately 230 feet vertically over 1,000-foot distance). In 2007, the airport installed a 405- × 150-foot (123- × 45-m) EMAS. This is on a built-up overrun area in which the slope has been moderated to about 1 percent (Figure 11.3). A PSA Airlines Canadair CRJ-200 on takeoff from Charleston on January 19, 2010, rejected takeoff at high speed and reached the EMAS area while still traveling at 50 knots. It was stopped 130 feet (40 m) into the area. The aircraft received substantial landing gear damage, but there were no injuries, and the runway was reopened within 6 hours.

SMSs can be used more mundanely too, say, to monitor seasonal issues that influence technical operations, such as the frequency of fog, incidence of snowfall or ice, or even bird observations over seasons.

Figure 11.3   EMAS installation at Charleston Airport, West Virginia. (*Source:* Charleston Airport.)

In such cases, the airport is expected to apply risk-assessment techniques to justify that appropriate measures are taken in operational procedure.

An SMS is not a process that will ensure safety. It has to be applied sensibly for it to assist in justifying what safety procedures should be applied and when or even why some procedures cannot be justified. Every airport technical service operative has to appreciate that this emerging and overarching regimentation is being imposed to ensure appropriate public safety in airport operations, and it must be seen to be a tool, not just an overhead, that does ensure appropriate due diligence in the way that services are operated.

11.3   Air Traffic Control

Fundamental Changes

Air traffic control (ATC) capability was already under scrutiny in the 1970s, in that it was questionable that existing paradigms could ever provide capacity that would be consistent with demand. The concern was both in terms of peak movement capacity and the minimization of environmental impact. The latter has become increasingly relevant over subsequent time.

In 1983, ICAO established a special committee on the Future Air Navigation System (FANS), and while the name implied that its terms of reference were limited to the navigation function, it was charged with developing very wide-ranging operational concepts for ATC. The newer-generation systems have since evolved under the title *air traffic management* (ATM). The FANS report was published in 1988 and laid the basis for the industry's future strategy for ATM through digital communication, navigation, and surveillance (CNS) using satellites and data links.

Traditional ATC systems use analogue radio systems for aircraft CNS, and their ground-service methodology is still more tactical (i.e., tradition has been "first come, first served") than strategic (i.e., planning and implementing according to capacity-sensitive criteria). This balance is being tipped, with digital advances making it more favorable to prioritize between strategic and tactical functions, moving more effort into strategic concepts being implemented that will lessen the demands on tactical process implementers.

The FANS concept has led to implementation changes across communications navigation and surveillance operations. *Communications* technology, which was solely voice-only radiotelephony-based messages, has become a hybrid voice- and digital-data-based communication technology using ground stations and satellite-based relay of data and messages. This affects all aircraft operations.

Navigation technology has become more involved with the integration of satellite-based navigation (satnav) systems. Historically, flight crews determined their position by the best perceived systems available (the usual choice being between inertial or radio-based systems) and regarded the system's probability of error as a function of the equipment. As soon as satellite-based navigation was possible, because it introduced a worldwide system capability, crew selection of which system to use was made more problematic. The FANS-assisted solution has been to introduce systems on aircraft that monitor navigation error probability and select the navigation technology most appropriate to the requirements of the current phase or mode of operation. This is done by declaring a required navigation performance (RNP) for the operation and managing sensor integration to attain required performance capability.

Surveillance operations have undergone a transition from being radar-based, supplemented by occasional voice reports to radar-based location detection, supplemented by automatic digital reports. The reports are conveyed from aircraft to the ground using an automatic dependent surveillance (ADS) system.

These improvements to CNS allow new procedures. In oceanic regions (Atlantic and South Pacific), where radar surveillance is not available, lateral track operations are capable of meeting safety requirements with separations of 30 nautical miles (55.6 km). These were formerly 120 nautical miles (222.2 km) and 60 nautical miles (111.1 km).

Aircraft can be spaced in trail at 30 nautical miles (55.6 km). Formerly, a time-based separation criterion was used that rarely reduced the in-trail separation below 60 nautical miles (111.1 km). In overland airspace, where there is still surveillance radar support, while ADS report position is not so crucial, the ability to add further control-related data—principally, knowledge of the three-dimensional vector—can enhance the improvement in decision-making support, with the intent to assess compromise of environmental and service efficiency clearance impacts when control actions are communicated to aircraft.

The ICAO FANS final report was released in 1991, and an implementation plan was available in 1993. Since then, ATM system plans for ground-system improvements have evolved in all leading nations. The introduction of ADS and Controller Pilot Data Link Communications (CPDLC) in aircraft commenced most immediately and is being followed through with the emergence of newer ground-based systems that will introduce new capability incrementally and possibly over an indefinite time period, but certainly spanning a number of decades.

Function of ATC

The primary purpose of ATC is the prevention of collisions between aircraft in flight and also between aircraft and any obstructions either moving or stationary on an airport. Additionally, it is concerned with promoting an efficient flow of air traffic. *Efficient flow* has tended to mean using up to the maximum capacity available in airspace, accepting that as the capacity limit is approached there will be an increasing level of delay.

It has been common practice in most countries that the central administration and management of air traffic services was vested in a governmental or quasi-governmental agency. This was usually a civilian organization but might be the military authority in some countries. Since the late 1980s, there has been a movement to "partial privatization" of some organizations, and the term *air navigation service provider* (ANSP) has become a widely accepted nomenclature for the whole sector.

In the case of a civilian organization, for example, U.S. legislation contained in the Federal Aviation Act of 1958, Title III, Section 307, provides the Federal Aviation Administration (FAA) with powers relating to the "Use of Airspace, Air Navigation Facilities, and Air Traffic Rules." The FAA is also the central authority for issuing airman[2] certificates, including those for air traffic controllers. It is not, however, responsible for the economic regulation of the air transport industry.

---

[2]A comprehensive term for holders of pilot, flight engineer, and other specialized qualifying licenses, including air traffic controller, aircraft dispatcher, and so on.

In the United Kingdom, National Air Traffic Control Services became National Air Traffic Services (NATS) when it was made part of the Civil Aviation Authority (CAA), on its establishment in April 1972, and although it remains an independent organization, it manages ATC developments that are governed through a coordinated program addressing regional as well as national demands on airspace.

Because repayable capital investment was only available through the public-sector borrowing requirement, privatization was proposed in 1992. NATS was reorganized into a Companies Act company in April 1996 and became a wholly owned subsidiary of the CAA, and a proposed public-private partnership for NATS was announced in June 1998 and enshrined in the Transport Act of 2000. Government chose the Airline Group (five U.K.-registered airlines) as the preferred partner in March 2001, and the transaction was completed in July 2001 with the sale of 46 percent of the share value to the Airline Group and the transfer of shares with a value of 5 percent to staff. The U.K. government retained the balance, but regarded the company as free of Treasury control. A financial restructuring of NATS, involving £130 million of additional investment (£65 million each from the government and BAA plc) was implemented in 2002, however, to reduce the debt accrued from the downturn of operational income following the events of September 11, 2001, and BAA plc took a 4 percent shareholding, reducing the Airline Group's holding to 42 percent. Debt was further reduced by a £600 million bond issue successfully completed in October 2003.

CAA is still the designated U.K. agency set up by the act of Parliament (Civil Aviation Act of 1971) with specific powers relating to ATC as set out in Air Navigation Orders. As in the United States, this is also the agency responsible for the issue of airman licenses, but unlike the FAA, it is also responsible for the economic regulation of the air transport industry.

In West Germany, the air traffic services agency is the Deutsche Flugsicherung (DFS), formally the Bundesanstalt für Flugsicherung (BFS), and is still 100 percent government owned. At most airports, therefore, ATC and its associated departments will be under the immediate management of a central government agency official and not a member of airport management. There are only a few exceptions to this where ATC staffs are employees of the airport authority, although licensed by the government authority. Under European law, it is possible for an organization such as DFS to take a stake in the partially privatized air traffic organizations of other countries, such as the U.K. NATS.

In addition to airports, a large part of any air traffic service is responsible for en route airspace, exercising control from regional centers—in the United States, air route traffic control centers (ARTCCs), and in Europe, air traffic control centers (ATCCs). Although most of these personnel are governmental employees, there are some areas of the world where this service is contracted out to specialist organizations. In all cases, however, governments maintain ultimate policy control in line with the central concept of national sovereignty over airspace.

International ATC Collaboration

There has been considerable change over time in the way that the air traffic service function has been viewed internationally. Whereas it was always a "national" service, the advantage of imposing pan-national service-quality attributes has been long recognized. In Europe in 1960, Eurocontrol was formed as a unit funded by national governments to address the fundamental operational issue of the control of aircraft in the physically restricted volume of airspace over Belgium, the Netherlands, and Luxembourg (Benelux countries). This resulted in the successful establishment of an ARTCC at Maastricht, in the Netherlands, that had control authority over the Benelux region and the upper airspace regions of northern Germany [at that time the Federal Republic of Germany (West Germany)]. An attempt to spread the application of this philosophy at Karlsruhe, in southern Germany, was not as successful, and this ARTCC was later placed into German ownership.

Eurocontrol is responsible to the European Commission (EC). The organization occupies adjacent offices in Brussels and through European Union (EU) funding is able to establish research and development capability, and eventually further operational capability, to address emerging service-provision requirements. In the past this has included the Central Flow Management Unit (CFMU) at Brussels. Since the late 1970s, the CFMU has worked with airline and airspace providers to conduct strategic evaluations of demand and airspace-sector capacity up to a year ahead and to ameliorate traffic delays through consultation on routings and schedules.

The coalescing of the new concept of long-term strategic flight planning (often referred to as *flow control*) with the preexisting strategic and tactical elements of ATC led to the establishment of air traffic management (ATM) research programs. Initially, these were national and independent because they tended to address national airspace hot spots. It was clear when implementations were considered that the establishment of a comprehensive ATM capability in any nation was never likely to be attainable without wide promulgation of similar and compatible practices. Several fundamental issues had to be resolved in order for the European and North American ATM concepts to bear their ultimate fruit, and the establishment of common means of conducting future air operations was achieved by the ICAO FANS Committee, which reported in 1991.

ATM concepts that are compatible with FANS requirements are managed by national and international research teams. In Europe, the international team is within Eurocontrol and is based in Brussels. It is called the *Single European Sky Airspace Research* (SESAR) Program. Similarly, the FAA is managing the *Next Generation* (NextGen) program, which will affect the whole of the continental U.S. airspace, and in Australia, the Airservice Australia organization has an ATM program called Astra.

These are examples of the impact of globalization requirements on air transport services, with higher levels of service quality (e.g., minimized delay, moderated fuel-burn penalties, and improved airspace capacity) being addressed in a coordinated manner. Major stakeholders, such as Eurocontrol, are also active throughout neighboring regions to ensure that even if capacity is not a significant issue yet, they adopt implementation policies that will ensure shared strategic objectives. As this book goes to press, full implementation of the Single European Sky (SES) program has been undertaken as a key responsibility by the European Aviation Safety Authority (EASA). While individual nations retain their own authorities, EASA is expected to supplant the national authorities and will commence by taking control of mandatory aviation personnel licensing and airspace implementation.

One significant change in Europe is the introduction of new airspace regions, each one called a *functional airspace block* (FAB). These are not based on state boundaries and can be regarded as an extension of the Benelux region that Eurocontrol pioneered. The first designated block was across the United Kingdom and Ireland, but whereas the legislation has been in place for some years, as of 2012, the FAB had not yet operated as a nonsovereign airspace region. The FAB concept requires analysis of operations across regions and the apportioning of responsibilities to configurations of airspace that respect control boundaries. Criteria for the definition of these involve analysis of hand-over workloads, the containment of airspace blocks that serve interacting major airports, and physical constraints such as water masses and mountainous areas. In Europe, the FABs should replace flight information regions (FIRs), reducing the total number of ARTCCs by about 50 percent (Figure 11.4). There are many political challenges to this kind of development.

A major benefit that should accrue from ATM developments is the implementation of "free flight" services. The ultimate goal is to offer aircraft a straight track from departure point to destination. This will be compromised still by terminal-area procedures at airports, which must respect the runway direction(s) in use. In some countries there is already considerable progress with continuous-decent-approach (CDA) operations allowing considerable descent and landing-fuel savings. Also reduced vertical separation minima (RVSM)

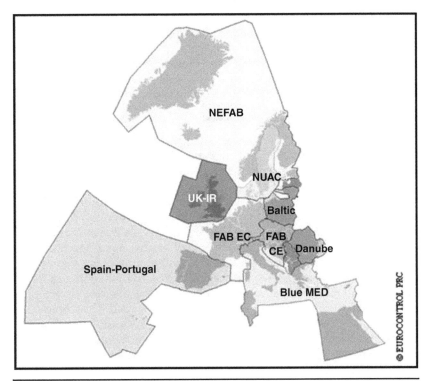

FIGURE 11.4   Planned FAB configuration for Europe. (*Source:* Eurocontrol.)

are being introduced in areas of the upper airspace, and by reducing the once-obligatory 2,000-foot separation minimum between opposing directions traffic above FL290 (flight level equivalent to 29,000 feet) to 1,000 feet, there is consistent vertical separation throughout controlled airspace and an approximate 100 percent of additional cruise-level capacity. This benefit has arisen largely from the improvement in on-board sensing of vertical position on modern aircraft and is not directly attributable to ground systems improvements.

Emerging from research into implementation in recent years has been an ATM-originated concept called *collaborative decision making* (CDM). Airports can be affected by operational circumstances that affect capacity, for example:

- Weather (short term)—passing storm
- Weather (longer term)—snow accumulation
- Incident/accident—blocking runway/taxiway
- Work in progress—adding/servicing infrastructure
- Capacity deficiency in the terminal (passenger handling, etc.)

The concept of CDM is to use the latest information technology to conduct data capture throughout the ground-based operational infrastructure and to share this information "collaboratively" with the ATM system. The airport "capacity" becomes a time-dependent, or dynamic, quantity in the airspace-management process, and this can allow traffic-handling processes to be adjusted to accommodate short-term capacity dips (based on actual data) or longer-term capacity variations (based on predictive data). The first CDM airports in Europe have been Brussels and Munich, and in January 2012, Zurich and Athens were close to full implementation, with 17 major airports in the process of adopting CDM procedures and 9 further airports having entered the development process.

This summary of ATC/ATM developments has seen a considerable change from when the preceding edition of this book was produced, and it should serve as a reminder that this is a dynamic and, in the current era, very fast-changing area of development. Much of what has happened to steer the air traffic system from being a conglomerate of relatively stilted but basically "safety first" organizations into a fit for the future interoperable states that are now visible suggests that the tenets of the past have been well served. The whole commercial aviation business is approaching the wide-scale implementation of a technology-based transition that will have such vast influence that it will be referred to in almost every aspect of technical operation arising subsequently in this chapter. In the majority of nations this has been possible without any privatization of air traffic services, but the appetite for such change remains very strong, and whatever is recorded here is merely a snapshot.

Flight Rules

There are three sets of flight rules depending on the circumstances listed:

> *General flight rules.* Observed by all aircraft in any class of airspace.
> *Visual flight rules (VFR).* Observed by aircraft flying in weather conditions equal to or above prescribed limits.
> *Instrument flight rules (IFR).* Observed by aircraft in weather conditions below VFR limits and/or in Class A airspace.

General Flight Rules

As the name implies, these rules refer to the conduct of flight in such general matters as the safeguarding of persons and property on the ground, avoidance of collision, right-of-way rules, and aircraft navigation lights. Details are listed in the various regulatory documents of each country. For example, in the United States, this is Part 91 of the *Federal Aviation Regulations*; the appropriate regulation in the United Kingdom is achieved by means of the *Air Navigation: Order and Regulations*.

Visual Flight Rules

In addition to observing the general flight rules of the air, each flight has to be conducted according to either visual flight rules (VFRs) or instrument flight rules (IFRs). In the case of VFRs, the flight is conducted on a see-and-be-seen basis in relation to terrain and other aircraft. It is therefore necessary for a pilot to have certain minimum weather conditions known as *visual meteorologic conditions* (VMCs). Anything worse than these conditions is referred to as *instrument meteorologic conditions* (IMCs). General international usage is VFR and IFR conditions.

The weather criteria for visual flight (this can refer also to an IFR flight making a visual approach) are intended to provide to pilots an adequate opportunity to see other aircraft or obstructions in time to avoid collisions. For this reason, limits for lower and slower aircraft are less than for those flying at higher speeds, and some allowance is also made for areas where traffic is less dense (i.e., in uncontrolled airspace).

ICAO sets out VFR minimums for the various classes of airspace, which countries by and large have adopted with some slight variations to suit their own circumstances.

Instrument Flight Rules

When visibility and/or proximity to clouds are less than the quoted VFR limits (VMC), flight has to be conducted under IFRs. With respect to flight under IFRs in controlled airspace, the rules require that ATC must be notified of flight details in advance by what is known as an *ATC flight plan*. Thereafter, the flight has to conform to the plan or to any other instructions issued by ATC. To do this, a continuous watch has to be maintained on the appropriate radio frequency by the pilot and reports made, as required, to ATC regarding the aircraft's position. Another of the rules requires that an instrument flight must be conducted at a minimum height of 1,000 feet (300 m) above the highest obstacle within 5 miles (8 km) of the aircraft's position, except while landing or taking off.

Elsewhere in the legislation of all countries there is a requirement that an aircraft must be suitably equipped for the type of flight being undertaken and the flight crew suitably qualified for that flight.

Flights may, of course, be conducted under IFRs outside controlled airspace and therefore not receive a specific level assignment from ATC. In order to provide some safeguard for such flights and for VFR flights, there are basic rules for a simplified form of vertical separation that is self-administered by pilots. Under this system, the level to be flown depends on the magnetic track being followed by the pilot. This system ensures vertical separation that increases with increasing altitude or flight level (FL), resulting in a separation of

2,000 feet (600 m) for an aircraft in the upper airspace intervals of 4,000 feet (1,200 m) in the "semicircular" system. In the United Kingdom, a quadrantal split of tracks is used for the lower airspace [i.e., 3,000 feet (900 m) above mean sea level (amsl) to 19,500 feet (6,000 m) with a similar "semicircular" split of flight levels above 19,500 feet (5,000 m)]. In the United States, IFR traffic below FL180 will be assigned altitudes on whole thousands of feet (e.g., 4,000, 5,000, 6,000, etc.) depending on the magnetic track being followed, whereas VFR traffic operates 500 feet above or below the whole thousands (e.g., 2,500, 3,500, 4,500, etc.) to provide 500 feet of separation between IFR and VFR traffic. There is consideration of this practice above/below 18,000 feet being implemented worldwide, with consultation in progress in mid-2012. It is anticipated that the implementation of global ATM systems will harmonize airspace classifications and the rules applied within them throughout the world.

Classes of Airspace

Owing to the variations in density of air traffic and the constraints imposed by weather conditions, ATC authorities apply more stringent control in some areas than in others. As a result, there are several different classes of airspace that relate to the varying level of provision of air traffic services. Those in the vicinity of busy airports, for example, are designated for an intensive level of control, whereas other areas with only light traffic may not have any control at all; this would be *uncontrolled* airspace. The basic geographic division of airspace is that which occurs at national boundaries. Within the national boundaries there might be one or more flight information regions (FIRs). The profusion of FIRs in Europe (about 50) is brought about mainly by the comparatively large numbers of national frontiers. For air traffic purposes, the airspace within the continental United States is divided across 20 flight advisory areas. Each is the responsibility of an ARTCC.

There are occasions when FIR boundaries between countries do not coincide with national geographic boundaries. In these cases, boundaries are mutually agreed on between the nations involved, as occurs over the high seas or in airspace of undetermined sovereignty (e.g., the rim of the North Atlantic, where the United States, Canada, and Portugal agree on the boundaries between the New York Oceanic FIR, Gander Oceanic FIR, and Santa Maria Oceanic FIR). These remain the case while an overriding development, the *functional airspace block* (FAB), has emerged.

Those parts of an FIR that are uncontrolled are sometimes referred to as the *open FIR* because no restrictions in terms of ATC are placed on aircraft in these areas, and no separation is provided by ATC. It is usual, however, for an air traffic service to be available on a notified frequency via flight information service. It might be broadcast at notified times or supplied on request to pilots by ground radio stations. Flight information service can, of course, also be provided by any control unit, including airport or approach control. The basic operational units of airspace are *control zones* and *control areas*, the difference being that zones start from the surface to a given altitude, whereas control areas are controlled airspaces extending upward from a specified limit above the earth, including airways. *Controlled airspace*, as defined by ICAO Annex 2, *Rules of the Air* (2005), is an "airspace of defined dimensions within which air traffic control service is provided to IFR flights and to VFR flights in accordance with the airspace classification."

Owing to the differing requirements for IFR and VFR operations in various locations, ICAO notifies airspace as any of seven classes, A to G. The intention is that worldwide it will be sufficient to know the classification of airspace to know exactly what the ATC conditions are in respect of IFRs and VFRs and the type of air traffic service available. However, not all countries have implemented all classes in a seamless manner. Table 11.1 provides a summary of ICAO airspace class definitions.

The controlled airspace found in the vicinity of busy airports generally consists of a surface area and two or more layers often resembling an upside-down wedding cake. In a typical example, the layers surrounding a surface area might have lower levels gradually lifting in steps from 7,000 to 8,000 to 10,000 feet. Figure 11.5 illustrates this effect as it relates to the London terminal maneuvering area (TMA) and shows airspace over the region and extending to ground level to serve the airports at Heathrow, Gatwick, Stansted, Luton, London City, and Southampton.

In the United States, such airspace includes primary civil airports such as Atlanta Hartsfield, Boston Logan, Chicago O'Hare International, Los Angeles International, Miami International, Newark International, New York JFK, New York LaGuardia, San Francisco International, Washington National, and Dallas–Fort Worth International, among others; all are now classified as Class B airspace. Class A airspace in the United States is generally the airspace from 18,000 feet amsl up to and including FL600, in which only IFR operations are authorized. Secondary airports (U.S. definition) in the United States are designated Class C airspace.

Separation Minima

The criteria used by ATC to determine the required spacing between aircraft to achieve safety are known as *separation minima*. These are specific criteria relating to vertical or horizontal distances or times established between aircraft. In the vertical plane, IFR aircraft are separated by being required to fly at different altitudes or flight levels[3] so that they are 1,000 feet (300 m) apart up to FL290 and 2,000 feet

---

[3]A flight level is measured at a preset barometric altimeter pressure setting. This is either 1,013.2 hectopascals or 29.92 inches of mercury. These setting are used worldwide.

Class	Type of Flight	Separation Provided	Service Provided	Speed Limitation*	Radio Communication Requirement	Subject to an ATC Clearance
A	IFR only	All aircraft	Air traffic control service	Not applicable	Continuous two-way	Yes
B	IFR	All aircraft	Air traffic control service	Not applicable	Continuous two-way	Yes
B	VFR	All aircraft	Air traffic control service	Not applicable	Continuous two-way	Yes
C	IFR	IFR from IFR, IFR from VFR	Air traffic control service	Not applicable	Continuous two-way	Yes
C	VFR	VFR from IFR	1) Air traffic control service for separation from IFR; 2) VFR/VFR traHic information (and traffic avoidance advice on request)	250 kt IAS below 3,050 m (10,000 ft) AMSL	Continuous two-way	Yes
D	IFR	IFR from IFR	Air traffic control service, traffic information about VFR flights (and traffic avoidance advice on request)	250 kt IAS below 3,050 m (10,000 ft) AMSL	Continuous two-way	Yes
D	VFR	Nil	IFR/VFR and VFR/VFR traffic information (and traffic avoidance advice on request)	250 kt IAS below 3,050 m (10,000 ft) AMSL	Continuous two-way	Yes

Class		Separation	Service	Speed limitation	Radio communication requirement	Subject to ATC clearance
E	IFR	IFR from IFR	Air traffic control service and, as far as practical, traffic information about VFR flights	250 kt IAS below 3,050 m (10,000 ft) AMSL	Continuous two-way	Yes
	VFR	Nil	Traffic information as far as practical	250 kt IAS below 3,050 m (10,000 ft) AMSL	No	No
F	IFR	IFR from IFR as far as practical	Air traffic advisory service; flight information service	250 kt IAS below 3,050 m (10,000 ft) AMSL	Continuous two-way	No
	VFR	Nil	Flight information service	250 kt IAS below 3,050 m (10,000 ft) AMSL	No	No
G	IFR	Nil	Flight information service	250 kt IAS below 3,050 m (10,000 ft) AMSL	Continuous two-way	No
	VFR	Nil	Flight information service	250 kt IAS below 3,050 m (10,000 ft) AMSL	No	No

*When the height of the transition altitude is lower than 3,050 m (10,000 ft) AMSl, FL 100 should be used in lieu of 10,000 ft.

TABLE 11.1 ICAO Classification of Airspace

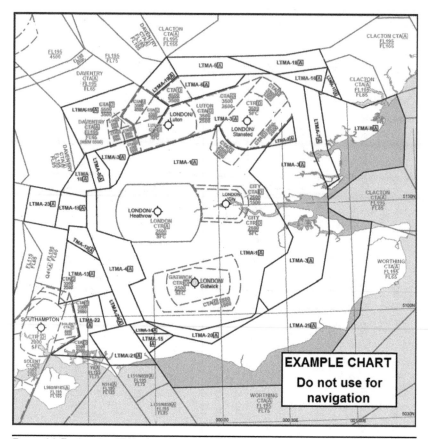

EXAMPLE CHART
Do not use for
navigation

Figure 11.5   London TMA plan. (*Source:* UK AIP.)

(600 m) apart from FL290 upward. Horizontal separation is classified into three groups: lateral, longitudinal, and radar. Aircraft are *laterally* separated if their tracks diverge by a minimum angular amount with reference to a radio navigational aid, for example, 15, 30, or 45 degrees, or if they report over different geographic locations and thereafter they continue to move farther apart.

ATC prescribe intervals of time or distance between aircraft to provide *longitudinal* separation to those at the same altitude or departing from the same airport. When ATC is able to see for itself on radar where aircraft are in relation to each other and, thus, is not dependent on aircraft position reports, it is are able to substantially reduce these intervals.

Typically, a 15-mile (24-km) longitudinal separation may be reduced to a 3-mile (5-km) *radar* separation. The widespread use of radar has been a major factor in helping to increase airspace capacity.

In relation to arriving aircraft, the radar separation of 3 miles (5 km) may need to be increased to take account of the phenomenon known as *wake turbulence*. This is the turbulence created behind an aircraft by vortices shed from the lifting surfaces, particularly a wide-bodied aircraft, commonly referred to as a *heavy jet*, and the associated hazard to a smaller, lighter aircraft following behind.

Operational Structure

Before any authority is able to provide positive ATC, especially in the present busy environment around most air carrier airports, it must establish a complex system of navigation aids and communications facilities to delineate the pattern of routings for arriving and departing aircraft and to provide information to pilots and controllers on the position and progress of flights.

The short-range ground-based radio aids used include nondirectional beacons (NDBs), VHF omnidirectional radio range (VOR) beacons, distance-measuring equipment (DME), and fan markers (FMs). For precision approach and landing guidance, there is the instrument landing system (ILS). The functioning of and information provided by each of these systems are presented in brief detail in the later section "Radio Navigation Services."

The improved technology now being used in aircraft equipment provides better instrumentation and enhanced instrument flight capabilities. There is a requirement for this to be matched on the ground by ILS installations and associated ground aids capable of providing for automatic landing in fog. Landing requirements are categorized according to two criteria, as shown in Table 11.2.

Runway visual range (RVR). The maximum distance in the direction of takeoff or landing at which the runway or the specified lights or markers delineating it can be seen from a position above a specified point on its centerline at a height corresponding to the average eye level of pilots at touchdown.

Decision height. A specified height at which a missed approach must be initiated if the required visual reference to continue the

Category	DH (ft)	RVR (m)
I	200	800
II	100	400
IIIA	—*	200
IIIB	—*	50

*No decision height applicable.

TABLE 11.2 Decision Heights (DH) and Runway Visual Ranges for Precision Approach Runways

approach to land has not been established. All air transport aircraft are operated into airports in compliance with company weather minima. These may conform to specific guidance from government or be established by the company and then passed on to the government agency.

They are expressed in terms of visibility and decision height or cloud base. When aircraft are approaching and landing in minimum weather conditions, ATC requires especially precise information on the positions of all aircraft. To assist with this, ATC relies in the first instance on primary surveillance radar, which generally gives good coverage out to some 30 miles (48 km) from the airport. Such radars do not, however, give two other vital pieces of information: identification of the radar echoes and height. To provide this information requires secondary surveillance radar (SSR). This is so vital to ATC in busy terminal areas and other types of controlled airspace that its carriage became mandatory in controlled airspace in the 1960s. It has effectively supplanted primary surveillance radar as the main sensor used by ATC to present position and related information on aircraft, and it is through the use of more refined SSR technology and equivalent developments that ATM will gather and disseminate information in the future.

Finally, once the aircraft is on the ground in fog conditions, ATC requires a "mapping" capability. The use of an SSR transponder to track vehicles and aircraft in airport maneuvering areas has been considered widely for many decades, and Global Positioning System (GPS) tracking capability added interest as the newer technology has emerged. The amelioration of runway incursion risk was addressed by considering systems that will detect movements within a protected zone around an active runway and adjacent areas and using computer-based monitoring will provide a conflict-detection capability to ATC tower staff.

There are suppliers now of advanced surface movement guidance control systems (A-SMGCS) that will combine the information from SSR, GPS, and surface movement radar (SMR) and provide conflict detection and resolution assistance to ATC staff. They target delivery of a capability that will lead to a reduction in accidents, savings in taxi time, delivery of increased movement throughput, fewer delays owing to weather, and an improved situational awareness in all operations for ATC staff.

The main functions of A-SMGCS are defined by ICAO as

- *Surveillance.* Providing controllers (and eventually pilots and vehicle drivers) with situational awareness (a surveillance display showing the position and identification of all aircraft and vehicles)

- *Control.* Providing conflict detection and alerting on runways (this is being expanded gradually to address the whole aerodrome movement area)

- *Routing*. Proposing, through manual or automatic control, the most efficient route to designate for a declared aircraft or vehicle
- *Guidance*. Giving pilots and drivers indications that enable them to follow an assigned route.

Implementation of A-SMGCS has been through a four-level program roughly conforming to adding functionality in the order just presented, with large airports introducing the new capability almost as some as it is available, and lower-level systems proving to be attractive in terms of cost and capability for smaller airports as they seek to upgrade from simpler radar surveillance systems. The implementation of data sharing from the ground to aircraft is in research, with a view ultimately to providing a runway-incursion warning on a head-up display where the imagery is superimposed on the outside world in a manner that assists with situation awareness for the crew.

Operational Characteristics and Procedures

The characteristics of an airport with the full range of ATC equipment just described will clearly differ greatly from those of the smaller general aviation (GA) airfield. At the GA airfield, the balance of traffic will lean toward relatively small aircraft, possibly including training flying, mainly under VFR. In the United States, however, non–precision instrument approaches are available at hundreds of GA airports, most of which do not have ATC towers. While accidents are not significantly greater in these operations, one of the impacts of ATM-essential CNS technologies is that these may be "remote tower–operated" in the future and a nonlocal ATC service provided with fidelity equivalent to larger airports.

ATC services are adjusted to suit the differing characteristics of airports. Some locations may have the full range of air traffic services, including aerodrome, approach, and radar, whereas others may have only an airport. The two main categories of ATC units associated with airport operations are aerodrome control (tower) and approach control.

Aerodrome Control

This is the more familiar part of ATC carried out from the glass-topped part of the tower building that affords controllers a panoramic view of the airport and its immediate surroundings. There has been a tendency for tower structures to become higher as airports have expanded and better sightline visibility has been needed across a wider area. The extension of terminal buildings and other ground facilities has resulted, on occasion, in critical points on an airfield, that is, runway touchdown point, apron, taxiway, and parking areas, being hidden from the view of the tower. Remote television cameras with display units in the tower have assisted in solving such problems. Airport staff responsible for stand/gate assignment can face similar problems of finding a suitable vantage point, and some tower build-ings or subsidiary ground-control towers make provision for them. The essential characteristic of airport control is that it principally con-trols only that which can be seen from the tower itself—the "visual" control room.

Control therefore is exercised over aircraft moving in the maneu-vering area or flying in the immediate vicinity of the aerodrome either in a specified traffic pattern known as the *circuit* or, in the United States, the *pattern* or in the final stages of approaching to land or departing. Because it controls aircraft movement on the airport, airport control also exercises authority over all other traffic, including vehicles using the same operational areas. Hence, vehicles using these areas must carry radio equipment to provide immediate voice communications with the tower and in the future may be mandated to carry ATC transponder devices.

At some airfields that are not dealing with many air transport operations or a large amount of GA activity, there may not be a control tower. All users (ground vehicles in the maneuvering area included) need to be familiar with the internationally agreed light signals used by ATC, as set out in Table 11.3. In the United States, of the more than 5,000 airports open for public use, many of which have heavy concen-trations of GA traffic, fewer than 700 have ATC towers. Most of these airports without a control tower, however, operate a unicom radio, which cannot be used for ATC but over which pilots may announce their positions and intentions for others to hear. ATM developments are requiring all aircraft to carry transponders, and some implemen-tations perceive operations of this nature carrying affordable CNS capability that will give every GA pilot a real-time in-cockpit display of all local movements.

Color and Type of Signal	On the Ground	In Flight
Steady green	Cleared for takeoff	Cleared to land
Flashing green	Cleared to taxi	Return for landing (to be followed by steady green at proper time)
Steady red	Stop	Give way to other aircraft and continue circling
Flashing red	Taxi clear of landing area (runway) in use	Airport unsafe—do not land
Flashing White	Return to starting point on airport	
Alternating red and green	General warning signal—exercise extreme caution	

TABLE **11.3** Aerodrome Control Light Signals

At airports with high volumes of traffic, it is necessary to divide airport control into *ground* and *air*. Ground control is responsible for all movement in the maneuvering areas *except* the runways and approaches to the runways. Air control (in the United States usually referred to as *local control*) is responsible for the runways and immediate approaches/turnoffs. This results in departing aircraft coming under ground control once clear of the stand/gate until the holding point prior to moving onto the runway itself. The split of control responsibility in an airport control tower at busy airports calls for extreme vigilance, especially at night and in fog. On October 8, 2001, at Milan (Linate), Italy, a McDonnell Douglas MD-87 carrying 110 passengers and at the point of rotation on takeoff collided with a Cessna Citation CJ2 business jet carrying four people, the latter aircraft having taxied onto the active runway and been unseen in fog. All 114 people on board the two aircraft were killed, as were four on the ground when the airliner hit the baggage-handling portion of the airport terminal. The airport had no operational SMR and, thus, ATC could not observe the traffic movements on the airport.

At busy airports, ground control also will be responsible for implementing any departure flow-control regulations that might be in force. This usually is originated by a central control unit, which increasingly will address national airspace needs while also addressing simultaneously the intranational implications of a clearance on airspace capacity utilization further afield. Traditionally, the implementation of flow control has meant that aircraft may be delayed on the ground. In such circumstances, ground control has provided *startup* clearances once a *slot* has be notified to be available that will allow the departing aircraft to join the general flow of air traffic over the area. In most of the cases where there are separate frequencies for ground and air control, they are so busy that it has become necessary to provide a third communications channel devoted solely to passing clearances to aircraft, or *clearance delivery*. In common with all elements of split responsibility, coordination among all elements of aerodrome control is vital, particularly between air and ground control to ensure that there is no obstruction to the landing and departing aircraft on the runways. ATM developments are intent on addressing the less than optimal procedures that have evolved from long-standing procedures by formalizing data exchanges. These may be largely electronic and nonverbal between aircraft and ATC systems throughout an element of the ATM systems regarded as part of the collaborative decision-making (CDM) process and only apparent on voice radio when the movement is actually underway.

Approach Control

Approach control has similar problems of coordination. It deals with IFR traffic approaching the airport and departing IFR traffic once it is handed over by aerodrome control, usually almost immediately after takeoff. It is common practice worldwide for approach control to also handle arriving and departing VFR traffic. The area of responsibility of approach control extends out typically to a distance of some 20 miles (32 km) from the airport. Traditionally, approach controllers have been situated in the same building as aerodrome control, but in a separate room, but increasingly it is being accepted that the best place for this function to be performed is in the ARTCC unit. This is especially true when approach control may be responsible for more than one airport.

The use of radar, both primary and secondary, has proved an essential aid to dealing with the difficult conflicts of departing climbing traffic and arriving descending traffic. In this regard, the European practice is to employ compulsory IFRs in busy traffic areas around airports, even in VMCs. Furthermore, it is not the practice to mix VFR and IFR traffic in the same airspace. The reasoning behind this is that the critical factor for safe visual separation by pilots themselves is not so much prevailing weather in terms of visibility and distance from cloud but the pilots' ranges of vision from the flight deck.

At all busy IFR airports, the dominant factors for approach control are the established instrument procedures. These take several forms, but all have the common factor of providing pilots and controllers with known, predictable patterns that will be flown by arriving and departing aircraft. Instrument approaches may be flown using any of the short-range radio aids already mentioned and, in some few cases, using surveillance-radar-approach (SRA) guidance.

An SRA approach and landing chart, for Leeds Bradford, U.K., runway 14, is shown in Figure 11.6. This defines the plan view of the path to be flown in the upper part of the diagram and the heights in the lower vertical cross section. There are specific notes on local issues for pilots unfamiliar with local conditions, such as on the example chart a warning of the possibility that the ground-proximity warning system (GPWS) might be triggered. In this case, the ground profile could cause the warning, although the flight path may not be unsafe. Examples of this occurring are less frequent as enhanced GPWS (EGPWS), sometimes incorporated in terrain alerting and warning systems (TAWSs), has become more widespread in the best-equipped aircraft types.

At particularly busy airports, procedures also have been standardized for departing and arriving routes:

- Standard instrument departure (SID)
- Standard instrument arrival route (STAR)

One of the obvious advantages of SIDs and STARs is that their use reduces very considerably the load on radio frequencies as a result of the "shorthand" descriptions that can be used for the complicated patterns of tracks and altitudes to be flown.

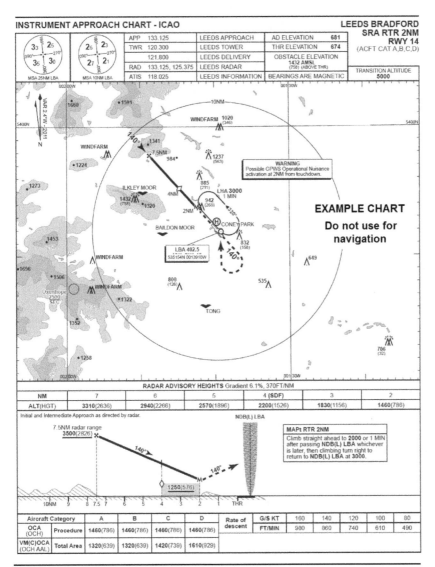

Figure 11.6 Aerodrome SRA chart. (*Source:* UK AIP.)

ICAO recommendation is that the system of designators shall permit the identification of each route (whether arrival or departure) in a simple and unambiguous manner. An example of a SID in plain language is BRECON ONE, which will have a coded designator, in this example BCN 1. The latter can be entered by the crew of an appropriately equipped airliner to recover the routing information electronically and then initiate automatic flight control guidance.

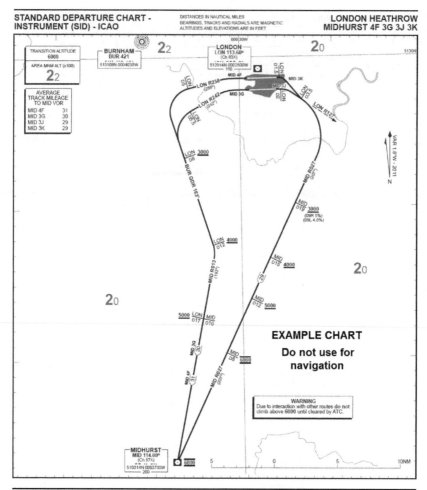

FIGURE 11.7   London Heathrow–Midhurst SID. (*Source:* UK AIP.)

The pattern to be flown for one London Heathrow procedure is illustrated in Figure 11.7 and can be transmitted to the pilot as "Midhurst Departure." A suffix, either 4F, 3G, 3J, or 3K, is added by the crew or automatically if appropriate data have been stored in the aircraft's database already to select the one route specific to the runway in use.

11.4   Telecommunications

The provision and maintenance of suitable aviation communication and navigation equipment and facilities are worldwide requirements for civil aviation and, as such, are another of the technical services subject to international agreement and standardization through ICAO. Details are contained in Annex 10 to the ICAO Convention: *Aeronautical Telecommunications* (in two volumes). The standardization of communications equipment and systems is dealt with in volume 1, whereas volume 2 deals with communication procedures. International aeronautical telecommunications services are formally classified as

- Fixed services
- Mobile services
- Radio navigation services
- Broadcasting services

All these are the responsibility of Member States of ICAO, although some of these facilities may be provided by commercial companies. In the Unites States, there is Aeronautical Radio, Inc. (ARINC), which has progressed from providing airlines with voice communications to a comprehensive data network service, including an aircraft communications addressing and reporting system (ACARS). ARINC is airline-owned, as also is another airline cooperative effort, the telecommunications network Societe International de Telecommunications Aeronautique (SITA), based in Europe. The main difference between the commercial and governmental agencies is that the governmental agencies restrict the traffic they will accept to certain types of messages, essentially those concerned with the safety of civil air transport. Thus, the government circuits are in frequent use for the transmission of flight-plan messages and urgent operational information between Member States, whereas the commercial channels may be used for passing company messages (e.g., requirements for crew transport, catering, supplies, etc.). Precise formats for various types of messages exchanged via the telecommunications and other governmental communications channels have been agreed on at the international level, and they are usually coded so that problems of language differences are much reduced.

Coding has the additional advantage of lending itself to computer techniques, and it is possible for teleprinter/teletype messages to travel around the world passing through several ground stations en route by means of automatic switching exchanges without the need for human intervention. This remains the main international nonurgent communication link in some countries, but a move away from teleprinter to electronic display and eventually into packaged messages similar to Internet e-mail is under development, and its acceptance is limited only by technical implementation difficulties.

Fixed Services

Fixed-service communication fills the need for a rapid means of point-to-point ground communications between fixed points (either by cable or radio link) to pass messages relating to safety and the regular, efficient, and economical operation of air transport and general aviation. The basis of the worldwide service is the Aeronautical Fixed Telecommunications Network (AFTN). It is in effect a dedicated network confined to the following categories of messages:

- Distress messages and distress traffic
- Flight safety
- Meteorologic
- Flight regularity
- Aeronautical administration
- Notices to Airmen (NOTAM) distribution
- Reservations
- General aircraft operating agency

The first six in the preceding list are broadly classified as Category A traffic; the last two, as Category B. Generally, Category B traffic is not accepted on government-operated AFTN circuits. Priority for all AFTN messages is indicated by a two-letter priority code in the following order: SS, DO, FF, GG, JJ, KK, and LL. Distress messages, for example, would bear the highest-priority indicator, SS, and "general aircraft operating agency" messages the lowest, LL. There is also a prescribed format for AFTN messages, and this is set out in detail in Volume 3, Chapter 8 of Annex 10.

Mobile Services

In the context of telecommunications, the term *mobile* refers to the service being provided for aircraft (moving vehicles), although the facilities provided by the individual government agencies are primarily fixed installations on the ground. Mobile service covers two vital aspects of aircraft movement:

- Communications
- Navigation

The major users of the communications facilities for air/ground and vice versa contact are aircraft, but to a very small extent they also may be used by ground vehicles moving on the airport (e.g., vehicles towing aircraft or conducting airport pavement inspections). The vast majority of communications facilities in the mobile service are devoted to radio voice communications.

There is still a very limited need for direction finding (DF) because aircraft with GPS receivers can provide this information from the stored aerodrome position. A DF facility is retained in many aerodromes both as a backup and as a coarse indication of relative position of a calling aircraft. This enables a controller to establish a radar or visual contact based on the indication it provides.

A mobile unit might be used to conduct a runway inspection before an aircraft arrives/departs, and on a small airport, the sole occupant then can assume the role of air traffic controller. Such a person usually will carry handheld equipment, such as a thermometer and an anemometer, to determine and transmit air temperature and wind data. In the latest ATM programs, this kind of operation, which should be integrated formally into the strategic demand model used by the regional ATC authority, is being addressed under "remote tower" research projects. This is likely to be the last area into which, and even if ever, ATM reaches.

At the level of individual airport facilities, the main concern is with the voice communications used by aerodrome and approach control, and for this purpose, very high frequencies (VHF) are used to avoid the risk of interference. The band of frequencies 118.0 to 136.0 MHz is used, and this provides up to 714 channels with 25-kHz spacing. A 100-kHz block is used to protect either side of the international emergency frequency of 121.5 MHz. These frequencies and others in the VHF range 30 to 200 MHz are characterized by a line-of-sight reception range. Representative reception distances versus aircraft heights are 39 nautical miles (72 km) at 1,000 feet, 55 nautical miles (101 km) at 2,000 feet, 122 nautical miles (226 km) at 10,000 feet, and 200 nautical miles (370 km) at 25,000 feet. At greater ranges, remote receivers/transmitters are used with microwave or cable links back to the ATC unit involved. In cases where it is not possible to provide the remote facilities, for example, ocean or desert areas, then a different frequency band, high frequency (HF), 3 MHz, (3,000 kHz) to 30 MHz (30,000 kHz), is used to provide longer-range propagation.

"Families" of such frequencies are provided by governmental agencies on a worldwide basis for international flights in such areas as the North Atlantic, Europe, the Mediterranean, and the Pacific. These are all part of the ICAO Major World Air Route Area (MWARA) HF Network. Typical frequency groups for the European/Mediterranean region are 2,910, 4,689, 6,582, and 8,875 kHz. Although developments in technology allowed closer spacing of frequencies, HF radio effectiveness has been limited by message congestion, which occurs in unavoidable periods associated with the peak-hour characteristics of long-haul air transport operations.

Satellite communications have been making inroads over many years and have been supplemented by automatic dependent surveillance (ADS), which is a digital satellite data link that brings unprecedented potential to revolutionize long-haul operations safety though improved communications and surveillance capability. It is an onboard system that gathers information, in a manner similar to the ATC transponder, and transmits a message at regular intervals to a satellite receiver. This is achieved using microstrip antennas that are flush to the skin and mounted on the fuselage crown. They are of such minute depth that they are barely deeper than a coat of paint and achieve adequate directional and signal-strength capability to alleviate any need to use a steerable or dishlike antenna.

In ADS broadcast mode (ADS-B), each aircraft's incoming data stream is retransmitted not just to the ground but omnidirectionally and, thus, is receivable by all other aircraft across a vast area (potentially half the earth's surface, but practically less, yet still several thousands of miles). This means that not just the ground stations feeding ARTCC units provide call-sign, three-dimensional-position, and vector data (track and ground speed) for the display of aircraft on a synthetic situation display, but they also supply the same information to all aircraft. It is now possible for an aircraft in midocean regions to have an indication on a navigation map display of the relative positions of all traffic within a considerable distance.

This development is pivotal to ATM developments that are introducing CNS capabilities of a similar fidelity worldwide so that oceanic and remote regions of the world are no less or better served than anywhere else. The implementation has to comply with ICAO FANS requirements, and this means that the service is not regarded as meeting safety-criticality requirements until it has appropriate redundancy.

In fact, ADS-B is not dissimilar in nature to the SITA-provided operational support systems that convey digital messages to and from aircraft and have been available in digital radio-network-based form since the 1970s. This is commercial service that is not approvable under ICAO FANS safety criteria, but it is by now a very substantial worldwide satellite-link communication system categorized as the Aircraft Communications Addressing and Reporting System (ACARS).

The capacity of these newer-generation communications systems became clear on June 1, 2009, when an Air France Airbus A330-200 (Flight AF447) flying from Rio de Janeiro to Paris crashed into the Atlantic Ocean killing all 216 passengers and 12 flight crew. The Air France flight operations center in Paris received messages from the ACARS between 02:10 and 02:14 UTC that included 6 failure reports and 19 warnings. The warning messages provided sufficient data to determine that the autopilot and autothrust system had disengaged, that the traffic conflict advisory system (TCAS) was in fault mode, and that the flight control system mode had changed from "normal law" to "alternate law." The aircraft coordinates indicated that the aircraft was at 2°59′N 30°35′W, and one of the two final messages indicated that the aircraft was descending at a high rate. The ACARS data provided the airline operations team in Paris with information very soon after the incident began.

Spoken radiotelephony (r/t) messages are still vital and must be as precise and succinct as possible. Certain standard phraseology and abbreviations therefore are used in air/ground exchanges. Even so, misunderstandings can arise, as was so tragically demonstrated at

Tenerife in 1977 with the collision and destruction of two B747 aircraft—one just about to become airborne.

Depending on the volume of traffic at a particular airport and any adjoining airports, there may be as many as six or seven different channels used by ATC for various communications purposes. A typical international airport has separate frequencies for each of the following:

- Information service
- Approach control
- Aerodrome
- Ground control
- Clearance delivery
- Helicopters

The channels/frequencies used by ATC serving a particular airport are not necessarily operated from the airport itself. It is becoming normal practice for larger airports to have their associated approach controller staff integrated within the local ARTCC so that they can coordinate through the sharing of a common database in the local computer system. This has happened with regard to six airports in the London area (i.e., Heathrow, Gatwick., Stansted, London City, Luton, and Biggin Hill), and there have been equivalent combinations to assist coordination in such locations as the San Francisco Bay and New York areas in the United States.

The coalescing of facilities in this manner does pose serious failure-case issues, but the architecture of modern ATC facilities is based on massive redundancy. In the ultimate case, clusters of workstations have the capacity to carry out all the computing needs of the ARTCC. It is feasible to consider the transfer of all control to/from adjacent ARTCCs so that a catastrophic disruption at one unit will be survivable. This could invoke the functional airspace block (FAB) concept and involve the transfer of right of control of airspace between adjacent sovereign states.

Radio Navigation Services

There are internationally agreed standards for radio navigation equipment laid down by ICAO. The technology used is not what ICAO specifies, but rather it is the minimum navigational performance specification (MNPS) of equipment used in certain applications. Traditionally, this specification addressed each application according to the relationship between surveillance and communications capabilities, with the most extreme case being "remote" area operations—over oceans and deserts—where surveillance was virtually nonexistent. In most other en-route operational scenarios, the quality of radar surveillance coverage often was very different, whereas the radio navigation system capability was similar, with the best radar surveillance coverage available in busy airspace regions. Additionally, in the region of airports, and especially for landing aircraft, there was particularly precise radio navigation guidance.

The oldest type of radio aid in use is the *nondirectional beacon* (NDB). It is akin to the nautical lighthouse, radiating a nondirectional signal to which the aircraft receiver can determine a direction relative to the aircraft's heading. The onboard compass is the reference used, therefore, to determine the bearing to an NDB. As a navigation aid it, therefore, offers no precise track guidance or selection. Because it is in the low-frequency band (200 to 1,750 kHz), it is also subject to interference in bad weather. It is easy to install, needs little power, and is not sensitive to local terrain, and it is also cheap to buy and maintain. The NDB is still very widely available, especially as a "locator" beacon for guidance toward precision navigation aids such as ILS. However, it is not recognized as a primary source of navigation information by ICAO, and airspace service providers cannot promulgate IFR operations with NDBs unless there is an alternative navigation and/or surveillance backup.

The basic short-range navigation aid, found in the vicinity of airports and across almost all populated areas of the world, is the *VHF omnidirectional range* (VOR) *beacon.* (Although called a *range*, there is no distance function—it is a legacy from the early days of radio engineering when many beacon-based systems were called *ranges*.). This employs a radio transmission that can be interpreted by the receiver unit as a precise track to or from the beacon's location. This is usually regarded as accurate to ±1 degree because it is unnecessary to promulgate operations with greater navigation requirements. A pilot can select a *radial* (specified by a number between 1 and 360). The number will correlate with the magnetic directions associated with the beacon's position, and the radio aid guides the crew, or autopilot, to fly to or from the VOR station. Instrumentation allows the pilot to determine deviation from the selected radial and to monitor automatic track capture of the autopilot.

VOR beacons provide directional guidance only and are usually colocated as ground stations with *distance-measuring equipment* (DME) that provides an accurate distance from the facility. Usually this system presents information on a display integrated with the VOR instrument, thus allowing the actual position of an aircraft relative to the beacon location to be determined and monitored.

VOR beacons operate in a band close to the VHF communications band of frequencies, between 108.0 and 118.0 MHz. DME operates at much higher frequencies, between 960 and 1,215.0 MHz, which is within the group of ultrahigh frequencies (UHF). The DME is an interrogation and response system that has limited capacity and therefore exhibits an apparent reduction in range capability in heavy-traffic areas.

Radio navigation guidance for approach and landing at an airport requires a precision-approach aid. The *instrument landing system* (ILS) became the universal precision-approach aid before 1950, and it has been developed to the stage where it now can provide a pilot with a "blind" landing capability. The ILS radiates two sets of intersecting signals (four in total). One pair is aligned to the horizontal plane and is called the *localizer*, and it provides azimuth/centerline guidance to the runway. The installation is a large fencelike system at the far end of the runway being approached and is often the most significant overrun obstacle at an airport. It is a frangible structure. The second pair of signals (on a higher radio frequency than the localizer) is called the *glide path* and is aligned upward from the touchdown point. This pair of signals provides vertical guidance to the runway with respect to a preset approach angle. Typically, the approach is at a 3-degree slope (roughly 300 feet/nautical mile decent on approach), but slightly lower and considerable higher slopes are used, the latter at airports surrounded by high terrain or where there are noise-sensitive areas beneath the approach path. It is rarely greater than 4.5 degrees, but in exceptional circumstances, 7-degree approaches have been used.

The pilot, if flying a manual approach, uses horizontal and vertical needle indicators on the ILS instrument to fly a precise approach and landing profile. It is more usual in modem airliners for the ILS signals to be fed into the aircraft's flight-guidance system for an automatic (coupled) approach. Distance information on the approach is best provided by a DME located at the midpoint between touchdown points and calibrated to show distance from touchdown to aircraft approaching from either runway direction. In the older process, marker beacons use upward-pointed and low-power radio transmitters. Passage over the beacons is indicated in the cockpit by an audio signal and a flashing light—blue, amber, and white for outer (OM), middle (MM), and inner marker (IM), respectively. Usually the OM is 4 to 5 miles (6–8 km) from touchdown, the MM is situated 0.5 to 0.9 miles (0.8–1.4 km) from touchdown, and a small number of ILS installations also have an IM situated 1,500 to 1,700 feet (457–548 m) from touchdown. Their positions are indicated on aerodrome charts, the variability of distance being unavoidable because of ground obstacles and terrain.

Two different frequency bands are used for each ILS installation. The localizer transmits on VHF between 108.1 and 111.9 MHz. The glide path operates on UHF between 328.6 and 335.4 MHz. These are paired such that the operator has only to know the ILS localizer frequency, and the glide-path receiver will be tuned automatically to the correct frequency. Therefore, only the localizer frequency is ever quoted on charts.

The *microwave landing system* (MLS) was devised as an ILS successor. It offered wider-angle and higher-accuracy signal coverage and was adopted as an approved navigation aid by ICAO in 1978. Although planned to replace ILS, the system was not widely adopted, and that was largely because of the perceived potential for satellite navigation systems raising the possibility of overtaking MLS. All avionics suppliers have provided a multimode receiver (MMR) capability for modern airliners for almost 20 years. This means that MLS always has been catered for and is routinely available but rarely used. Some clusters of MLS-equipped airports were implemented in the United States, especially in mountainous areas, where ILS has propagation limitations but where regularity of operations demands a high-integrity approach aid. These have been supplemented by the Global Navigation Satellite System (GNSS; see next section), and experience will determine whether MLS is regarded as redundant to needs or an essential part of a failure-tolerant low-visibility landing guidance system. The most well-used MLS is installed at London (Heathrow), and since 2009, it has been cleared for use in category 3 weather conditions, where experience from date gathering during monitored approaches will be used to determine the likelihood of needing to maintain development of the system as a part of landing systems in the future.

Satellite Navigation

It is difficult to dissociate satellite communication and navigation because much of the navigation information that traditionally has been passed by radio voice links has been superseded by satellite-based communication systems such as ADS-B and ACARS. A predominant source of the position (and vector) information sent through these systems is derived from satellite navigation sensors on the aircraft. It is the latter element that is regarded as constituting a satellite navigation (satnav) system.

Global Positioning System (GPS) is a system synonymous widely in the public domain with the way that satellite navigation has developed, but it was simply the first all-encompassing system (in that military-only systems such as Transit were around in the 1960s). There are more systems being introduced into service. GPS emerged from a series of U.S. military satellite-navigation developments able to offer very precise position fixing, but its utility for civilian users was to have been diminished by the imposition of signal signatures that limited the benefits of GPS to civilian users. It was said in the late 1980s that GPS would provide a 50 percent probability of position error of 5 m (15 ft) in the military (precision) mode and 200 m (650 ft) or better in the civil (coarse) mode.

Conflict in the Middle East in late 1990 led to an acceleration of deployment, and the military commanders were so keen to equip ground vehicles and troops that civil receivers were used, so for a while the precision code was accessible to civilian users. At about the same time, the ICAO FANS vision of the attributes of a future navigation system were produced, and these showed that GPS was able to meet all the criteria with one exception—military ownership of the satellites meant that there was no assurance on necessary systems integrity.

In the early 1990s, the U.S. Department of Defense submitted failure-case analyses to ICAO that hinged largely on describing the, until then, highly classified way in which the monitoring of satellite performance was conducted.

GPS needed a "constellation" of 18 satellites, 6 spaced in three orbits, with about a 12-hour orbital period, to be fully operational because a user has to be able to receive a signal from at least four satellites at any point in time to determine precise position (in three dimensions). Over time, the satellite constellation has grown from 18 to 24. In GPS, the satellite transmissions are synchronized using a high-precision clock on each satellite. Periodic monitoring is conducted on each satellite at very frequent intervals (a few hours apart) by ground-control stations.

An aircraft (or any earth-based) receiver determines position by interpreting the time relationship of incoming radio pulses from at least four satellites. The process is complex and completed using specially produced microelectronics that interpret the identity and time-tag data from each satellite, which are compared with ephemeral data (a description carried over from astronavigation star charts) and effectively treat each satellite in a global constellation of "radio stars" to determine position three-dimensionally in space. This can be related to the Earth's surface and presented as latitude and longitude and height above the Earth.

Two significant GPS-related system developments, called *wide-area augmentation system* (WAAS) and *local-area augmentation system* (LAAS), also were proposed. These were applications based on a technique called *differential GPS* (D-GPS) developed principally for surveyors. This uses a GPS receiver located at a precisely known position and determines a measure of the error in the GPS-derived measurement of position. In WAAS and LAAS, this was to be performed at airports, and the error data were to be communicated by radio data link. (It is necessary to measure the error from different combinations of satellites and to provide a tabulation of errors so that a user can correlate position with the appropriate satellite combination.) WAAS was put into trial in the late 1990s, certification was expected to be straightforward, and LAAS was expected to be approved in time for full use in 2010. The latter system was to be so precise that it could do the job expected of MLS—hence, the loss of interest in the earlier system. In the event, WAAS certification was much more difficult than anticipated, and the program slipped, but WAAS precision equaled that planned of LAAS. The precise implementation is still undecided, but a D-GPS system developed from WAAS and capable of replacing all preexisting radio navaids is entirely feasible, although this will not meet ICAO FANS specifications because a dissimilar redundancy monitoring system is required for civil aviation use.

The United States took the lead in satellite navigation, and the Soviet Union started a project called *Glonass* in the 1980s. This has been sustained, albeit with less tempo, by Russia, and it provides satellite navigation from an independent source. Europe has opted to develop a third system, called *Galileo*, that will become operational toward the end of the current decade (by 2020) as satellites are launched. There is a Chinese system in prospect as well, and the European Space Agency has developed a geostationary "spot" navigation system called *EGNOS*.

This chapter has looked at systems already in service, and as GPS has begun to show its capability in the last decade or so, the scene has started to change very suddenly. There is little evidence that the use of radio navigation systems will diminish suddenly, but their numbers must be expected to recede, and a navigation environment will evolve that exhibits the essential redundancy and control-authority requirements demanded of service providers, perhaps over 20 to 25 years.

Since the mid-1990s, there has been a policy, initially in the United States but now embraced worldwide, to produce Global Navigation Satellite System (GNSS) charts for airports. These are numerous in almost every national aeronautical information publication (AIP) nowadays and can be expected to become the primary guidance approach and departure charts at airports. Initial approval for IFR applications is still to fly "overlay" approaches, whereby the GNSS procedures are identical to existing non–precision approach procedures that use VOR beacons, VOR beacons/DME, NDBs, or NDBs/DME for position fixes, and then approval is granted in stages up to the point where IFR approaches can be flown with GNSS chart guidance. The aircraft equipment and installation must have an approved and operational alternate means of navigation for IFR operations to be approved.

The ATC service at airports, therefore, can offer a range of possible approach-guidance options. The most well-equipped and, therefore, usually the busiest airports offer options that range from GNSS or ILS/DME, usually with radar surveillance. Where MLS will fit into future operations, if at all, is still to be determined. With ATC sequencing of arrivals, the capacity of such airports is very high. A small commercial airport may offer precision-approach radar (PAR) "talkdown" only, which limits capacity greatly because the ATC service has to devote time exclusively to one aircraft, perhaps throughout 10 minutes of sequencing and approach, and a general aviation airport may not have any IFR approach capability.

Broadcast Services

A great deal of information relating to air navigation is required by aircraft in flight or about to depart. Such information on weather and airport and radio aids serviceability is of particular importance. Since the requirement is universal, the telecommunications agency of each country makes available suitable broadcast facilities and is required by international agreement to publish details of the frequencies used and times of broadcasts. These channels are separate from those used for normal control purposes, and no acknowledgment is required from aircraft receiving the broadcasts. There is an increasing tendency for information to be prerecorded or produced electronically using synthesized speech created from a database. While initially crude and monotonic, modern systems produce a result that is often almost indistinguishable from a real-time reader.

Automatic Terminal Information Service (ATIS) has for many years been the most common type of broadcast concerned with airport operations. It is a recorded broadcast on either the voice facility of a nearby VOR beacon or a discrete radio frequency of its own. Each broadcast is individually identified by a phonetic letter of the alphabet commencing the day with "alpha" then "bravo" and so on. Details given in each broadcast include surface wind (magnetic direction) and prevailing weather. The purpose of the broadcast is not only to inform but also to reduce the amount of traffic on the vital control frequencies.

Thus, on initial voice contact with the destination airport, the pilot advises that the ATIS broadcast has been received by quoting its identifying letter—thus, "Information sierra received." The controller then knows that is it is not necessary to pass this information on the control frequency. At some airports there might be two ATIS broadcasts: one for arriving aircraft and one for departing aircraft. Contents of ATIS messages are repeated continuously until such time as a change takes place in any of the items reported. The message then is changed and assigned a new letter identification. This service is likely to be discontinued as more data-link-based systems are introduced.

VOLMET is volume meteorologic information for aircraft in flight and comprises both reports of actual weather conditions at specified airports and also landing forecasts. Broadcasts are made via the "mobile" service on both VHF and HF. High frequencies (HF) are assigned for broadcasts to North Atlantic flights.

There is no intention to remove these services, but a more cost-effective means of serving crew requests for meteorologic data is through digital data links, such as ADS-B and ACARS. The former is usable for such an application, but it has to be lower-priority messaging content. The existing VOLMET will be retained while there is a need for redundancy, and in many scenarios, it is forecast to be retained almost indefinitely. Message formats are being retained, and details of the information sent in this way are given in the following section.

11.5 Meteorology

Function

Aviation meteorologic services are provided by governmental organizations in all ICAO Member States, and their services are organized to conform with ICAO Annex 3. Some countries employ their military to produce aviation-related weather products, but most use the civil meteorologic organization.

World Area Forecast System

The World Area Forecast System was established by ICAO and the World Meteorological Organization (WMO) in 1982 with the purpose of providing worldwide aeronautical forecasts in a standardized form. Currently, there are two world area forecast centers (WAFCs), one in Washington (United States) and the other in London (United Kingdom).

The main task of the WAFCs is to provide significant weather forecasts as well as upper-air forecasts (grid-point forecasts) in digital form and on a global basis. These forecasts are disseminated by a satellite-based system. The two WAFCs are designed to back up each other and produce the same products for different areas. The products generated by the WAFCs are described below.

Meteorologic Observations and Reports

Meteorologic observations are vital to forecasting, and reports are generated by meteorologic services worldwide. In context with aviation, four types of routine observations of surface weather have been established by ICAO and are produced in either hourly or half-hourly intervals at many airports and partly at other geographically relevant sites.

- Aviation routine weather report (METAR)
- Aviation selected special weather report (SPECI)
- Local routine met report (met report)
- Local special met report (special)

Additionally, the 3-hourly surface synoptic observations (SYNOP) reports are used for a wide array of products; primarily, they are the source data for most weather-model applications. While not specifically used for aviation, they are frequently included in meteorologic briefings where other data are not available.

METAR and SPECI

The METAR is the most common surface weather report for aviation purposes. It includes a time-tagged observation of the current weather situation at the observation station and a 2-hour forecast for the same location. It is therefore current weather information and a short-term forecast. Usually METARs are disseminated at half-hour intervals, although some stations only produce METARs hourly.

The METAR includes the following data:

- Station identification and time of observation
- Surface wind direction and speed (direction in *true* north)
- Visibility

- Runway visual range (RVR) when appropriate and available
- Present weather
- Cloud amount and type [only cumulonimbus and towering cumulus (CB and TCU)]
- Temperature and dew point
- QNH or atmospheric pressure above mean sea level (amsl)
- Supplementary information
- Trend forecast
- Remarks when applicable (national dissemination only)

There can be different units of measurement for items such as wind speed (e.g., knots or miles per second), pressure information (e.g., hectopascals or inches of mercury), and temperature/dew point (e.g., degrees Centigrade or degrees Fahrenheit) in various parts of the world.

SPECI reports use the same format and data as METAR reports and are issued at stations where only hourly METARS are produced.

A sample METAR produced at Amsterdam Schiphol Airport is

EHAM 240455Z 18002KT 9999-RA FEW017 SCT035 BKN047 12/11 Q1014 TEMPO 6000

The decoding tables for METAR and SPECI messages are presented at Table 11.4.

Local Routine Met Report (Met Report) and Local Special Met Report (Special)

Local routine met reports consist of more detailed information made available at shorter time intervals. They will include items such as

- Individual surface wind measurements at different locations on the airport
- Complete RVR readouts for all runways
- Individual cloud amounts and types for each runway direction
- Atmospheric pressure Q at field elevation (QFE) at each runway threshold

Additional information, as agreed on by the airport operator and the meteorologic organization, can be included in these messages. The reason for local routine met reports is to disseminate weather information that will allow optimization of operations at the airport itself. Some information included in this report is made available via data link in resolutions between 60 and 120 seconds (e.g., wind, temperature, and barometric pressure), and additional data are updated by the METAR report. Should weather phenomena change beyond

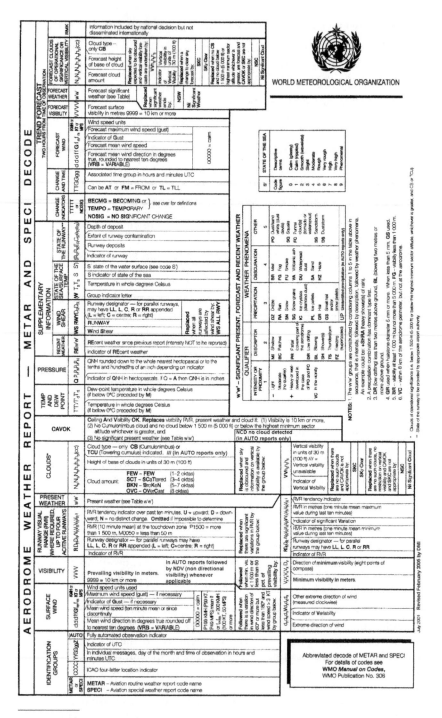

Source: WMO

TABLE 11.4 METAR and SPECI Decode

predefined threshold values within the time interval between messages, the local routine met report will be amended with the issue of a special meteorologic report.

Dissemination of airport met reports via ATIS or VOLMET will take their data from these local routine reports. ATIS and VOLMET transmissions will be updated if there are local routine met reports or local special met reports, but they will not include 1- or 2-minute values.

Aircraft Observations and Reports

Meteorologic data obtained from aircraft in flight provide valuable information about weather conditions in places where either no surface or upper-air observations are available or no observation is possible. Generally, pilot reports (PIREPs) are the best known such report, where an air crew gives a weather report, often together with other routine reports, to ATC. Recently, new technology has allowed expansion to this up to the point where aircraft automatically relay certain data in real time over special communication channels such as ACARS.

Pilot Reports (PIREPs)

The pilot report is a weather observation by an air crew relayed to ATC via either voice communications or ACARS data link. It contains the following elements:

- Message identifier (UA or UUA) for routine or urgent messages
- Position and flight level of the observing aircraft
- Time of report
- Aircraft type (important for reports on icing or turbulence)

Possible weather data include

- Wind direction and speed [true north or magnetic north (United States)]
- Cloud cover
- Icing (light, moderate, or severe)
- Turbulence (light, moderate, or severe)
- Temperature
- Visibility
- Remarks

An example message is

    UA /OV OAK104035/TM 0412/FL070/TP B737/TB LGT-MOD

This is a Boeing 737 near Oakland (OAK) reporting light to moderate turbulence at FL070 at time 0412 UTC.

Aircraft Meteorologic Data Relay (AMDAR)

AMDAR is an initiative by both the World Meteorological Organization (WMO) and ICAO to exploit data available from the sensors used
on board most modern airliners. The system is intended to automatically relay relevant meteorologic data such as wind direction, speed,
temperature, and pressure. While intended to supplement sounding
data, AMDAR potentially allows for a huge amount of data in critical
areas as well as such areas where no soundings are available.

Terminal Airport Forecasts

Terminal airport forecasts (TAFs) are generated for regional and
international airports by the meteorologic services of each country.
Usually TAFs are provided for airports also providing METARs. They
are normally issued four times a day, and the forecasts cover between
9 and 30 hours ahead. They use the same code and terminology as
METAR but with a few additional code groups to define time frames
and certain data items.

Table 11.5 is a complete decode table for TAF messages, and an
example message for Amsterdam airport is

EHAM 231708Z 2318/2424 24014KT CAVOK BECMG 2321/2324
20009KT BECMG 2400/2403 7000 -RA BKN045 TEMPO 2406/2410
4000 RA SCT015 BKN020 BECMG 2407/2410 22016KT 9999 SCT008
BKN012 PROB40 2409/2413 BKN008 TEMPO 2412/2416 26028G40KT
6000 SHRA BKN010 SCT015CB BECMG 2413/2416 28022G35KT NSW
SCT015 BECMG 2417/2420 23008KT

This example provides a 30-hour TAF that is valid between 18 UTC
on the twenty-third of the month to midnight of the twenty-fourth
and shows the passage of a weather system over Amsterdam airport
starting out with clear (CAVOK) conditions and gradually worsening
to possible IMC and strengthening winds.

Significant Weather Forecasts and Charts

Significant weather forecasts (SIGWXs) and charts (SWCs) as well as
upper-air forecast charts are generated and disseminated by the
World Area Forecast Centers (WAFCs) in London and Washington.
They are then produced and locally distributed by the national meteorologic authorities. Where local distribution is not available, the
charts provided by the WAFCs are usually used.

Significant weather forecasts are produced four times a day and
based on the relevant model runs at 00, 06, 12, and 18 UTC. They provide forecasts up to 30 hours from the relevant model data for high
levels (FL250 to FL630) globally and medium levels (FL100 to FL450)
more locally. Some state meteorologic authorities produce low-level
SIGWX charts. These do not always conform to the ICAO standards but,
nevertheless, are useful for low-level operation such as general aviation.

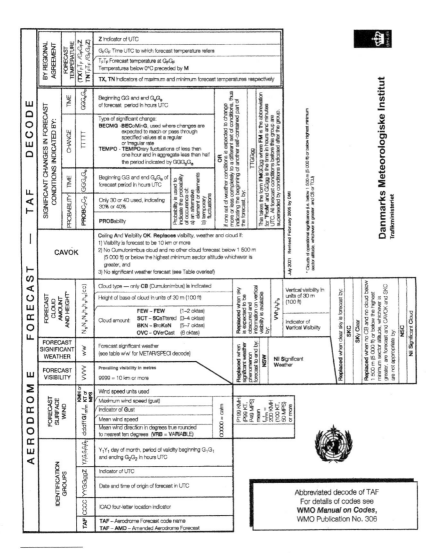

Source: WMO

TABLE 11.5 TAF Message Decode

Significant weather forecasts include the following elements:

- Information about tropical cyclones
- Squall lines
- Moderate or severe turbulence
- Moderate or severe icing
- Sand and dust storms
- Cumulonimbus clouds and thunderstorms

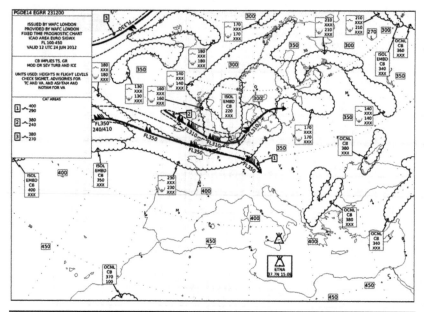

FIGURE 11.8　Synoptic weather chart. (*Source:* Aviation Meteorological Services.)

- Volcanic ash information on active volcanoes
- Jet streams
- Flight level of the tropopause
- Information about the position of nuclear accidents

An example significant weather chart is given in Figure 11.8.

Upper-Air Grid-Point Data Forecasts

Grid-point data forecasts are generated four times a day for fixed time points in 6-hour intervals from 6 to 36 hours after the relevant synoptic data on which they are based. They provide in-flight wind and temperature forecasts at standard flight levels 50 (850 hectopascals), 100 (700 hectopascals), 140 (600 hectopascals), 180 (500 hectopascals), 240 (400 hectopascals), 300 (300 hectopascals), 340 (250 hectopascals), 390 (200 hectopascals), 450 (150 hectopascals), and 530 (100 hectopascals). Additional data produced from the same source include the tropopause flight level and temperature, plus humidity data and the geopotential heights of forecast levels.

As with the significant weather forecasts, state meteorologic authorities produce upper-air charts for use by their customers. Where locally produced charts are not available, those issued by the WAFCs are used. A sample chart is shown in Figure 11.9.

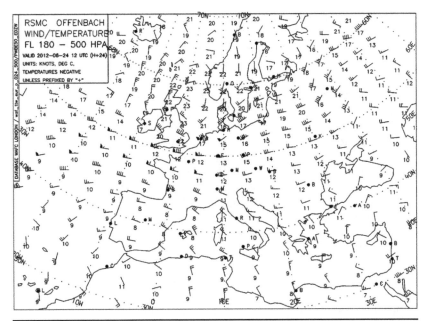

FIGURE 11.9 Airspace (FL 180) wind chart. (*Source:* Aviation Meteorological Services.)

It is with these forecasts that current electronic flight-planning systems generate flight-plan routes between departure and destinations points that provide for a minimum-time, minimum-fuel, or otherwise optimized schedule. The data are available for download from the WAFCs either in Washington (no charge) or in London (with a charge).

SIGMETs/AIRMETs

Significant meteorologic information (SIGMET) and airmen's meteorologic information (AIRMET) messages are weather advisories directed at pilots concerning phenomena that may jeopardize the flight safety of aircraft flying in the area covered by the advisory. SIGMETs advise of severe phenomena such as icing, turbulence, volcanic ash, thunderstorms, and other convective phenomena, as well as mountain waves and radioactive contamination from sea level to FL600. AIRMETs warn of less pronounced phenomena and are restricted to lower airspace up to FL245.

SIGMETS are disseminated worldwide, but AIRMETS usually have a lower dissemination range. Usually it is within the country of origin, but in many cases by agreement or via briefing platforms they are still received and read in neighboring areas. AIRMETs and SIGMETs are published using numbering from 1 for the first message in a day in ascending order until the end of the day.

Some examples of the codes used in SIGMET/AIRMETS include

EMBD TS	Embedded thunderstorms
FRQ TS	Frequent thunderstorms
SQL TS	Squall-line thunderstorms
OBSC TSGR	Obscured thunderstorms with hail
EMBD TSGR	Embedded thunderstorms with hail
FRQ TSGR	Frequent thunderstorms with hail
SQL TSGR	Squall-line thunderstorms with hail
TC	Tropical cyclone
SEV TURB	Severe turbulence
SEV ICE	Severe icing
SEV ICE FZRA	Severe icing due to freezing rain
SEV MTW	Severe mountain wave
HVY DS	Heavy dust storm
HVY SS	Heavy sandstorm
VA	Volcanic ash
RDOACT CLD	Radioactive cloud

SIGMETs with special purposes such as volcanic ash activity and radioactive clouds are published with a different header and outside the usual dissemination chain.

The codes for AIRMETs are similar to those for SIGMETs, but include moderate (MOD) or isolated (ISOL) or occasional (OCNL) phenomena. Some samples can be examined below:

EGJJ SIGMET 01 VALID 240500/240900 EGJJ-EGJJ CHANNEL ISLANDS CTA SEV TURB AND WS FCST AND OBS BLW FL030 NC=

LBSR SIGMET 01 VALID 240340/240740 LBSR-LBSR SOFIA FIR FRQ TS OBS AT 0340Z NW OF LINE N4340 E02438-N4219 E02227 TOP ABV FL330 MOV E 10KT NC=

LSAS AIRMET 1 VALID 240550/240800 LSZH- LSAS SWITZERLAND FIR/UIR ISOL TCU OBS S OF ALPS MOV NE INTSF =

Should a SIGMET or AIRMET become obsolete, it will be cancelled by the issuing met authority.

Weather Information Support for General Aviation

Some countries produce specific forecasts for general aviation, especially VFR operating aircraft. These products include locally produced low-level-significant weather charts (France, Austria, and Germany). These are only locally distributed ICAO products produced in large parts of Europe; in particular are the general aviation forecast (GAFOR) and general aviation meteorologic information (GAMET) messages, which cater primarily to the needs of aircraft operating under VFR.

GAMET messages are combined text/chart products that indicate wind direction and speeds and freezing level, as well as weather phenomena such as turbulence and significant precipitation will be in then-valid SIGMET/AIRMET messages. Where necessary, further and weaker phenomena are also included. GAFOR messages are combined text/chart products that describe flight conditions on selected routes or areas by a classification of four states of usability:

- X-ray   Closed   Ceiling and/or Visibility below VFR limits
- Mike   Marginal   Ceiling and/or Visibility marginal for VFR (1,000–1,500 feet /2–5 km)
- Delta   Difficult   Ceiling and/or Visibility difficult (1,500–2,000 feet AGL/5–8 km)
- Oscar   Open   Ceiling and Visibility above 2,000 feet AGL/8 km

A GAFOR is valid from the time it is issued and includes three time periods of 2 hours each. In a GAFOR valid between 06 and 12 UTC, a classification of XMD means that route/area is closed (X) between 06 and 08 UTC; marginal (M), between 08 and 10 UTC; and difficult (D), between 10 and 12 UTC.

GAFORs and GAMETs provide a general aviation pilot with a concise overview of the conditions he or she may expect for a given area/route. Both products are also used by search and rescue (SAR) providers and are in some countries provided on a 24-hour basis, but by others only during daylight.

Climatologic Information

ICAO requires Member States to provide climatologic information about those airports for which meteorologic information is provided. Climatologic bulletins provide a climatologic overview of the airport over a number of years and are for use by interested parties, such as airport managers and airlines.

Services for Operators and Flight Crew Members

Meteorologic information for flight planning as well as in-flight updating is provided by the relevant meteorologic authority. In additional to all previously described products, the services include

- Ground-based weather radar
- Satellite imagery
- Aerodrome warnings

Information is obtained either from briefing personnel at the airport or via telephone, and increasingly it is also available via automated briefing systems. Recent years have seen a distinct decline in face-to-face service at general aviation and even larger airports. This has led to reductions in the staff numbers employed, and even observations nowadays are being created at centralized meteorologic centers rather than directly at the airports.

There is a trend toward providing as much data as possible directly into aircraft in flight via ACARS or other data links. While some states still specify a personal briefing as obligatory (Germany), most pilots today rely on either meteorologic service–provided automated briefing systems or other sources either within their organization or on the Internet.

Information for ATC, SAR, and AIS

It is the duty of the meteorologic service provider to establish a dedicated service for use of ATC, SAR, and Aeronautical Information Service (AIS). It is still recommended that every FIR and every airport should have its own meteorologic office, but this level of service is already relatively rare. ICAO allows, however, for alternative means of compliance by meteorologic watch offices, which then take care of whole areas rather than individual airports. It is also recommended that each FIR should have its own watch office, and SAR must have 24-hour access to meteorologic information as well as personal briefings.

Use of Communication

ICAO specifies the requirements and use of communications among meteorologic organizations, ATC, pilots, and other interested parties as well as the contents and composition of VOLMET broadcasts. VOLMET broadcasts are continuous transmissions of meteorologic data containing METARs of several airports and AIRMETs and SIGMETs via normally a VHF frequency. ATIS broadcasts are continuous transmissions of local met reports and local special reports via VHF of one specific airport plus AIRMET's and SIGMET's as applicable.

Trends in Meteorologic Services

In the last decade, the aviation meteorologic service in most nations has changed significantly from a one-to-one service provider to an increasingly automated organization where data provision is used more than direct communication. Even with all the technologic advantages meteorology has today, it is still a science that thrives on the expertise and experience of the human beings who provide the service. Meteorology remains a natural science where the most elaborate technologic solutions will only partly replace the experience of the humans who live the weather day by day. The rapid analysis and dissemination of data still will need to be complemented with expert analysis for forecasting to be as reliable as clients would wish. Airlines depend on accurate forecasting of fog density, not just its probable occurrence, in order to meet service-reliability targets without having to equip their whole fleets with appropriate equipment and maintain the currency of crews to perform operations with such systems and know that the experienced forecaster with exemplary knowledge of the locale is the best source of reliable information they will find.

11.6 Aeronautical Information

Scope
The complexities of civil aviation are such that it is almost impossible to conduct a flight of any kind, even a short GA flight, without having recourse to a considerable amount of aeronautical information such as ATC requirements (including airspace restrictions), airport layout, hours of operation, and availability of fuel. The requirement is multiplied many times over if an international flight is involved. Most states have acknowledged this and comply with the standards laid down by ICAO regarding an international format for production of aeronautical information. It is interesting to note, however, that a state may delegate the authority to provide such service to a nongovernmental agency, and there are several instances where this has been done. It then would be the responsibility of that agency to comply with the specifications for such a service as laid down by ICAO.

These specify that an aeronautical information service is responsible under international agreement for

- The preparation of an aeronautical information publication (AIP)
- The origination of NOTAMs
- The origination of aeronautical information circulars (AICs)

In preparing an AIP, there is a laid-down format and content that comprises the following parts listed by specified three-letter identifiers:

Part 1: GEN   General
Part 2: ENR   En route
Part 3: AD   Aerodrome

An amendment service is provided in each country to keep its AIP up-to-date. Most operational units will accept amendments in the form of replacement pages and maintain a regularly updated readable copy. However, throughout the world there has been a move toward disseminating the data electronically.

The evolution of an electronic aviation data (EAD) standard that replaces the AIP now allows dissemination across the Internet, although in many nations it is necessary to register as a user before one can gather information. Information is still presented in the time-honored AIP format (usually as PDF files), but the fundamental data are in a database format that can be accessed by real-time applications. An example of this is an aircraft flight-management system (FMS) that will use information to compute essential operational parameters, such as takeoff and landing distances required, and then compare them with the database declarations of distances available and thus approve or recommend operational changes.

Urgent Operational Information

NOTAMs are urgent notices for the attention of flight crews and operations personnel. They are given the fastest possible circulation. Most NOTAMs will refer to information already published in the AIP, and as such, they are the usual means of transmitting the urgent AIP amendments noted earlier.

NOTAMs are distributed electronically. This replaces the previously used methods of teleprinter or fax, and while these are still the media used in remote areas, most nowadays will get very rapid electronic dissemination worldwide. Less urgent supplements are posted out, but again, the electronic media will accommodate them, with a lesser delivery priority. Equivalent distribution services are used to disseminate aeronautical information regulation and control (AIRAC) NOTAMs, which are used to give advance notification of changes in facilities, radio aids, and so on that have been planned in advance.

NOTAM Code

With the very considerable amount of aeronautical information that is distributed, it is still desirable to condense it as far as possible, and this is achievable by using a NOTAM code. An advantage of using a code is that it overcomes language difficulties, but one disadvantage is that small errors can render the whole message inaccurate or even meaningless. Increasingly, as electronic systems offer so much more capacity than did the distribution methods they replace, NOTAMs are being written and distributed in plain language.

The NOTAM code is used to enable the rapid dissemination of information regarding the establishment, condition, or change of

- Radio aids
- Lighting facilities
- Aerodromes
- Dangers to aircraft
- Search and rescue facilities

Availability of Information

International recommendations regarding the use of aeronautical information are that it should be made available in a form suitable for the operational use of

- Flight operations personnel, including flight crews
- Services responsible for preflight information
- Air traffic service units responsible for flight information

The extent of the information made available at specific airports will depend on the stage lengths of flights operating out of that airport. Thus, the information made available at large airports would include international information, whereas a regional airport with domestic-only services would receive only domestic information within a certain range. This does not prevent special requests being made for aeronautical information for areas not normally covered at a certain airport.

Although international recommendations do not specifically call for the provision of information/briefing rooms at airports for the distribution of aeronautical information, many governmental agencies and airport authorities provide such facilities. These usually include suitable reference material such as the AIP and display charts as well as information bulletins, NOTAMs, and so on.

Major airlines frequently will use the information obtained from the AIS together with their own operational information to publish company NOTAMs.

11.7   Summary

International agreements made through ICAO have laid down the standards and recommended practices for each of the technical services described in this chapter. Individual countries have shown a remarkable consistency in adopting these principles with only minor differences. When such differences do exist, they are required to be published and well documented in both national and international publications, including the various ICAO annexes to the convention. Implementation of these services on a regional basis is documented in the various air navigation regional plans published by ICAO.

As new technologies are developed, there will be inevitable changes, but with a total membership approaching 200, clearly these changes will not be accomplished overnight. So far the system of international standards negotiated through the aegis of ICAO has proved up to the task of satisfying the essential technical and safety requirements of the air transport industry.

References

Eurocontrol. 2009. *European Air Traffic Management Master Plan*, 1st ed.; available at www.eurocontrol.int/sesar/gallery/content/public/docs/European ATM Master Plan.pdf.

Federal Aviation Administration (FAA). 2012. *NextGen Implementation Plan*, March 2012 (annually updated); available at www.faa.gov/nextgen/implementation/plan/.

International Civil Aviation Organization (ICAO). 2006. *Procedures for Air Navigation Services: Aircraft Operations*, Vol. II, 1st ed. (Document 8168 Ops/611). Montreal: ICAO.

International Civil Aviation Organization (ICAO). 2005. Annex 2: *Rules of the Air*, 10th ed. Montreal: ICAO.

International Civil Aviation Organization (ICAO). 2010. Annex 3: *Meteorological Service for International Air Navigation*, 17th ed. Montreal: ICAO.

International Civil Aviation Organization (ICAO). 2007. Annex 10: *Aeronautical Telecommunications*, Vol. I: *Radio Navigation Aids*, 6th ed., July 2006; Vol. II: *Communication Procedures*, 6th ed., July 2001; Vol. III: *Communication Systems*, 2nd ed., July 2007; Vol. IV: *Surveillance and Collision Avoidance Systems*, 4th ed., July 2007; Vol. V: *Aeronautical Radio Frequency Spectrum Utilization*, 2nd ed., July 2001. Montreal: ICAO.

International Civil Aviation Organization (ICAO). 2001. Annex 11: *Air Traffic Services*, 13th ed. Montreal: ICAO.

International Civil Aviation Organization (ICAO). 2010. Annex 15: *Aeronautical Information Services*, 13th ed. Montreal: ICAO.

CHAPTER 12

Airport Aircraft Emergencies

12.1 General

At any particular airport, a number of different types of emergencies might occur, including an aircraft emergency, a building fire, or other major disruptions, such as a major spillage of flammable or poisonous liquids or nonaccidental emergencies caused by attacks, bomb scares, or other terrorist activities. This chapter concentrates on the first category, the aircraft emergency, which is peculiar to airports and aviation and could involve loss of life on a scale that is rightly termed disastrous. Air travel is not a particularly hazardous mode of transportation; indeed, commercial air passenger transportation has a safety record that is bettered only by the railroads. Nevertheless, every airport operator must recognize the possibility that an aircraft accident on the airport or in its vicinity can take place. For airports serving air carriers, this imposes a special responsibility to plan for the saving of a large number of lives through the provision of competent firefighting and rescue services, recognizing and even hoping that during the life of the airport they will never be employed to the limits of their capability.

12.2 Probability of an Aircraft Accident

Aircraft accidents are unlikely occurrences. A Boeing study established that the statistical probability of an accident for a commercial aircraft flown by a Canadian or U.S. operator was one accident in approximately every 6 million departures (Boeing 2009). The same study found that the accident rate for the rest of the world was approximately four times that level. Depending on the type of commercial flight, the accident rate varies considerably, as can be seen from Table 12.1, which demonstrates the high safety record of medium-sized and large passenger aircraft in comparison with small

Flight Type	Fatalities per Million Flight Hours
Airline: Scheduled and nonscheduled operating under *FAR*, Part 121 Regulations	4.03
Commuter airline: Scheduled operating under *FAR*, Part 135 Regulations	10.74
Commuter airplane: Nonscheduled air taxi operating under *FAR*, Part 135 Regulations	12.24
Private general aviation: Operating under *FAR*, Part 91 Regulations	22.43

Source: Boeing 2009.

TABLE 12.1   Aircraft Accident Fatality Rates

air taxis and general aviation (NTSB 2011). This low accident rate is largely due to the very demanding performance standards set by the aircraft certification bodies, which use a scale of probabilities, as shown in Table 12.2. The reliability of modern air transport aircraft

Frequency	Probability	Examples
Frequent Likely to occur often during the life of the aircraft	↑ 10^{-3}	
Reasonably probable Unlikely to occur often but may occur several times in the life of the aircraft	↑↓ 10^{-5}	Engine failure
Remote Unlikely to occur to each aircraft in its life but may occur several times during the life of a number of aircraft of the type	↑↓ 10^{-7}	Low speed overrun Falling below the net takeoff flight path Minor damage Possible passenger injuries
Extremely remote Possible but unlikely to occur in the total life of a number of aircraft of the type	↑↓ 10^{-9}	High speed overrun Ditching Extensive damage Possible loss of life Hitting obstacle in net takeoff flight path Double engine failure on a twin
Will not happen	↑↓	Aircraft destroyed Multiple deaths

TABLE 12.2   Probabilities of Aircraft Failure by Type

has largely removed aircraft failure as a cause of accidents; more than 60 percent of air transport accidents are directly attributable to some form of human error.

Even with low accident probabilities, aircraft disasters do occur at airports, often with a large loss of life. On March 27, 1977, at Norte Los Rodeos in Tenerife, a KLM B747 crashed while taking off, colliding with another B747 of Pan American back-taxiing on the runway. In all, 555 passengers and 25 crew members were killed, making this the worst air disaster in civil aviation history. A routine operating day had been converted into a day on which the airport administration had to cope with an aviation disaster of unprecedented scale. While casualty levels of this scale are unusual, with the size of modern airliners, accidents involving more than 100 persons are more common.

12.3 Types of Emergencies

The International Civil Aviation Organization (ICAO) classifies aircraft emergencies for which rescue and firefighting services might be required into three categories (ICAO, 1991):

Aircraft accident. When an aircraft accident has occurred either on or in the vicinity of an airport, air traffic control (ATC) at the airport will alert the airport rescue and firefighting service (RFFS), giving details of the time and location of the accident and the type of aircraft involved. Other appropriate organizations, such as the local fire department, are notified in accordance with the airport emergency plan (see Section 12.9).

Full emergency. When an aircraft approaching an airport either is or is suspected to be in danger of an accident, the rescue and firefighting service is called to predetermined standby positions for the approach runway and is given details of the type of aircraft, number of occupants, type of trouble, runway to be used, estimated time of landing, and location and quantity of any dangerous goods on board. In accordance with the procedure laid down in the emergency plan, the local fire department and other organizations are also alerted.

Local standby. When an aircraft has or is suspected of having some defect, but the trouble is not sufficiently serious to cause any difficulty in landing, the rescue and firefighting service is alerted to its predetermined standby positions for the approach runway and is given all essential details by ATC. Table 12.3 indicates the relative frequencies of the various categories of emergencies for two large airports in recent years. It can be seen that whereas aircraft accidents are a relatively rare occurrence, other forms of emergencies are much more frequent.

Emergency Type	LIS (2010)
Aircraft accident	0
Full emergency	2
Local standby	24
Total number of operations	142,671

TABLE 12.3   Frequency of Aircraft Emergencies

12.4   Level of Protection Required

Not surprisingly, the level of rescue and firefighting protection depends on the size of the largest aircraft using the airport and the frequency of operation. Ten different levels of protection are designated by ICAO (ICAO 1991); the category into which the airport is assigned is determined by the dimensions of the longest aircraft that uses the airport. The categories are shown in Table 12.4. Based on the airport category determined from this table, the amount of extinguishing agents to be carried on rescue and firefighting vehicles is obtained from Table 12.5 and the minimum number of vehicles from Table 12.6.

The quantities shown in Table 12.5 are computed from the amount of liquid required to control an area adjacent to the aircraft fuselage (the so-called critical area) in order to maintain tolerable conditions for rescue of the occupants.

Airport Category	Aeroplane Over-all Length	Maximum Fuselage Width
(1)	(2)	(3)
1	0 up to but not including 9 m	2 m
2	9 m up to but not including 12 m	2 m
3	12 m up to but not including 18 m	3 m
4	18 m up to but not including 24 m	4 m
5	24 m up to but not including 28 m	4 m
6	28 m up to but not including 39 m	5 m
7	39 m up to but not including 49 m	5 m
8	49 m up to but not including 61 m	7 m
9	61 m up to but not including 76 m	7 m
10	76 m up to but not including 90 m	8 m

TABLE 12.4   Airport Categorization

Airport Category	Foam Meeting Performance Level A		Foam Meeting Performance Level B		Complementary Agents		
	Water (L)	Discharge Rate Foam Solution/ Minute (L)	Water (L)	Discharge Rate Foam Solution/ Minute (L)	Dry Chemical Powders or (kg)	Halons or (kg)	CO_2 (kg)
(1)	(2)	(3)	(4)	(5)	(6)	(7)	(8)
1	350	350	230	230	45	45	90
2	1,000	800	670	550	90	90	180
3	1,800	1,300	1,200	900	135	135	270
4	3,600	2,600	2,400	1,800	135	135	270
5	8,100	4,500	5,400	3,000	180	180	360
6	11,800	6,000	7,900	4,000	225	225	450
7	18,200	7,900	12,100	5,300	225	225	450
8	27,300	10,800	18,200	7,200	450	450	900
9	36,400	13,500	24,300	9,000	450	450	900
10	48,200	16,600	32,300	11,200	450	450	900

TABLE 12.5   Minimum Usable Amounts of Extinguishing Agents

Airport Category	Rescue and Fire-Fighting Vehicles
1	1
2	1
3	1
4	1
5	1
6	2
7	2
8	3
9	3
10	3

TABLE 12.6   Minimum Number of Vehicles

Extinguishing agents are of two major types: *principal agents*, which are used for the permanent control of fire, and *complementary agents*, which have a high capability to "knock down" a fire, but which provide no exposure or reflash protection. Modem principal extinguishing agents provide a fire-smothering blanket. ICAO recommends the use of

1. *Protein foam (foam meeting performance level A).* This is mechanically produced foam capable of forming a long-lasting blanket.

2. *Aqueous film–forming foam (AFFF) (foam meeting performance level B).* This is effective on spill fires, providing faster extinguishing than protein foams. However, the liquid film over the fuel surface is destroyed by high temperatures. AFFF foams are not suitable for fires involving large masses of hot metal.

3. *Fluoroprotein foam.* This is a development from protein-base foams. The addition of a fluorocarbon to protein foam cuts down the amount of pickup of fuel on the surface of the foam bubbles. Although more expensive than protein foam, it is more suited as an extinguishing agent for fires where there is some depth of fuel.

Care must be taken to select a complementary extinguishing agent that is compatible with the principal agent. Although complementary agents do not have any significant cooling effect on liquids and other materials involved in the fire, they act rapidly in fire suppression and can ensure that a fire does not get out of control before permanent control can be achieved with the principal agent. Several complementary extinguishing agents are available:

- Carbon dioxide
- Dry chemicals
- Halocarbons

It is obviously important that there is a sufficient reserve supply of principal and complementary agents to ensure replenishment of the rescue and firefighting vehicles so that after an accident, continued cover can be given to subsequent aircraft operations. It is recommended that a minimum reserve supply of 200 percent of the quantities of the agents shown in Table 12.5 should be maintained on the airport for the purpose of vehicle replenishment.

A rescue and firefighting service should have as its operational objective a response time of not more than three minutes, preferably not more than two minutes, to any part of the movement area in ideal visibility and surface conditions. This sort of performance level was achieved in the crash of the Eastern Airlines B727 at JFK airport on June 24, 1975. The first Port Authority appliance reached the aircraft two minutes after the alarm and three minutes after the crash. It took two minutes to control the main fire and three minutes to extinguish it. There were 11 survivors, but 107 passengers and 6 crew members died. The New York City Fire Department was notified within four minutes of the crash, and the first units arrived within eight minutes.

Response time must be considered in conjunction with evacuation times. In a trial conducted by Douglas, 391 occupants were evacuated from a DC10 in 73 seconds in the dark using only emergency lighting

(United States 1977). This must be seen as an ideal time, achieved by subjects in an unshocked condition, with no fear of danger of fire and with all escape chutes in operation. In the Continental Airlines DC 10 crash at Los Angeles in March 1978, the evacuation of 183 passengers and 14 crew took five minutes, not the 90 seconds demonstrated in the Federal Aviation Administration's (FAA's) aircraft certification program.

In November 2011, a Boeing 767 operated by Lot Polish Airlines, landing on a flight from Newark, New Jersey, to Warsaw, suffered a failure of the hydraulic undercarriage gear and was forced to make a "wheels-up" landing. The runway had been hosed previously with water and foam by the RFFS. No fire occurred, and within 1½ minutes of the aircraft coming to a stop, the cabin crew reported that all passengers had been evacuated from the passenger cabin. All 231 passengers and crew survived without major injury.

12.5   Water Supply and Emergency Access Roads

Water for aircraft rescue and firefighting purposes can come from either the airport water supply or natural water supplies within the airport area. It is desirable that the airport water supply is provided in apron and service areas and in the vicinity of administration areas. Firefighting vehicles are more easily replenished if the supply is extended to hydrants spread about the movement area where this is economically feasible. Natural surface water from rivers, lakes, streams, and ponds can be used only if the firefighting vehicles are adequately equipped to pick up and pump such supplies.

Although rescue and firefighting vehicles should have some all-terrain ability, reduction of the response time to an accident and the subsequent evacuation of casualties are made easier by the provision of emergency access roads to various areas in the airport and to areas beyond the airport boundary, especially in the final approach areas and the clearways designated for takeoff. ICAO recommends particular attention to providing ready access to approach areas up to 3,300 feet (1,000 m) from the threshold. It is common for airport operators to assume that provided that obstruction clearances are observed in the approach and takeoff areas, the airport's responsibilities have been met. The crash of a DC9 at Toronto on June 26, 1978, is an example of the very severe problems of rescue where terrain beyond the runway is virtually impassable. Rescue vehicles were able to reach the site in the clearway beyond the runway threshold only through the quick thinking of an airport bulldozer driver who immediately cleared a temporary road through an otherwise impassable wooded ravine.

12.6   Communication and Alarm Requirements

An aircraft emergency is an unplanned event, and accidents are very rare occurrences. Consequently, a reliable rescue and firefighting operation can be achieved only with a defined chain of command linked with effective communications. For each airport, the individual requirements for the communications network are likely to be different; however, in general, there must be provision for the following:

Direct communication between the emergency activation authority (usually ATC) and the airport fire station. This should be in the form of a two-way radio network and a direct telephone line not passing through an intermediate switchboard. The satisfactory operation of the equipment should be continuously monitored and arrangements made for 24-hour maintenance. The fire station itself should have alarm and public-address systems to alert the crew to emergencies and to permit general crew briefing of the details of the emergency. It is usual to have a device that silences alarm bells while broadcasts are being made over public-address systems.

Communication between rescue and firefighting vehicles and both ATC and the airport fire station. Overall control of vehicles in the operational area must, for safety reasons, be under the direction of ATC, and entry into the active areas can be made only with its permission. Logistically, this requires vehicles to be fitted with a two-way radio communication system. Desirably, this is a multichannel discrete-frequency system that permits vehicles to contact ATC, the airport fire station, and each other. Portable radiotelephone equipment also should be supplied to the officer in charge of the rescue and firefighting operation at the accident/incident site to permit the officer to move away from his or her vehicle without losing contact with the common-frequency radio link. It is essential that the vehicles and ATC have unbroken communications en route to the accident and on site.

Other communication and alerting facilities. At a large airport, in the event of an emergency, a number of different parties are required to be informed and to take action. They include (ICAO 1991):

- ATC
- Airport rescue and firefighting services
- Airport security
- Airport management
- Airline station managers
- Military units (at joint-use airports)

- Local fire departments
- Medical services
- Local police

Simultaneous notification of interested parties can be achieved by the use of *series* or *conference* circuits in the emergency communications system. These require a trained response of strict communications discipline to ensure prompt and uninterrupted transmission of information and messages.

12.7 Rescue and Firefighting Vehicles

The overall level of protection recommended for airports has been covered in Section 12.4 from the viewpoint of the number of rescue and firefighting vehicles and the amount of extinguishants. Clearly, as traffic grows at an airport, its needs for protection change. Therefore, the carrying and discharge capabilities of the rescue and firefighting vehicle fleet should be based on the needs over the short- and medium-term future during the economic life of the vehicles. A forward-looking policy in the commissioning of vehicles permits a greater degree of standardization and overall long-term savings. Once commissioned, vehicles must be protected and maintained in a manner that permits immediate availability should an emergency arise. This implies not only the routine maintenance customary for all vehicles but also maintaining the fitted equipment such as pumps, hoses, nozzles, turrets, two-way radios, searchlights, and floodlights, all of which are necessary for adequate functioning of the vehicle. Providing the day-to-day protection for the vehicles must take account of emergency operational requirements, which necessitate that access to the movement area be unobstructed, that the vehicle running distance to the runways be as short as possible, and that crew members have the widest possible view of flight activity. In airports where water rescue vehicles may be required, it is normal to locate the craft on the airport and to provide launching sites so that the boats can be brought into action with a minimum response time.

The rescue and firefighting vehicle is designed to carry out the principal attack on an aircraft fire. Its design and construction should make it capable of carrying a full load at high speeds in all weather over difficult terrain. The recommended characteristics of the vehicle are shown in Table 12.7. The off-road performance of the major vehicle is a primary factor in equipment choice. It should have traction and flotation characteristics that accord with the terrain in which it is to be used. The minimum ground clearance and the angles of approach and departure must be adequate to permit the vehicle to cross depressions and slopes that could be obstacles to movement. Moreover, the

	Rescue and Firefighting Vehicles up to 4,500 L	Rescue and Firefighting Vehicles over 4,500 L
Monitor	Optimal for categories 1 and 2.	Required
	Required for categories 3 to 10	
Design feature	High discharge capacity	High and low discharge capacity
Range	Appropriate to longest airplane	Appropriate to longest airplane
Hand lines	Required	Required
Under-truck nozzles	Optional	Required
Bumper turret	Optional	Optional
Acceleration	80 km/h within 25 s at the normal operating temperature	80 km/h within 40 s at the normal operating temperature
Top speed	At least 105 km/h	At least 100 km/h
All-wheel-drive capability	Yes	Required
Automatic or semiautomatic transmission	Yes	Required
Single-rear-wheel configuration	Preferable for categories 1 and 2	Required
	Required for categories 3 to 10	
Minimum angle of approach and departure	30°	30°
Minimum angle of tilt (static)	30°	30°

Source: ICAO.

TABLE 12.7   Suggested Minimum Characteristics for Rescue and Firefighting Vehicles

design of the equipment should be such that operation is simple in order to avoid delay and confusion should an accident occur. Moreover, the design should minimize crew requirements for operation. A modem major vehicle is shown in Figure 12.1. Table 12.8 shows the manual and power tools to be carried to the accident site.

Many airports have approaches and departures over water. Water rescue vehicles must be chosen with similar care. They must have as high a speed as is practicable to reach possible accident sites, and the power unit must be reliable and capable of delivering maximum power

Equipment for Rescue Operations	Airport Category			
	1–2	3–5	6–7	8–10
Adjustable wrench	1	1	1	1
Axe, rescue, large non-wedge type	—	1	1	1
Axe, rescue, small non-wedge or aircraft type	1	2	4	4
Cutter bolt, 61 cm	1	1	1	1
Crowbar, 95 cm	1	1	1	1
Crowbar, 1.65 m	—	—	1	1
Chisel, cold 2.5 cm	—	1	1	1
Flashlight/hand lamps	2	3	4	8
Hammer, 1.8 kg	—	1	1	1
Hook, grab or salving	1	1	2	3
Saw metal cutting or hacksaw, heavy duty, complete with spare blades	1	1	1	1
Blanket, fire resisting	1	1	2	3
Ladder, extending (of over-all length appropriate to the aircraft types in use)	1	1	2	3
Rope line, 15 m length	1	1	2	3
Rope line, 30 m length	—	—	2	3
Pliers, 17.8 cm, side cutting	1	1	1	1
Pliers, slip joint 25 cm	1	1	1	1
Screwdrivers, assorted (set)	1	1	1	1
Snippers, tin	1	1	1	1
Chocks, 15 cm high	—	—	1	1
Chocks, 10 cm high	1	1	—	—
Powered rescue saw complete with two blades; or—pneumatic rescue chisel complete—plus spare cylinder, chisel, and retaining spring	1	1	1	2
Seat belt/harness cutting tool	1	2	3	4
Gloves, flame: resistant pairs (unless issued to individual crew members)	2	3	4	8
Breathing apparatus and spare cylinder	One set per fire fighter on duty			
Oxygen inhaler	—	1	1	1
Hydraulic or pneumatic forcing tool	—	1	1	1
Medical first aid kit	1	1	2	3
Tarpaulin	1	1	2	3
Fan for ventilation and cooling	—	1	2	3
Protective clothing	One set per fire fighter on duty			
Stretcher	1	2	2	2

TABLE **12.8** Rescue Equipment to be Carried on Rescue and Firefighting Vehicles

FIGURE 12.1   A major all-terrain RFFS vehicle. (*Courtesy of Oshkosh.*)

in a minimum time under low-temperature and high-humidity conditions. The design of the vehicles must be related to the environment in which they are to operate (e.g., mudflats, tidal areas, and ice), and they must be capable of carrying the flotation equipment necessary to effect a rescue. It is not uncommon to find that airports with approaches over water provide a significantly lower level of rescue and firefighting capability for accidents in water than for those provided on land. This is unreasonable because the passengers and crews of aircraft using a facility expect a similar level of protection regardless of the airport into which they are flying or the particular approach they happen to be using.

12.8   Personnel Requirements

In determining the number and deployment of personnel required for the rescue and firefighting services, ICAO recommends use of the following criteria:

1. The rescue and firefighting vehicles can be staffed such that they can discharge both principal and complementary extinguishing agents at their maximum designed capability and can be deployed immediately with sufficient personnel to bring them into operation.

2. Any control room or communications facility related to the rescue and firefighting service can continue operation until alternative arrangements are made under the airport emergency plan.

All personnel must be fully trained and familiar with their equipment and duties. Should an accident occur, they must be able to perform their duties without any limitation that could come from physical disability, and they must be mentally and intellectually capable of performing the responsibilities of a rescue and firefighting team. It is common from an economic standpoint, especially at small airports, to assign full-time rescue and firefighting personnel to other duties. These duties include fire-prevention inspections, fire guard functions, bird scaring, grass cutting, and frequently at small airports, apron-related functions such as baggage loading and apron equipment handling. In the event of a disastrous accident, the airport emergency plan will provide for alerting all personnel who can act in a support role to the full-time rescue and firefighting crew.

12.9   The Airport Emergency Plan

International aviation agreements, as set out in Annex 14 of the *International Convention on Civil Aviation* (ICAO 2010), require that each airport establish an emergency plan commensurate with the level of aircraft operations and other activities at the airport (ICAO 1991). Individual countries have their own regulations, such as *Federal Aviation Regulations* (*FAR*), Part 139 in the United States, which to some degree parallels the international requirement. The purpose of the airport emergency plan is to minimize the effects of the emergency, particularly with respect to the preservation of life and maintaining aircraft operations by establishing a coordinated program between the airport and the surrounding community. In addition to setting up an agreed-on and recognized structure of command during the emergency, the plan should include a section of instructions to ensure immediate response of rescue and firefighting services, law enforcement, medical services, and other persons and agencies both on and off the airport.

A comprehensive airport emergency plan considers the following:

1. *Preplanning before an emergency.* Defining organizational authority, testing, and implementing the plan.

2. *Operations during an emergency.* Defining what must be done at each stage and the structure of responsibilities as the emergency progresses.

3. *After the emergency.* Handling matters not usually having the urgency of preceding events but necessarily defining the transition of command and operations back to normality (ICAO 1991).

The emergency plan should cover an on- or an off-airport incident. Normally, the airport authority will be in command for an on-airport incident. A mutual emergency agreement among the airport and the jurisdictions of the surrounding community will define command and responsibilities in the case of an off-airport incident. The purpose of an airport emergency plan is to ensure the following:

1. Orderly and efficient transition from normal to emergency operations

2. Delegation of airport emergency authority

3. Assignment of emergency responsibilities

4. Authorization of key personnel for actions contained in the plan

5. Coordination of efforts to cope with the emergency

6. Safe continuation of aircraft operations or return to operations as soon as possible (ICAO 1991)

Even an on-airport accident might be of such a magnitude that the airport services will not on their own be capable of coping with the situation. It is therefore essential that the airport authority arrange mutual-aid emergency agreements with surrounding jurisdictions defining the responsibilities of each party, for example, in the provision of ambulances, additional firefighting staff and equipment, and medical personnel.

The agreements should

• Clarify the responsibility of each involved agency

• Establish an unambiguous chain of command

• Designate communications priorities at the accident site

• Designate an emergency transportation coordinator and indicate the organizational structure of emergency transportation facilities

• Predetermine the authority and liability of all cooperating emergency personnel

• Prearrange for the use of rescue equipment from available sources

A recommended outline of an airport emergency plan is given in Table 12.9. The airport and community agencies to be included in the emergency plan are air traffic services, rescue and firefighting services, fire departments, police, national aviation security services, airport authority, medical services, hospitals, aircraft operators, government authorities, communication services, airport tenants, transportation authorities, civil defense, military harbor patrol or Coast Guard, clergy, public information office, and other mutual-aid agencies (ICAO 1991). The flow of control will be different in the case of an accident depending on whether the site is on- or off-airport. Figures 12.2 and 12.3 are the ICAO-suggested flow-control charts for these respective cases.

ICAO further recommends that an emergency plan be tested by full-scale emergency exercises using all facilities and associated agencies at intervals not exceeding two years, with partial exercises each year and tabletop exercises every six months. The exercises should be followed by a full debriefing and critique in which all involved organizations participate. In fact, full emergency drills of this scale are not as common as ICAO recommends. For example, in

Section 1. Emergency Telephone Numbers
This section should be limited to essential telephone numbers according to site needs, including • Air traffic services • Rescue and firefighting services (departments) • Police and security • Medical services • Hospitals • Ambulances • Doctors, business/residence • Aircraft operators • Government authorities • Civil defense • Others
Section 2. Aircraft Accident On the Airport
• Action by air traffic services (airport control tower or airport flight information service) • Action by rescue and firefighting services • Action by airport authority • Vehicle escort • Maintenance • Action by medical services • Hospitals • Ambulances • Doctors • Action by aircraft operator involved • Action by emergency operations center and mobile command post • Action by government authorities • Communications network (emergency operations center and mobile command post) • Action by agencies involved in mutual aid emergency agreements • Action by transportation authorities (land, sea, air) • Action by public information officer(s) • Action by local fire departments when structures involved • Action by all other agencies
Section 3. Aircraft Accident Off the Airport
• Action by air traffic services (airport control tower or airport flight information service) • Action by rescue and firefighting services • Action by local fire departments • Action by police and security services

TABLE **12.9**   Example of Contents of Emergency Plan Document

- Action by airport authority
- Action by medical services
- Hospitals
- Ambulances
- Doctors
- Action by agencies involved in mutual-aid emergency agreements
- Action by aircraft operator involved
- Action by emergency operations center and mobile command post
- Action by government authorities
- Communication networks (emergency operations center and mobile command post)
- Transportation authorities (land, sea, air)
- Action by public information officer
- Action by all other agencies

Section 4. Malfunction of Aircraft in Flight (Full Emergency or Local Standby)

- Action by air traffic services (airport control tower or flight information service)
- Action by airport rescue and firefighting services
- Action by police and security services
- Action by airport authority
- Action by medical services
- Hospitals
- Ambulances
- Doctors
- Action by aircraft operator involved
- Action by emergency operations center and mobile command post
- Action by all other agencies

Section 5. Structural Fires

- Action by air traffic services (airport control tower or flight information service)
- Action by rescue and firefighting services (local fire departments)
- Action by police and security services
- Action by airport authority
- Evacuation of structure
- Action by medical services
- Hospitals
- Ambulances
- Doctors
- Action by emergency operations center and mobile command post
- Action by public information officer
- Action by all other agencies

TABLE 12.9   Example of Contents of Emergency Plan Document (*Continued*)

Section 6. Sabotage Including Bomb Threat (Aircraft or Structure)
• Action by air traffic services (airport control tower or airport flight information service) • Action by emergency operations center and mobile command post • Action by police and security services • Action by airport authority • Action by rescue and firefighting services • Action by medical services • Hospitals • Ambulances • Doctors • Action by aircraft operator involved • Action by government authorities • Isolated aircraft parking position • Evacuation • Searches • Handling and identification or luggage and cargo on board aircraft • Handling and disposal of suspected bomb • Action by public information officer • Action by all other agencies
Section 7. Unlawful Seizure of Aircraft
• Action by air traffic services (airport control tower or airport flight information services) • Action by rescue and firefighting services • Action by police and security services • Action by airport authority • Action by medical services • Hospitals • Ambulances • Doctors • Action by aircraft operator involved • Action by government authorities • Action by emergency operations center and mobile command post • Isolated aircraft parking position • Action by public information officer • Action by all other agencies
Section 8. Incident on the Airport
An incident on the airport may require any or all of the actions detailed in Section 2. Examples of incidents the airport authority should consider: fuel spills at the ramp, passenger loading bridge, and fuel storage area; dangerous goods occurrences at freight-handling areas; collapse of structures; and vehicle-aircraft collisions.

TABLE **12.9**   Example of Contents of Emergency Plan Document (*Continued*)

Section 9. Persons of Authority—Site Roles
To include, but not limited to, the following, according to local requirements: • Airport chief fire officer • Airport authority • Police and security officer in charge • Medical coordinator • Off-airport • Local chief fire officer • Government authority • Police and security—officer in charge
The on-scene commander will be designated as required from within the prearranged mutual-aid emergency agreement.
Experience indicates that confusion in identifying command personnel in accident situations is a serious problem. To alleviate this problem, it is suggested that distinctive colored vests with reflective lettering and hard hats be issued to command personnel for easy identification. The following colors are recommended:

Red	Chief fire officer
Blue	Police chief
White (red lettering)	Medical coordinator
International orange	Airport administration
Lime green	Transportation officer
Dark brown	Forensic chief

An on-scene commander should be appointed as the person in command of the overall emergency operation. The on-scene commander should be easily identifiable and can be one of the persons indicated above or any other person from the responding agencies.

Source: ICAO 1993.

TABLE 12.9   Example of Contents of Emergency Plan Document (*Continued*)

many countries, there are regulations that all airports must have a written airport disaster plan, but frequently the regulation does not require that the plan be tested. The first full-scale U.S. air carrier airport that used an air transport aircraft to increase the reality of its disaster drill was Oakland International Airport in May 1971. Since then, many full-scale drills have been held that have emphasized the three C's of disaster planning: command, communication, and coordination.

Command

Following through an on-airport accident, it can be seen that the captain is in command of the aircraft while in the air and immediately after the crash. Command changes as the airport fire trucks arrive and

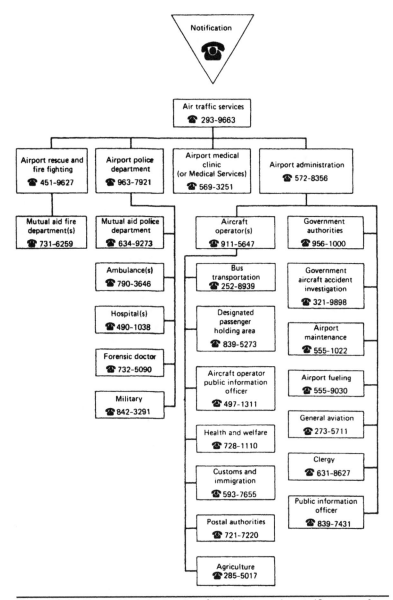

FIGURE 12.2 Flow-control chart—aircraft accident on airport. (*Courtesy of ICAO.*)

the flight crew evacuates passengers. In the United States, the highest-ranking fire officer assumes command on the disaster site (or the designated airport operations officer, depending on the chain of command set up in the airport emergency plan). In some countries, command is assumed by the most senior police officer. This situation holds until all fires are stabilized and all

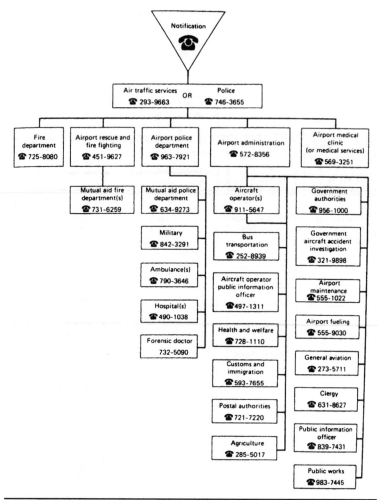

FIGURE 12.3   Flow-control chart—aircraft accident off airport. (*Courtesy of ICAO.*)

casualties are treated and dispatched to hospital. The accident site then remains under the control of the airport operations officer until the arrival of the accident investigation team; in the United States, this is the National Transportation Safety Board (NTSB). The "persons in charge" of the various agencies should wear distinctive-colored vests, as indicated in the emergency plan and shown in Table 12.8. Until the arrival of a medical officer, a paramedic should take control of removal of casualties to the triage area.[1] Ideally, the medical coordinator designated under the emergency plan will be a trauma-trained doctor.

---

[1]Triage is the sorting and classification of casualties to determine the order of priority for treatment and transportation.

Passengers are evacuated initially to the triage area, which is at least 300 feet (100 m) upwind of the crash. Here casualties are examined and tagged according to the severity of injury; the tags represent the priority for transportation and care:

Priority I: Immediate care—red rabbit symbol
Priority II: Delayed care—yellow turtle symbol
Priority III: Minor care—green ambulance symbol
Priority 0: Deceased—black cross symbol

The layout of an accident site is shown in Figure 12.4.

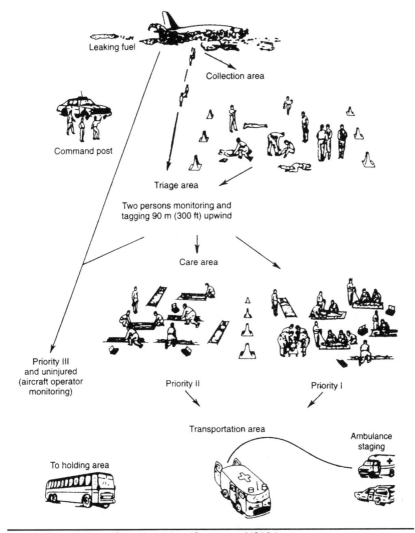

FIGURE 12.4 Layout of accident site. (*Courtesy of ICAO.*)

Communications

Communications from the accident site are achieved through the command post at the accident site. At well-equipped airports, this can be a field trailer specially equipped with radio, radiotelephone, loudspeakers, elevated platform, and floodlighting. Experience has shown that there is usually little problem with communication between the fire and police departments and the airport authorities; the main difficulties have been found to lie among the medical coordination, hospitals, and ambulances. Clear-channel radio networks, such as the Los Angeles County hospital administrative radio system, have been found to be particularly effective.

Coordination

Coordination of the many agencies and individuals to be involved in the case of an emergency or an accident requires planning, patience, and teamwork. It is essential that everyone involved, on- and off-airport, know what his or her responsibilities are when an emergency is decla-red. San Francisco International Airport, which has carried out many simulated disaster drills, has a separate building designated as a disaster building. In Jeddah, Saudi Arabia, it is part of the Haj Terminal. Converted aircraft cargo trailers are used to transport disaster supplies and have been fitted out as a mobile command post and communications trailers.

12.10   Aircraft Firefighting and Rescue Procedures

The principal objective of a rescue and firefighting service is to save lives in the event of an aircraft accident (ICAO 1990). Because of the very large amounts of inflammable material present at an aircraft accident, the likelihood of a fire is high. It is important to bear in mind that it is a characteristic of aircraft fires that they reach lethal intensity in a very short time from outbreak. Coupled with the fact that the occupants of a damaged aircraft have suffered the shock of a high-speed impact, this means that effective rescue from a survivable crash is possible only if

- The firefighting and rescue team is well trained and familiar with equipment and procedures.
- The equipment is effective for the purpose.
- The accident site is reached in time.

The details of firefighting are complex and beyond the scope of this text. However, a basic understanding of the techniques of rescue should exist. First, rescue must be seen to include the protection of escape routes from the aircraft. Although the rescue of occupants is the primary objective, the overall requirement is to create conditions where survival is possible and rescue might be carried out. The prompt arrival of the RFFS vehicle at the site of a crash is most vital to achieving the rescue of victims. After four minutes of exposure to fire, aircraft windows will melt, and within one further minute, the cabin temperature will be sufficiently high to ensure no survivors. Therefore, it is often necessary to concentrate first on fire suppression prior to any rescue. Failure to suppress a fire or blanket a fuel spill might endanger many lives. Second, the rescue of those unable to escape without aid might be a time-consuming process. During this period, additional fire-security measures will be necessary, as well as possibly the delivery of ventilation air into the fuselage.

Seen in this context, the RFFS vehicle will be used to suppress any fires and put down a precautionary blanket over the critical areas, especially in fuel-wetted areas. The first vehicle provides protection against ingress of fire into the fuselage when the aircraft doors and windows are opened. With the arrival of other vehicles, there is usually sufficient fire-suppression capacity to permit the crew of the first vehicle to perform a number of functions:

- To form rescue teams to enter the aircraft to assist aircraft occupants who need help to evacuate
- To provide firefighting equipment within the aircraft for internal fires
- To provide lighting and ventilation within the fuselage

An accident of disaster proportions is something that few rescue and firefighting teams actually face. Therefore, adequate training and proper planning are essential so that when suddenly confronted with the reality of an accident, a crew is capable of coping with a very fast-moving situation that can develop rapidly from an accident into a disaster. Two examples of crashes with B727s in April 1976 serve to outline the problems of unpreparedness that result from insufficient training.

On April 5, 1976, an Alaska Airlines Boeing 727 crashed at Ketchikan International Airport. There was a full-time crash firefighting and rescue service under the control of the airport manager. However, it had no formal training in firefighting techniques or in operating the two crash trucks. The airport manager had neither the training to direct firefighting activities nor the experience and personnel to do it. Firefighting activities by the airport crew at the accident site, therefore, were described as minimal. The city fire department arrived at the site, but ran out of foam and hose in 20 minutes. The airport had not been inspected for almost 12 months by the FAA certification program.

On April 27, 1976, an American Airlines Boeing 727 crashed on takeoff from an aborted landing at St. Thomas, Virgin Islands. The crash firefighting and rescue crew responded immediately but did not take all available emergency equipment; airpacks and some proximity suits were left behind. The RFFS crew chief was the port authority duty officer. Consequently, no one was left to take overall charge of the operation. Flight operations were resumed shortly after the crash, but with no runway inspection and no crash fire and rescue cover for these operations. Subsequently, it was found that a direct line from the airport to the outside emergency service had, unknown to the FAA, been removed, leaving only a commercial line.

In comparison with these two accidents, a potential disaster was averted by a well-trained and organized force when a DC-10 Continental Airlines aircraft crashed after an aborted takeoff at LAX on March 1, 1978, with 183 passengers and 14 crew members. The left main gear collapsed as the aircraft entered consolidated ground at the runway end at about 65 mi/h (105 km/h). A fire, fed by ruptured wing tanks, was attacked by firefighters from a satellite station only 90 seconds after the crash. The vehicle from the main airport fire station 2.5 miles (4 m) away arrived after four minutes. Six exits were opened, and evacuation took five minutes. More than 100 persons were still on board when the first foam tender arrived. Forty-three persons were injured, but only two died, these being individuals who, against the advice of the cabin attendant, used the left overwing exit and jumped down into the fire. The rapid intervention of the airport rescue and firefighting service saved many lives as well as instruments and fittings from the aircraft interior then worth more than US $10 million.

Since these accidents, civil aviation authorities increasingly have made national accident training requirements more stringent and RFFSs are now far more professional than previously. On August 2, 2005, Air France 358 crashed after landing in turbulent weather at Toronto Lester Pearson Airport with 309 passengers and crew. Overshooting the end of the runway, the aircraft came to a stop and immediately caught fire. The RFFS crew was on site at the crash scene 52 seconds after the crash alarm sounded. Evacuation took place within 90 seconds of the aircraft coming to a stop, even though one chute failed to deploy and another failed through puncture. All passengers and crew survived; the aircraft was destroyed by the subsequent fire.

12.11   Foaming of Runways

Protein foam (performance level A) has been applied to the runway in a number of emergency landings in the hope that the landing could be carried out in a safer manner. Fluoroprotein foam and aqueous film–forming foam are not suitable for such an operation owing to their short drainage time. Foaming usually has been used for wheels-up landings or aircraft with defective nose gears. In some cases, foaming operations were successful; in others, the purpose was not accomplished, mainly owing to the aircraft missing or overrunning the blanket.

Theoretically, there are four principal benefits from foaming that bear some examination (ICAO 1990):

1. *Reduction in aircraft damage.* There is no firm evidence that the minimum damage attained in a number of well-executed landings on foam-coated runways would have been greater without foaming. Controlled emergency landings also have been carried out with minimum damage on dry runways. ICAO indicates that too many other variables are involved to draw valid conclusions from the small sample involved. The practice, however, is still widespread.

2. *Reduction in deceleration forces.* In the case of a wheels-up emergency landing, the reduced runway friction can be an advantage. Where the main gear is down, braking performance is only slightly worse than on a wet runway.

3. *Reduction of spark hazard.* Under the conditions of an emergency landing, aluminum alloys do not appear to be capable of generating sparks capable of igniting aircraft fuel. Magnesium alloys, stainless steel, and other aircraft steels can produce such sparks, and it is found that in the majority of tests, a properly laid foam blanket did suppress sparks. Titanium friction sparks cannot be suppressed effectively.

4. *Reduction of fuel spill hazard.* Vapors above the foam blanket are unaffected by the application. However, fuel released over the blanket will fall through the foam and will spread under it, with a reduction of burning area. A more intense fire might occur locally.

Other factors must be considered before deciding to foam a runway:

- Whether there is sufficient time to lay a foam blanket (this usually takes about an hour)
- The reliability on the landing techniques to be used: wind, visibility, pilot skill, visual and radio aids, aircraft conditions
- Adequacy of foam-making equipment
- Problem of cleanup to reopen operations
- Ambient conditions such as rain, wind, or snow
- Runway condition and slope

In cases where it has been decided to proceed with foaming the runway, the following criteria must be met:

1. The pilot must be fully informed on how the operation is to be carried out and what protection is to be provided.

2. The vehicles providing the foam must not decrease the minimum level of airport protection required.

3. The foam used should be additional to the minimum level of protection plus replenishment needs.

4. The positioning of the blanket should recognize that with a gear-up landing the aircraft contacts the runway much farther from the threshold than usual.

5. In reduced visibility, the pilot must be given a reference point to indicate where the foam starts.

6. Unnecessary personnel must be cleared from the area.

7. Prior to use, the foam should be aged about 10 minutes to permit drainage and surface wetting.

8. The foam blanket should be continuous.

9. The blanket depth is about 2 inches (5 cm).

10. After laying the foam, all personnel and equipment are removed to emergency standby positions.

12.12   Removal of Disabled Aircraft

Annex 14 states that participating countries to the international convention are required to establish a plan for removing disabled aircraft on or adjacent to the movement area and recommends the designation of a coordinator to implement the plan (ICAO 2009). The reason for an aircraft's immobility can range from a relatively minor incident, such as a blown tire or frozen brake, to a major accident involving disintegration of the aircraft itself.[2]

Whereas in the early days of aviation the task of removing a disabled aircraft was relatively simple, with large modem aircraft, the job has become technically difficult and expensive, requiring special equipment and organization. Obviously, if the aircraft is immobilized on a part of the airport where it interferes with operation, it is in the interest of the airport authority, other operators, and the traveling public that removal be carried out rapidly. However, where a large aircraft is concerned, the task is complex and potentially dangerous, and, therefore, it will not always be possible to clear the airport as rapidly as might be desired. With large, expensive aircraft, care must be taken to avoid further damage during the removal procedure itself. The responsibility for controlling the removal of an aircraft lies with the registered owner. With small aircraft, it is usually

---

[2]Except as specified in Annex 13, wreckage of aircraft after an accident must be left undisturbed.

possible, if necessary, for the airport authority to undertake removal with the agreement of the owner. With large aircraft, however, generally, the airport does not have the expertise or experience necessary for speedy and safe removal without secondary damage. This will require specialized teams, often involving not only the airline but also the aircraft manufacturer and the insurer. For each type of aircraft, special knowledge of safe jacking techniques and other lifting procedures is necessary (Paluzek 2009).

The main purpose of the disabled aircraft removal plan is to ensure the prompt availability of appropriate recovery equipment and expertise at any incident site. The plan should be based on the characteristics of aircraft types that normally can be expected to use the airport. Such a plan should include

- Itemization of equipment and personnel necessary, together with location and time required to get to the airport
- Necessary access routes for heavy equipment
- Grid maps of the airport to locate the accident site, access gate, and so on
- Security arrangements
- Arrangements for accepting specialized recovery equipment from airports in the technical pool
- Manufacturers' data on aircraft recovery for aircraft normally using the airport
- Defueling arrangements with resident oil companies
- Logistics of supplying labor and special clothing
- Arrangements for expediting the arrival of the investigator in charge

In addition to heavy lifting equipment and general recovery equipment, which can be covered by advanced agreements with local companies and organizations, specialized lifting equipment such as pneumatic lifting bags and jacks will be necessary for some problems with heavy aircraft (Table 12.10). The International Air Transport Association (IATA) has found it necessary to make such equipment available on a worldwide basis through the International Airlines Technical Pool (IATP). Around the world in widely scattered locations (Sydney, Paris, Mumbai, Tokyo, London, Johannesburg, Chicago, São Paulo, Honolulu, Los Angeles, and New York), four lifting kits are available in case of need. These kits, consisting of pneumatic lifting bags, large extension hydraulic jacks, and tethering equipment, are stored on pallets and are available for immediate shipment accompanied by skilled personnel to any accident location. They can be used by pool-member

Conditions	Typical Methods of Aircraft Recovery
Collapsed nose landing gear	Jacking and use of pneumatic lifting bags; hoisting with cranes and the use of specially designed slings or by pulling down on tail tie-down fitting
Collapsed or retracted main landing Gear, but nose landing gear intact and extended	Jacks, pneumatic lifting bags, or cranes
Collapsed main landing gear, one side only	Jacks, pneumatic lifting bags, or cranes
Collapse of all landing gear	Jacks, pneumatic lifting bags, or cranes
One or more main landing gear off pavement, no aircraft damage	Assuming that the aircraft has the landing gear bogged down in soft soil or mud, extra towing or winching equipment or use of pneumatic lifting bags usually will suffice for this type of recovery. It may be necessary to construct a temporary ramp from timbers, matting, etc.
Nose landing gear failure and one side of main landing gear failure	Jacks, pneumatic lifting bags, or cranes
Tire failure and/or damaged wheels	Jacks and parts replacement

Source: ICAO 1983.

TABLE 12.10   Typical Methods of Recovery of Heavy Aircraft for Various Conditions of Damage

airlines and nonmembers on a fee basis. It is estimated that the kits can be at any accident site within 10 hours and at most locations within 5 hours. The complexity of the recovery operation is exemplified by Figure 12.5, which shows the lifting of a large aircraft with airbags.

12.13   Summary

Disaster-scale accidents at airports seldom occur. Many airport operators will spend their entire careers at airports that will have no accidents where there is a large loss of life, although there will be many emergencies. Safety in aviation, however, requires a continuous alertness to the possibility that a serious accident could occur unexpectedly at any time and will require a rapid and expert response from the airport rescue and firefighting services and from aircraft recovery.

Figure 12.5   Lifting a damaged aircraft with airbags. (*Courtesy of AMS Systems Engineering.*)

The ability to respond promptly can be achieved only through preplanning and training.

Much of what has been covered in this chapter is applicable as much to the United States as to other countries. However, U.S. operators must comply with the FAA Airport Certification Program and should seek to follow Department of Transportation Order 1900.4, "Emergency Planning Guidance for the Use of Transportation Industry." Therefore, the FAA has prepared literature that covers U.S. airports (FAA 2009):

- Scope and arrangement of plans
- Arrangements for mutual assistance and coordination
- Functions including management responsibilities, personnel assignments, and training
- An overall checklist for plan content
- Guidance summaries concerning aircraft accidents and incidents, bomb incidents, structural fires, natural disasters, crowd control and measures to prevent unlawful interference with operations, radiologic incidents, medical services, emergency alarm systems, and control tower functions

The structure of U.S. requirements can best be exemplified by the FAA checklist shown in Table 12.11 (FAA 2009).

Airport Emergency Plan Review Checklist

Inspector: _____ Airport: _____ Date: _____

Incident & Action	Aircraft Incidents and Accidents	Bomb Incidents	Structural Fires	Fuel Farm Fire or Fuel Storage Areas	Natural Disasters	Hazardous Materials/ Dangerous Goods Incidents	Sabotage, Hijacking Incidents	Airfield Power Failure	Water Rescue Situations
A. Plans									
1. Largest air carrier type (c)(1)[7]	X[8]								
2. Identification of response agencies/ personnel	X	X	X	X	X	X	X	X	X
a) Hospitals/Medical Facilities	X								
(1) Name (c)(2)	X								
(2) Location (c)(2)	X								
(3) Telephone # (c)(2)	X								
(4) Emergency capacity (c)(1)	X								
b) Medical personnel (doctors, nurses, etc.)	X								
(1) Business Address (c)(2)	X								
(2) Telephone # (c)(2)	X								
c) Rescue squad, ambulance service, military installation, government agency	X	X	X						X
(1) Name (c)(3)	X	X	X	X	X	X	X		
(2) Location (c)(3)	X	X	X	X	X	X	X		
(3) Telephone # (c)(3)	X	X	X	X	X	X	X		

406

d) Law Enforcement (c)(6)	X	X	X	X	X	X
e) Rescue and firefighting	X	X	X	X	X	X
(1) Name (c)(3)	X	X	X	X	X	X
(2) Location (c)(3)	X	X	X	X	X	X
(3) Telephone # (c)(3)	X	X	X	X	X	X
f) Principle tenants (including air carriers and Control Tower)	X	X	X	X	X	X
(1) Name (c)(3)	X	X	X	X	X	X
(2) Telephone # (c)(3)	X	X	X	X	X	X
g) Equipment inventory	X	X	X	X	X	X
(1) Surface vehicles/aircraft to transport injured/deceased	X	X	X	X	X	X
(2) Hangars/buildings to accommodate uninjured, injured, and deceased (c)(5)	X	X	X	X	X	X
(3) Designated parking area—Bomb/ hijacking Incident		X				X
h) Removal of disabled aircraft agencies (responsibilities or capabilities)	X					
(1) Name (c)(7)	X					
(2) Location (c)(7)	X					
(3) Telephone # (c)(3)	X					
i) Agreements (c)(2,3,6)	X	X	X	X	X	X
j) Communications network (a)(1)	X	X	X	X	X	X

TABLE **12.11** General Checklist for Preparing an Emergency Plan

Airport Emergency Plan Review Checklist

Inspector: _____ Airport: _____ Date: _____

Incident & Action	Aircraft Incidents and Accidents	Bomb Incidents	Structural Fires	Fuel Farm Fire or Fuel Storage Areas	Natural Disasters	Hazardous Materials/ Dangerous Goods Incidents	Sabotage, Hijacking Incidents	Airfield Power Failure	Water Rescue Situations
k) Plan Coordination (g)(1)	X	X	X	X	X	X	X		
l) Plan development (g)(2)	X	X	X	X	X	X	X		
m) Training (airport personnel only) (g)(3)			X	X	X	X	X		
n) Annual review (g)(4)		X	X	X	X	X	X		
o) Triannual full scale exercise (h)	X								
B. Procedures	X	X	X	X	X	X	X		
1. Injured and uninjured accident survivors	X	X	X	X	X	X	X		
a) Marshaling (d)(1)	X								
b) Transportation (d)(1)	X								
c) Care (d)(1)	X								
2. Removal of disabled aircraft (d)(2)	X								
3. Emergency alarm systems (d)(3)	X	X	X	X	X		X		
4. Airport/Control Tower emergency action coordination	X	X	X	X	X	X	X	X	

5. Notification of support agencies	X					
a) Hospitals/Medical facilities (e)	X	X	X	X	X	X
b) Medical personnel (e)	X	X	X	X	X	X
c) Rescue squad, ambulance services, military installation, government agency (e)	X	X	X	X	X	X
d) Crowd control agencies (e)	X	X	X	X	X	X
e) Disabled aircraft removal agencies (e)	X					
6. Water rescue (f)	X					
a) Sufficient water rescue vehicles (f)	X					

REMARKS

[7]Note: The letters and numbers contained in parentheses on the checklist are references to the appropriate requirements found in Part 139.325.

[8]"X" indicates actions related to specific incidents.

TABLE 12.11 General Checklist for Preparing an Emergency Plan (*Continued*)

409

References

Boeing. 2009 (July). *Statistical Summary of Commercial Jet Accidents, Worldwide Operations 1959–2008*, Seattle, WA: Boeing Commercial Airplanes.

Federal Aviation Administration (FAA). 2009. *Airport Emergency Plan* (Advisory Circular AC150/5200-31C). Washington, DC: FAA.

International Civil Aviation Organization (ICAO). 1990. *Airport Services Manual*, Part 1: *Rescue and Fire Fighting*, 3rd ed., including Amendments. Montreal, Canada: ICAO.

International Civil Aviation Organization (ICAO). 1991. *Airport Services Manual*, Part 7: *Airport Emergency Planning*, 2nd ed. Montreal, Canada: ICAO.

International Civil Aviation Organization (ICAO). 2009. *Airport Services Manual*, Part 5: *Removal of Disabled Aircraft*, 4th ed. Montreal, Canada: ICAO.

International Civil Aviation Organization (ICAO). 2010. "Aerodromes," Annex 14, *International Convention on Civil Aviation*, Vol. 1, 5th ed. Montreal, Canada: ICAO.

National Transportation Safety Board (NTSB). 2011. *NTSB Accidents and Accident Rates by NTSB Classification, 1991–2010*, Washington, DC: NTSB.

Paluzek, J. 2009. "Boeing Assistance in Airplane Recovery," *Aero Magazine*, fourth quarter 2009.

United States. 1977. Aviation Safety, Hearings before the U.S. House of Representatives, 95th Congress, July 1977, Washington, DC, U.S. House of Representatives.

CHAPTER 13

Airport Access

13.1 Access as Part of the Airport System

Until a few years ago, it was customary for airport operators to consider that the problem of getting to the airport was chiefly the concern of the urban or regional transportation planner and the surface transport operators. But congestion and difficulties in accessing airports have, as will be seen, very strong implications on their operations. Therefore, the airport administrator has an unavoidable vital interest in the whole area of access and accessibility, perhaps one of the most difficult problem areas to face airport management. An administration might have to watch severe deterioration in its own operations owing to problems outside the limits of the airport itself, conditions over which the airport operator appears to have less and less direct control.

Figure 13.1 is a conceptualized diagram indicating how potential outbound passengers and freight traffic through an airport will be subject to capacity constraints at the various points in the system; a similar chain operates in reverse for inbound traffic. It is important to realize that should any of the potential constraint areas become choke points, throughput is reduced, delay will occur, and there is actual disruption of flow. Lack of access capacity is far from being a hypothetical occurrence. Several of the world's major airports already face severe capacity constraints in the access phase of throughput. Using direct traffic-estimation methods, urban transport planners can show that some of the most severe access problems can occur at airports set in the environment of large metropolitan areas, if these airports depend largely on road access. In fact, three of the world's largest airports, Los Angeles, Chicago O'Hare, and London Heathrow, have for some time displayed severe symptoms of access congestion. In Los Angeles, in the 1970s, before the double decking of the landside circulation road, the environmental capacity of the airport was declared from a determination of the landside access capacity. In 1980, the Los Angeles

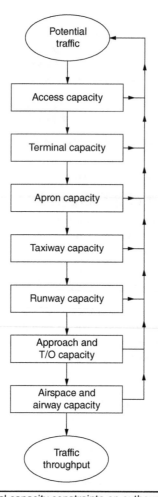

Figure 13.1 Sequential capacity constraints on outbound airport throughput.

Department of Airports proposed that the total number of aircraft operations should be determined as follows:

$$MTAO = \frac{365 \times 0.90 \times RCAP \times PPV}{ASOP \times CHTF \times ANPO} \qquad (13.1)$$

which is a reformulation of

$$AEDT = \frac{MTAO \times ASOP \times ANPO}{365 \times PPV} \qquad (13.2)$$

and

$$CHTF \times AEDT = RCAP \times 0.90 \qquad (13.3)$$

where AEDT = average number of vehicles entering the central ter-
                    minal area in the prior six months
        ANPO = average number of annual passengers per actual air
                    operation in the prior six months
        ASOP = actual number of air operations divided by the pro-
                    posed number for the prior six months
        CHTF = critical-hour traffic factor: the three-hundredth highest
                    hour of vehicular traffic during the prior 12 months
                    divided by the average number of vehicles entering
                    the central terminal area daily
        MTAO = Maximum Takeoff and Approach operations
          PPV = average number of air passengers per inbound vehicle
        RCAP = entering central terminal area roadway capacity in
                    terms of vehicles per hour
          0.90 = constant

This procedure was an attempt to ensure that the scheduled airside
activities would not impose unacceptable loads on a landside access
system. Although not adopted by the airport, in light of the double
decking of the access roads prior to the Los Angeles Olympic Games,
it is possible that similar procedures will have to be adopted by other
airports in the future.

   Chicago O'Hare, which in 2011 generated around half a million
person-trips per day, is principally served by the Kennedy Express-
way, which is one of the world's busiest freeways and is frequently
severely congested. As a consequence, as long ago as 1981, O'Hare
made plans to connect to downtown Chicago by linking to the urban
transit line. Ridership to the airport grew slowly and in 2011 had
reached approximately 21,000 passengers, workers, and visitors daily.
London Heathrow Airport has a severe access problem even though
slightly less than two-thirds of its passengers arrive by public transport or taxi. Access difficulties at the airport itself stem partly from the
configuration of three of its terminals, which are crowded into a small
"island" site between the two main runways and which can be
accessed by road only by a tunnel beneath one of the runways. To limit
the level of traffic into the terminal area, passengers are encouraged
by a pricing mechanism to use remote medium- and long-term parking.
There are also severe restrictions on taxis in the central terminal area.

   At nearly all airports, much of the access system in terms of the highways, the urban bus and rail systems, and taxis is outside the control of
the airport administrator, both financially and operationally. However,
the interface of the access system is very much within the administrator's control. A poor operational interface can discourage travelers from
using what otherwise would be an excellent and (from the airport's
viewpoint) desirable system. On the other hand, by imposing selective
operational constraints on a mode that is becoming less desirable (e.g.,
automobiles in a congested road access situation), significant, if not
large, changes in the traveler's modal choice can be brought about.

13.2   Access Users and Modal Choice

Airport passengers often, but not always, constitute the majority of persons entering or leaving an airport. Excluding individuals making trips as suppliers to the airport, the airport population can be divided into three categories:

- Passengers—originating, destined, transit, and transfer
- Employees—airline, airport, government, concessionaires, and such
- Visitors—greeters, senders, sightseers, and such

All but the transit and transfer passengers make use of the access system. There is no single figure for the division of the airport population among these categories. The split varies considerably among airports and depends on such factors as the size of the airport; the time of day, week, and year; the airport's geographic location; and the type of air service supplied. Large airports with large based-airline fleets have extensive maintenance and engineering facilities. Hong Kong, Atlanta Hartsfield, and London Heathrow airports had between 60,000 and 76,000 employees on site in 2010. Most of these were airline employees. Airports serving international rather than domestic operations tend to attract large numbers of senders and greeters. Similar numbers of visitors are not found at most U.S. airports, which serve mostly domestic and business flights, although there are exceptions, such as JFK, which provides service to the ethnic flights. Table 13.1 lists the spread of breakdowns of airport "populations" that have been found over time by a number of surveys. It can be seen that the range of values is very large. Because of the sources of the data contained in the table and the difficulty in getting homogeneous data, the reliability of one of the few available surveys of this type must be suspect.

Over the past 50 years, a number of superficial solutions have been proposed for the access problem, many of which have involved the use of some dedicated high-speed tracked technology to link the airport with the city center in an effort to reduce the demonstrated dominance of the automobile. These proposals fail to recognize that the reason that the automobile is widely used to access the airport is that except for a handful of very large metropolitan areas with dominant central business districts, air travelers do not for the most part begin or end their journeys in city centers. Table 13.2 lists the percentage of passengers originating from or destined to arrive in the central business district (CBD) of the city for a selection of airports in the United States and the United Kingdom. Concentrating on CBD-oriented access routes can solve only a part of the access problem. Table 13.3 indicates for a few selected U.S. and European airports the

Airport	Passengers	Senders and Greeters	Workers	Visitors
Frankfurt	0.60	0.06	0.29	0.05
Vienna	0.51	0.22	0.19	0.08
Paris	0.62	0.07	0.23	0.08
Amsterdam	0.41	0.23	0.28	0.08
Toronto	0.38	0.54	0.08	Not included
Atlanta	0.39	0.26	0.09	0.26
Los Angeles	0.42	0.46	0.12	Not included
JFK	0.37	0.48	0.15	Not included
Bogota	0.21	0.42	0.36	Negligible
Mexico City	0.35	0.52	0.13	Negligible
Curacao	0.25	0.64	0.08	0.03
Tokyo	0.66	0.11	0.17	0.06
Singapore	0.23	0.61	0.16	Negligible
Melbourne	0.46	0.32	0.14	0.18
U.S. Airports	0.33–0.56	—	0.11–0.16	0.31–0.42 (includes senders and greeters)

Source: Institute or Air Transport Survey 1979.

TABLE 13.1 Proportion of Passengers, Workers, Visitors, and Senders/Greeters at Selected Airports

general popularity of the car or taxi even in the presence of public transport (Coogan 1995; TRB 2000). The convenience of the auto mode is the principal reason for its popularity in countries where there is a high level of car ownership; the auto provides the most effortless connection between the origin or destination of the traveler and the airport unless both trip ends can be served by direct and easily accessible public transport.

In countries where the intercity public transport system (i.e., train and bus) is weak, highway-oriented private modes become essential for those accessing the airport. With the increased emphasis on public transport in urban areas since the 1970s, access patterns are only slowly changing in the United States. At Boston Logan Airport, for example, the combined use of rail and bus modes grew from 16 percent in 1970 to 18 percent by 2000 (TRB 2000). Some European airports, such as Zurich, Amsterdam, Frankfurt, Brussels, and London Gatwick, are connected to the intercity rail network. In theory, all such rail links provide direct connections to all parts of the country, and surface

	Distance from Airport to CBD (miles)	Percent of Passengers Oriented to CBD
United States		
New York (LGA)	6.5	46%
New York (JFK)	11.5	32% (to Manhattan)
Atlanta (ATL)	7.5	7%
Chicago O'Hare (ORD)	15.5	14%
Baltimore/Washington	10.0	14% (to central Baltimore)
Reagan Washington National (DCA)	2.0	33% (to central Washington)
Chicago Midway (MDW)	9.0	20%
Philadelphia (PHL)	6.3	14%
Denver (DEN)	7.5	20% (of nonresident business passengers)
United Kingdom		
London Heathrow (LHR)	16	29%
London Gatwick (LGW)	24	21%
Liverpool (LPL)	6	37%
Manchester (MAN)	8	11%
Glasgow (GLA)	6	28%
Birmingham (BHX)	7	25%
Newcastle (NCL)	6	17%

Source: ACRP and CAA.

TABLE 13.2   Percentages of Airline Passengers with Origin or Destination in the Central Business District

rail should be an attractive alternative to the car as the access mode. In reality, the form of the network can materially affect the efficacy of rail as a viable alternative. At some airports, connections are not good, and rail usage consequently is low. The Gatwick-Victoria link has proved successful, not necessarily for reasons of conventional network connectivity but more because Victoria Station in central London serves as a link to the city-wide Underground system; moreover, connections to other main-line rail stations and the West End are easily made by taxi. Other than Gatwick, Zurich Kloten was perhaps one of the first airports to be truly connected to an intercity rail network; the Swiss authorities originally hoped that eventually between 50 and 60 percent of access trips would be made by rail. In practice, the figure has fallen far short of this goal, attaining only 34 percent by 2000 (TRB 2000). A major problem that must be considered in all questions of access is the degree of coincidence between traffic peaks for urban transport, both road and rail, and airport passenger traffic. This is exemplified in Figure 13.2, which shows the combined arrival and departure passengers at San Francisco Airport and the vehicular traffic variations in the Bay Area. The variations in airport peaks are also related to the diurnal rhythms of urban life in

	Percent Car or Taxi	Public Transport Available
Oslo	37	Yes (bus and rail)
Hong Kong	40	Yes (bus and rail)
Tokyo Narita	40	Yes (bus and rail)
Geneva	55	Yes (bus and rail)
London Heathrow	58	Yes (bus and rail)
Munich	61	Yes (bus and rail)
Zurich	65	Yes (bus and rail)
London Gatwick	65	Yes (bus and rail)
London Stansted	68	Yes (bus and rail)
Paris Charles de Gaulle	69	Yes (bus and rail)
Frankfurt	70	Yes (bus and rail)
Amsterdam	70	Yes (bus and rail)
Brussels	74	Yes (bus and rail)
Paris Orly	77	Yes (bus and rail)
San Francisco	79	Yes (bus and rail via shuttle)
Boston Logan	82	Yes (bus and rail via shuttle)
Washington Reagan	83	Yes (bus and rail)
Los Angeles	87	Yes (bus and rail via shuttle)
Chicago Midway	88	Yes (bus and rail)
Chicago O'Hare	91	Yes (bus and rail)
Atlanta Hartsfield	92	Yes (bus and rail)
New York JFK	92	Yes (bus and rail via shuttle)
Baltimore/Washington	93	Yes (bus and rail)
Philadelphia	93	Yes (bus and rail)
St. Louis	94	Yes (bus and rail)
Clevelend	94	Yes (bus and rail)
Washington Dulles	94	Yes (bus and rail via shuttle)
LaGuardia	95	Yes (bus and rail via shuttle)

Source: ACRP.

TABLE 13.3 Access by Car or Taxi for Selected Airports

that airline schedules are often set to coincide with peak travel generated by the eight-hour work day (see Chapter 2). Consequently, the airport traveler competes with the urban dweller for road space and transit capacity during peak-hour periods. For the passenger using the automobile, taxi, and bus, this means delay through congestion; for those using urban and intercity rail systems, it means possible difficulties in finding seats and handling baggage in crowded facilities.

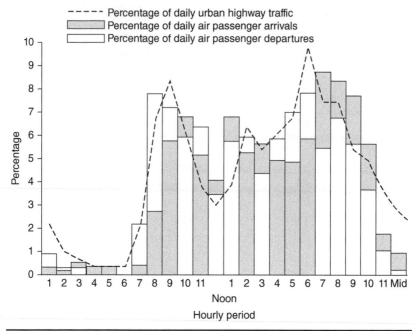

FIGURE **13.2**   Daily highway and air passenger traffic patterns in the San Francisco Bay Area.

Rail has been used successfully in connecting two major new European airports, Munich and Oslo. In both cases, the attracted ridership of the dedicated rail connection has been high. The new East Asian major airports, Shanghai, Hong Kong, Seoul, and Kuala Lumpur, are all well connected into the national rail system. The rail connection of Changi Singapore to the Singapore rail network is also a significant success, but the length of the network is necessarily small in the island republic.

Experiences in Europe of connecting directly into the high-speed rail (HSR) systems have shown varied success even in the same country. In France, while the use of the Trains à Grand Vitesse (TGV) has been very successful at Paris Charles de Gaulle, uptake of the mode at Lyons Airport has been poor. The linkage between Schiphol Amsterdam and Brussels, including Rotterdam and Antwerp (250 ki/h), has significantly expanded the catchment area of Schiphol Amsterdam to the potential passenger traffic of west Holland and Belgium.

13.3   Access Interaction with Passenger Terminal Operation

The method of operation of the passenger terminal and some of the associated problems of terminal operation depend partly on access in as much as this can affect the amount of time that the departing passenger spends in the terminal. Short dwell times in terminals require few facilities. For example, provincial domestic air terminals in Scandinavia often take the place of intercity rail and bus stations. Consequently, they are rightly designed and operated as very functional buildings and are relatively utilitarian facilities using half the space norms of many other European airports because passengers are not expected to spend much time in the terminal. Facilities where longer dwell times are expected must provide a high level of comfort and convenience (e.g., restaurants, bars, cafés, relaxation areas, shopping, post offices, and even barbers). Naturally, more terminal revenue to pay for such facilities can be generated in the longer dwell time. It is the departing passenger who places most demands on the airport terminal system. Departing dwell times depend chiefly on the length of access time, reliability of access time, check-in and security search requirements, airline procedures, and the consequences of missing a flight.

Length of Access Time

It is likely that the amount of time for a particular access journey is a random variable that is normally distributed about its mean value. It is reasonable to assume that the variance of the individual journey time about the mean is in some way proportional to the mean. This is shown conceptually in Figure 13.3a and b, where two access journeys of mean length t_1 and t_2 are shown with respective standard deviations of σ_1 and σ_2. If the access time t_1 is truly normally distributed about the mean, all but a negligible proportion (0.5 percent) is contained between $t_1 \pm 3\sigma_1$. As a result, if all but 0.5 percent of trips are to arrive a standard time K before scheduled time of departure (STD), then the cumulative curve (Figure 13.3c) shows that the average time spent in the terminal is $3\sigma_1 + K$ for the longer access time and $3\sigma_2 + K$ for the shorter access time. Since there is strong evidence that journey times are random, cumulative arrival patterns of the form of Figure 13.3c are observed frequently.

Reliability of Access Trip

The effect of reliability on departing terminal dwell times is shown in Figure 13.4. If there are two access trips each with the same mean trip time of t but with standard deviations of σ_A and σ_B, it can be seen that the mean terminal dwell time, under assumptions of normality and 99.5 percent arrivals by K minutes before STD, are $3\sigma_A + K$ and $3\sigma_B + K$, respectively. The effect on the cumulative curve is demonstrated to be a more gradual slope for low-reliability access. Access routes with very low reliability can result in very long average passenger dwell times in the terminal.

Check-in Procedures

Check-in requirements are not the same for all flights. For many long-distance international flights, check-in times are a minimum of one hour before scheduled time of departure, whereas for domestic and

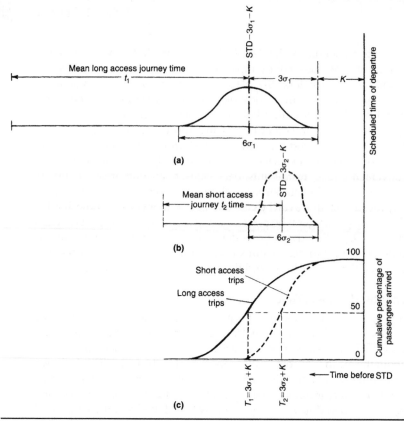

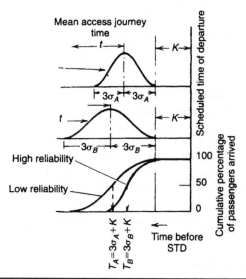

Figure 13.3 Comparison of passenger terminal dwell times for long- and short-access journeys.

Figure 13.4 Effect of reliability of access times on passenger terminal dwell times.

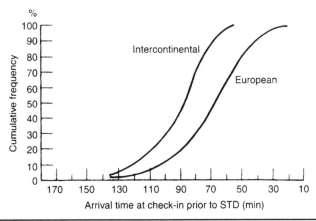

FIGURE 13.5   Effect of flight length on passenger terminal dwell times.

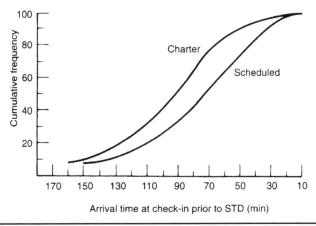

FIGURE 13.6   Effect of flight type on passenger terminal dwell times.

short-haul international flights, this is usually cut to 30 minutes.[1] The effect is a leftward shift of the cumulative arrival curve; an example of this can be seen on the passenger data from Manchester International Airport (Figure 13.5), with long-haul passengers spending an average of 22 more minutes in the terminal than short-haul passengers. Similar differences are often observed between check-in procedures for chartered and scheduled passengers. It is not unusual for a passenger on a charter flight to receive instructions to check in at least 90 minutes before the scheduled time of departure. Figure 13.6

---

[1]For international flights to the United States, Federal Aviation Administration (FAA) procedures in many parts of the world require a reporting time of at least two hours before the scheduled departure time.

shows observed differences in check-in behavior between charter and scheduled passengers at a European airport. Recent security measures have caused some European airlines to close the check-in 45 minutes before departure. The effect of longer closeout times is to increase passenger dwell time in the terminal prior to departure.

Consequences of Missing a Flight

Depending on the type of flight and the type of ticket, the passenger will have a very different attitude toward arriving after the flight has closed out and consequently missing the aircraft. This can be exemplified by considering a hypothetical trip maker making three different flights from Tampa International Airport. The first flight is on a normal scheduled ticket at full fare to Miami; the second is on a normal scheduled full-fare ticket to Buenos Aires; and the third is a special chartered holiday flight to London. The implications of missing the three flights are not at all the same. Should the passenger miss the first flight, there will soon be another flight, and there is no financial loss. In the case of the second flight, the ticket remains valid, but because the connections will now be lost and there might not be an alternative flight rapidly available, there is serious inconvenience and maybe some financial loss. Missing the third flight, however, could cause much inconvenience through a spoiled holiday and serious financial loss because the ticket is no longer valid. The passenger therefore will arrange his or her arrival at the airport in such a way that the risk of missing each flight is different.

Subconsciously, the risk levels might well be set at 1 in 100 for missing the first flight, 1 in 1,000 for missing the second, and 1 in 10,000 for missing the third. Figure 13.7 shows the variation in arrival time that could be expected for an access journey of 60 minutes mean duration with a standard deviation of 25 minutes at these risk levels, assuming for each a closeout time of 20 minutes before STD. It can be seen that the average time the passenger spends in the terminal is 59, 69, and 76 minutes, respectively.

In practice, the arrival patterns at individual airports are a mixture of all these factors. The variation between arrival times can be seen in Figure 13.8, which shows the cumulative arrival curves for four European airports. At the time these data were collected, prior to German reunification, Berlin Templehoff in West Berlin served a relatively small access catchment because the city was entirely surrounded by East Germany. Its access times were reasonably predictable, and most flights were short haul. Paris Charles de Gaulle and Schiphol Amsterdam both served a mixture of short- and long-haul flights, and access times were less predictable. London Heathrow also served short- and long-haul flights, but road access times varied a great deal and were subject to considerable congestion. The cumulative curves are a measure of the impact of these variables on terminal dwell times.

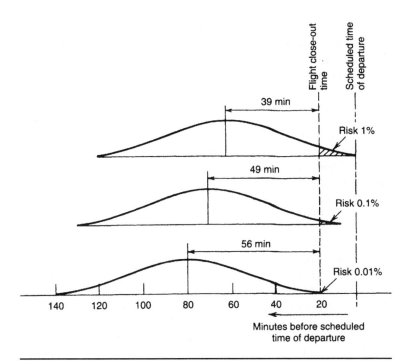

FIGURE 13.7 Effect of the risk of missing a flight on average passenger terminal dwell times.

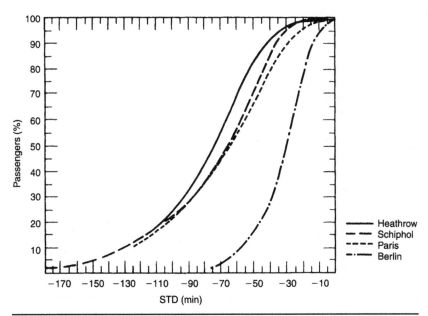

FIGURE 13.8 Check-in times for passengers prior to scheduled time of departure for European flights at four selected airports.

Research examining the effect of the length and reliability of access times has confirmed that unreliable access times can cause congestion in the check-in areas and long dwell times in the departure lounges (Ashford and Taylor 1995). While such dwell times may be favorable to developing commercial income for the airport, they also reduce the terminal's capacity to cope with flow.

The manner in which access is provided to the passenger becomes critical to the operation of airport terminals at many vacation locations. Airports such as Punta Cana in the Dominican Republic and Palma, Majorca, have large landside deliveries of passengers at times that have little to do with the scheduled time of departure of their flights. Vacation hotels, used by passengers on packaged holidays, are required to clear the accommodations of departing guests to make room for the next inrush of incoming guests. As a result, it is not uncommon for departing passengers to be delivered, by fleets of charter buses, to the departures area several hours before the scheduled time of departure, even before the check-in desks and baggage-drop facilities are open for their flights. The departing passengers are unloaded into the departures area, often before the next influx of passengers is delivered to the arrivals area, to be carried away by the same buses. This can cause severe overcrowding of departures facilities, long wait times queuing for processing, and a dramatic lowering of the level of service provided to passengers.

13.4 Access Modes

Automobile

In most developed countries, the private car is the principal method of accessing airports. This has been the case since the inception of commercial air transport, and the situation seems most unlikely to change in the foreseeable future. As a consequence, airports must integrate a substantial parking capability into their design and operation. Large U.S. airports, such as JFK and Chicago O'Hare, have extensive parking areas in locations both close to the terminal and remote.

As airports grow in size, it becomes difficult to provide adequate parking space within reasonable walking distance of the terminals. This is particularly true for largely centralized terminals such as Chicago O'Hare and London Gatwick, less so for decentralized designs such as Dallas, Paris Charles de Gaulle, and Kansas City. In the case of centralized operations, it is common to divide the parking areas into short-term facilities close to the terminal and both medium- and long-term parking areas often served by shuttle services. Normally, the pricing mechanism is a sufficient incentive to ensure that long-term travelers do utilize the remote parking areas. Serious internal circulation congestion can limit the airport's capacity if too many cars

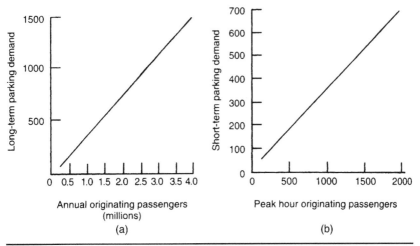

FIGURE 13.9 (a) Long-term parking demand related to annual originating passengers. (b) Short-term parking demand related to peak-hour originating passengers. (*La Magna et al. 1980.*)

attempt to enter the facilities close to the terminal, a condition that has caused problems with the operation of the Terminal 1 at Paris Charles de Gaulle airport, where parking is integral to the terminal, and access is via a tunnel under the apron. At London Heathrow, the constraints on space within the "central area," which contained Terminals 1, 2, and 3 up to 2010,[2] are so severe that short-term parking rates are very high and are set at approximately three times long-term rates to discourage cars from entering this central site. Parking charges at the space-starved London Heathrow Airport are two and a half times the rates at LaGuardia and JFK. Figure 13.9 and Table 13.4 show recommended criteria for providing long- and short-term parking that have been found useful in the United States and Canada (Whitlock and Cleary 1976; Ashford et al. 2011). Although parking requires considerable land area and space, there are substantial profits to be made from its provision. As airports increase in size, the relative importance of the contribution of the parking facilities to overall revenue also increases to about one-fifth of all revenues at the largest airports. At many major airports in the United States, car parking is almost as large a contributor to total revenue as the landing fees from the aircraft.

Major airports relying overwhelmingly on the automobile as the major access mode find that it is not solely in the matter of supplying

---

[2]In a major redevelopment of Heathrow, Terminal 2 was demolished in 2010, and the eastern portion of the old central terminal area was reconstructed. The three old terminals were replaced by Terminals 1 and 2 (see Figure 13.10).

Roads and Transport Association of Canada (smaller airports)	1.5 spaces per peak-hour passenger (short term) 900 to 1,200 spaces per million enplaned passengers (long term)	
FAA (nonhub airports)	1 space per 500 to 700 annual enplaned passengers	

Table 13.4 Parking Requirement Recommendations

	Curbside Access Length	Curbside Access Length per MPPA at Capacity
U.S. airports	—	100
London Heathrow (LHR)	3,100	103
LHR Terminal 1 only	1,500	102
LHR Terminal 2 only	1,000	116
LHR Terminal 3 only	600	59
Schiphol Amsterdam	1,200	66
Frankfurt	900	30

Source: Ashford 1982.

Table 13.5 Curbside Access Length per Million Passengers per Year at Declared Terminal Capacity

and operating car parking that this decision materially affects the operation of the passenger terminal. Use of the space-extensive car mode requires the provision of substantial lengths of curbside space in front of the terminal for dropping off and picking up passengers; de Neufville indicated that for U.S. airports this averages 1 foot per 3,000 annual passengers (de Neufville 1976; de Neufville and Odoni 2003) whereas for European airports the figure is 1 foot per 4,000 annual passengers (Ashford 1982; Ashford et al. 2011). Table 13.5 gives some examples of facility provision in the 1990s. It is common practice at newer large terminals to supply less curbside space and to require drop-off and even pickup within adjacent parking lots.[3] Large-volume terminals must be designed in one of two ways to accommodate this curb space requirement; either they must become long and linear, a form now familiar at Dallas–Fort Worth, Kansas City, and Terminal 2 of Charles de Gaulle Airport, or the space must be provided by separating departures and arrival traffic onto

---

[3]This has the double advantage to the airport of reducing the requirement of curb space and increasing income from parking fees.

separate floors, such as at London Gatwick, Singapore, Amsterdam, and Tampa. The first solution leads to a highly decentralized passenger terminal complex with possible difficulties in interlining, especially for baggage-laden international passengers. The second solution almost certainly will lead to the segregation of departing and arriving passenger flows throughout the terminal building, a desirable feature anyway on security grounds. In the early 1980s, LAX, which previously had operated a one-way, one-level access system, had to undertake a very expensive modernization scheme that involved double-decking the access road and substantial modification of the terminals to permit two-level operation. Only in this way could access and terminal capacity be brought up to airside capacity.

Strict policing of the way passengers are picked up or dropped outside the terminal can have a substantial effect on curb requirements. Table 13.6 shows substantial differences between the loading/unloading times and dwell times at four U.S. airports (La Magna, Mandle, and Whitlock 1980). Because the time need only be on the order of two to three minutes (the time to get out of the car, reenter, and move off), it would appear that vehicles are in fact actually parked for a short time rather than merely waiting. This time variation is emphasized when comparison is made with taxis, which use the curb front for drop-off and pickup only. Efficient utilization of the high-volume curbside access space requires an active presence of some form of traffic policing to keep vehicles moving (La Magna, Mandle, and Whitlock, and Whitlock 1980; TRB 1987).

Taxi

For the air traveler, the taxi is perhaps the ideal method of accessing the airport from all aspects except one—cost. In general, this mode involves the least difficulty with baggage, is highly reliable, operates from a real origin or destination, and provides access directly to the airport curbside. Unfortunately, it can be comparatively expensive, although not necessarily so if hired by a party of travelers or by an individual who otherwise would consider using a personally owned car and is likely to incur high parking charges for an extended parking stay. The airport operator normally has two principal interests with respect to taxi operations at the airport: (1) the balance of supply and demand and (2) the financial arrangements with taxi operators. It is likely that these two matters will warrant simultaneous consideration.

The airport has an interest in maintaining a reasonable balance of supply and demand of taxis at the airport. Taxis must be available at unsociable hours, such as at night when perhaps most other public transport is not operating, and during peak hours of operation, passengers should not have to wait an unreasonable amount of time for a taxi. Equally important, there should not be so many taxis within the airport terminal area that they cause a congestion problem.

	Miami		Denver		New York La Guardia		New York JFK	
	Unloading/ Loading Time (min)	Dwell Time	Unloading/ Loading Time (min)	Dwell Time	Unloading/ Loading Time (min)	Dwell Time	Unloading/ Loading Time (min)	Dwell Time
Departures								
Auto	1.3	3.0	1.0	2.3	0.6	1.2	1.2	2.5
Taxi	1.0	1.8	0.7	1.2	0.5	1.1	0.8	1.3
Arrivals								
Auto	2.8	4.3	2.9	4.2	1.2	2.4	1.6	3.3
Taxi	0.9	N/A	1.0	N/A	0.3	N/A	0.4	N/A

Source: La Magna 1980.

TABLE 13.6 Curbside Access Activity Times for Four U.S. Airports

To achieve these ends, the airport needs control. Many airports do not permit taxis to pick up a fare on airport property without a special license, for which the taxi operator must pay annually. The annual license fee gives this operator the privileged but controlled right to operate at such airports as Ronald Reagan Washington National and Jorge Chavez Lima, Peru. The license fee adds to the airport's income, and the airport operator can ensure that supply and demand are in reasonable balance. Some airports, such as Schiphol Amsterdam, previously did not charge a license fee but awarded and renewed licenses based on the performance of the operator. This system has been replaced by a controlled concession that prequalifies firms according to performance and selects on the basis of requests for proposals. The license can be withdrawn if the operator fails to supply sufficient vehicles or underperforms in any number of ways. In the United Kingdom, it is recent common practice for taxis to incur a charge for both a drop-off and a pickup at an airport. As airports become large, it is not unusual that they suffer from too many cruising taxis, which cause congestion on the terminal access roads. At a number of airports, this is controlled by a scheme that holds taxis in a parking area away from the central area until dispatched by radio communication to the required points in the terminal area.

Limousine

Limousine services, which are reasonably common in the United States and a number of other countries, are either minibuses or large automobiles that provide connection between the airport and a number of designated centers (usually hotels) in the city. The limousine company pays the airport operator in exchange for an exclusive contract to operate a service to provide access according to an agreed-on schedule. The actual form of service varies. In small cities, the limousine usually operates to only one central location; in larger cities, to designated multiple locations. In some very small cities, on paying a supplement to the standard fare, the limousine operates a multiple-origin-destination service, often to the home, and in that becomes very similar to a shared-taxi service.

Operationally, a limousine is similar to a bus, and where bus services are feasible, it is unusual to have limousines as well. Services that have multiple pickup and drop-off points in the urban area have gradually disappeared in most countries, their place being taken by a combination of bus and taxi services. From the airport operator's viewpoint, limousines require very few facilities. Because they use small vehicles, loading and unloading are simple and rapid. They can be carried out at the normal curbside. The only facilities necessary are signs to direct passengers to where they should congregate and wait to be picked up at the airport. For passengers, limousine service is usually relatively inexpensive, yet it gives a level of service that is very similar to that of a taxi, which can be up to five times as expensive. The contracts are lucrative to the limousine operator because passenger load factors are high, and therefore, the concessionary fees that go to the airport operator can be high in comparison with the cost of providing facilities. Because limousines are in fact a form of public transport, they relieve road congestion and the need for parking.

Rail

In the last 20 years, there has been a great deal of activity at large airports to move in the direction of providing more access by rail (TRB 2000, 2002). Airports as widely spread across the globe as Chicago O'Hare, JFK, London Heathrow, Hong Kong, Beijing, Singapore, and Seoul Incheon are just some of the airports that have added rail access routes. The rail access facilities fall into three categories:

Provision of a connection into an existing rail rapid-transit system—for example, Atlanta, Chicago O'Hare, Ronald Reagan Washington National, Paris Charles de Gaulle, and London Heathrow

Direct connection to an existing national intercity rail network—for example, Zurich Kloten, Schiphol Amsterdam, Frankfurt, London Gatwick, and Brussels

Dedicated link from airport to city center location or locations—for example, Munich, Oslo, Beijing, Incheon, and Shanghai

The connection of an airport to an existing urban rail rapid-transit system potentially overcomes a major problem of dedicated airport–city center links (i.e., that most travelers are not destined to the central city). The network provides the opportunity of traveling to many destinations in the urban area. Obviously, if the urban rapid-transit rail network is very limited in size, the attractiveness of the mode is likely to be low. The Heathrow Underground rail connection, completed in the late 1970s, was initially very successful for two reasons: the ease of connection at the airport terminal and the fact that there is direct connection to 250 stations on the London Underground network and easy connections to suburban and intercity rail lines. However, as air passenger traffic continued to increase, the mixing of air passengers encumbered by luggage and urban commuters became a severe problem (except for air passengers with very little luggage); further investment in rail access became necessary. Note also must be taken of the internal accessibility of the rapid-transit network of the urban area. However well the station at the airport end is designed, it is not possible to upgrade an entire rapid-transit network and eliminate stairs and long connecting corridors that are extremely difficult to negotiate with encumbering baggage. It may well be that only a few of the many stations on an existing city rapid-transit system are really accessible to some air passenger travelers.

Conventional rail gives access to the entire conventional rail network. Very few airports are served directly by intercity trains. Frankfurt, Amsterdam, Zurich Kloten, Paris Charles de Gaulle, and London Gatwick are among the best known. The success of a conventional rail line will depend on the level of connection it provides to the rest of the surface transport system. If connections are to a very limited network or a network on which very few trains are operated or to a poor surface bus and taxi system (as formerly with Paris Orly Rail), ridership will be low. The London Gatwick rail link, which carries one-fifth the airport's passengers, is successful because, at its town end, it links well to the London underground, the London transport bus system, and taxis. The line also carries a metropolitan train service, Thameslink, that serves a network of suburban and intercity stations both north and south of London.

Dedicated links, such as those built at the new airports in Munich, Oslo, and Incheon, can provide reasonably high-speed services with reliable access times to the city center. Generally, when assessing the viability of constructing and operating a dedicated link, the considerable cost of either purchasing right-of-way through an existing urban area or extensive tunneling must be taken into account, rendering the project economically unfeasible. By constructing a short spur line from the airport to the intercity rail network, the BAA was able to connect Heathrow to the center of London largely over existing rail right-of-way. British Rail and BAA jointly funded Heathrow Express, which provides direct nonstop conventional rail service to a downtown London mainline terminal. Figure 13.10 shows the layout of the airport rapid-transit station in relation to the three terminals in the central area (Terminals 1, 2, and 3). The more remote Terminals 4 and 5 have their own rapid-transit stops.

If rail service is to be successful for all three rail modes (i.e., urban rapid transit, conventional intercity rail, and dedicated links), it requires a compact connection at the airport end. The introduction of a bus or rail shuttle lowers the perceived level of service and results in a consequent failure to attract riders, as originally occurred, for example, at Paris Orly, Paris Charles de Gaulle, Boston Logan, JFK, and Birmingham.[4]

The access rail system and any system to which it connects must be able to accommodate storing of luggage on the trip. This has proved to be a problem during peak hours on the London Underground and on the Singapore Transit network, which includes Changi Airport.

Access time, provided that it is reasonable for the distance covered, is not extremely important to passengers, so the cost of supplying very high speeds may not be worth striving for. The Shanghai

---

[4]All these airports have subsequently made extensive improvements to the connection between the rail service and the airport terminal.

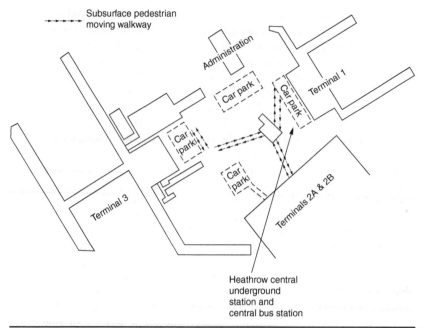

Subsurface pedestrian moving walkway

Administration

Car park

Car park

Terminal 1

Car park

Car park

Terminal 3

Terminals 2A & 2B

Heathrow central underground station and central bus station

FIGURE 13.10 Location of Heathrow Underground station (rapid transit) in relation to Terminals 1, 2, and 3 at London Heathrow Airport in 2012.

Maglev system covers the 19 miles (30 km) to the city center in an astonishing seven minutes, traveling at a top speed of up to 268 mi/h (431 km/h). Looking at other dedicated lines, this compares with 22 minutes for the 31-mile (50-km) trip for Oslo and 15 minutes for the 14-mile (22-km) journey on the Heathrow Express to Heathrow Central. Rapid-transit lines with multiple stops have much longer journey times from the airport to the CBD (e.g. the Chicago blue line, 45 minutes to the CBD, and the Heathrow underground, 50 minutes to Central London).

In both design and operation, the transfer onto the mode at the airport end must be easy. An example of poor design was the Ronald Reagan Washington National Airport rapid-transit station when originally opened. There was a considerable and inconvenient distance from the terminal to the station platforms, making the handling of baggage very difficult. This fault was corrected in the reconstruction of the airport landside during the early 1990s.

Where specialized links between the airport and the city are provided, it is essential that the town end of the line be well sited. Designers of facilities in Oslo, Brussels, Zurich, Munich, and Amsterdam have taken note of these problems and feed air passengers into the main intercity rail stations that are designed for passengers handling baggage and are linked to the urban network of taxis and buses.

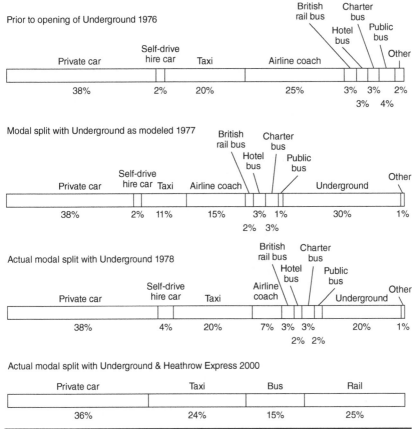

Figure 13.11   Effect of providing Underground (rapid transit) link on access modal split to London Heathrow Airport. (*Sources: London Transport, Civil Aviation Authority, British Airports Authority, TRB.*)

Frequently, the construction of a rail link does not materially increase the use of public transport at an existing airport; it simply results in a transfer from bus to rail. In 1978, the Heathrow Underground link was termed a success, but the traffic that accrued to it came mainly from other public transport modes, mainly airline coach (Figure 13.11). Car and taxi usage was not dramatically affected, actually growing slightly from 60 percent in 1976 to 62 percent in 1978. More than 30 years later, even with the additional provision of Heathrow Express, car and taxi still accounted for 60 percent of air passenger access trips.

Rail links seldom attract large percentages of airport employees. Because of the size of airports, employees' destinations at the airport can be a long way from the passenger terminal; also employees will not necessarily select a residential location that gives a good public transport link to the airport. The new Incheon International Airport is on a vast scale. Built to eventually handle 100 million annual passengers, the perimeter of the airport is 23 miles (37 km). Some workers using the rapid-access system to the passenger terminal find themselves a long distance from their own workplace at the airport. Because the airport operator probably would like to discourage large numbers of airport workers from using their cars for their work trip, it is usually necessary to provide a system of staff shuttles that feed areas remote from the passenger terminal. For example, many Schiphol Amsterdam–based staff arrive at the airport on trains and buses and use the staff shuttles from the passenger terminal to reach their workplaces.

Access journey time does not appear to be critical to air travelers, except in the very shortest hauls with competitive surface modes. The selection of an access mode is much more affected by the ability to cope with inconvenient and heavy baggage and the total cost to the traveling party. A relatively minor flaw in design, such as an inconvenient bus transfer, a long walk, or a bad flight of stairs that makes it difficult to use a mode with reasonable comfort, will result in low modal utilization. Equally, the cost of a shared-taxi ride to five or six fellow travelers may be more or less the same as the combined fares on public transport, but the journey, encumbered with luggage, will be considerably more manageable in a taxi.

Bus

Around the world, virtually all airports carrying reasonable volumes of passengers by scheduled and charter operators are connected by bus to the city center.[5] Normally this is arranged by a contract between the bus operator and the airport authority whereby the bus company usually pays the airport a concessionary fee or percentage for the exclusive right to provide an agreed-on scheduled service. Service is supplied to a number of points in large cities but perhaps to only one point in a small urban area.

At the larger European airports, buses are an extremely important mode of access. Many airports therefore emphasize bus access and supply sophisticated curbside bus bays and bus unloading arrangements, such as those at London Heathrow (Figure 13.12).

Buses become extremely important at airports serving many resorts. For example, in Malaga in Spain and Punta Cana in the Dominican Republic, the overwhelming majority of travelers are vacationers arriving from northern Europe and North America, respectively, who travel in organized groups and are brought to and from the airport by chartered buses. Because most passengers are

---

[5]This statement does not hold for the United States. Virtually all small airports and many medium-sized facilities have no bus transit or other timetabled bus service. Passengers and workers rely on personal automobile, limousine companies, and taxis to get to and from the airport.

Figure 13.12   Bus bays at London Heathrow in 2012.

leisure travelers, and because use of the personal auto is very low even among this group, at "vacation" airports, the landside access is designed almost entirely around the accommodation of bus passengers. Bus loading and unloading areas are designated and must be kept clear of taxis and automobiles. Bus parks are as important as car parks, and the airport operator has an interest in ensuring that the bus parks are kept operational and clear. Figure 13.13 shows a bus park that caters to chartered buses for vacation passengers.

Dedicated Rail Systems

In the area of airport access, nothing has caught the public imagination more than the concept of some form of futuristic high-speed,[6] tracked vehicle that will convey passengers from the airport to the town center unimpeded by surface-road or rail traffic. Many such schemes have been suggested and investigated at sites all over the world.

The realities of the economics of dedicated systems, especially high-speed systems, have been less sanguine. Monorail systems, such

---

[6]High speed in this context means technology capable of speeds of 180 mi/h (300 km/h), which are attained by modem European and Asian conventional trains (e.g., TGV).

Figure 13.13   Remote bus park for Caribbean resort airport. (*Courtesy: Punta Cana International Airport, Dominican Republic.*)

as that linking Tokyo Haneda Airport to central Tokyo, have had a mixed reception, taking many more years than anticipated to achieve financial viability. Studies indicate that dedicated systems are unlikely to reach economic viability at riderships of less than 10 million passengers per year with reasonable right-of-way costs. However, with city-center attractions on the order shown in Table 13.2 and estimated modal splits on the order of those shown in Table 13.3, ridership at the level of 10 million passengers per year is feasible only at the very largest airports. These are normally sited some considerable distance from the CBDs of large cities, requiring long stretches of very expensive right-of-way. Construction costs of surface or elevated dedicated lines in urban areas are very high, but these costs might be small in comparison with the cost of the purchase of central urban right-of-way. Normally, such schemes are possible only if some abandoned right-of-way becomes available, such as in Chicago or Atlanta. In 2004, the Chinese authorities opened a high-speed maglev line from Shanghai Pudong Airport to the Shanghai financial district. Capable of speeds of up to 268 mi/h (431 km/h), the train has come under criticism because it does not satisfactorily link into the existing transit network. Critics stated that the heavily subsidized train was a showpiece gamble to promote the construction of maglev interurban lines between the cities of Shanghai and Beijing and Shanghai and Hangzou. The economics of the project were not disclosed by the Chinese authorities.

High-speed tracked airport-access vehicles on dedicated rights-of-way are unlikely to be built anywhere in the world where the economics of access costs are correctly considered. High-speed links are unnecessary, save little time over trains operating nonstop at conventional speeds, are likely to cost half as much as the remote airport they purport to serve, and can move passengers only to and from the central city, where most travelers probably have no wish to go. Moreover, if they require public subsidy, they raise an ethical question as to whether the air traveler has any right to expect to travel to the urban area at a higher speed than any other traveler. Even so, it is likely that they will continue to receive a disproportionate amount of public and media interest.

13.5   In-Town and Other Off-Airport Terminals

Experience with in-town terminals with check-in facilities has been varied. Originally opened in 1957, when the West London Terminal serving Heathrow was closed, only 10 percent of passengers were using the facility. As well as being uneconomic, it was difficult to have reliable connections between the off-airport terminal and the airport owing to increasing road congestion on the airport access routes. Similarly, the Port Authority of New York closed its West Side Terminal because of poor usage. In contrast, at Victoria Station in central London, a very successful town check-in operation for the rail-served Gatwick Airport has been operated for more than 40 years. As a rule, airlines are not in favor of in-town check-in facilities. They find that the inherent duplication of staff leads to high operational costs that are untenable in the financial climate of the modem airline. There also may be considerable problems with baggage from the viewpoint of aviation security.

Examples of successful in-town airline bus terminals with no check-in facilities are more numerous. For example, in Paris, for many years Air France has had very successful bus terminals at Etoile, Les Invalides, and Montparnasse. For some time, San Francisco and Atlanta airports also had remote parking areas scattered throughout their regions and connected by shuttle minibus service to the airport. This park-ride-fly operation has met with reasonable success. They do not, however, involve baggage check-in.

Because of the availability of online ticketing and online check-in, there has been little recent development in remote check-in by the airlines. Throughout the United States, in areas serving the large airports, several specialist companies have started a remote bag check-in service, where the bag can be checked in at a location remote from the airport. The baggage company transports the bag to the airport and checks it onto the flight, arranging its passage through security onto the aircraft. Unlike the airline remote check-ins, this service entails a fee.

Since use of the in-town bus terminal is part of the modal-choice process, it depends on the factors that affect mode choice enumerated in Section 13.6. It is important to realize, when examining how off-airport terminals operate, that cost and time do not seem to be very sensitive factors in choice of access mode. An otherwise attractive service can perform poorly if the convenience level of the traveler is debased by long walking distances with baggage, frequent changes of level by stairs, crowded vehicles, and inadequate stowage space.

13.6   Factors Affecting Access-Mode Choice

The level of traffic attracted to any access mode is a function of the traveler's perception of three main classes of variables:

- Cost
- Comfort
- Convenience

Decisions in terms of these variables are made not only on the level of service provided by a particular mode but also on the comparative level of service offered by competing access modes. In addition to making an out-of-pocket price comparison, the traveler makes a decision based on the level of comfort and convenience provided by the various modes. The principal considerations are shown in Table 13.7.

Car	Bus	Rail
Ease of loading and unloading	Location of bus terminal	Location of terminal within airport
Distance baggage must be carried and difficulties such as stairs	Speed and reliability of service	Need to use shuttle bus to reach terminal at airport end
Ease of finding long- or short-term parking	Whether specialized express service of part of urban bus network	Difficulty in handling baggage
Access-route congestion and Travel-time reliability	Difficulty in handling baggage	Siting of station or stations in town relative to ultimate destination or to taxis, buses and other train terminals
Shuttle arrangements for long-term car parks	Siting of in-town terminal relative to ultimate destination or to taxis, buses, and trains	
Vulnerability of car to vandalism and theft		

TABLE 13.7   Factors Affecting Model Choice

Attribute	Rank
Ease of baggage handling	1
Convenience of transfer to check-in area	2
Expected access journey time	3
Comfort of mode	4
Parking space availability	5
Convenience of interchanges where more than one vehicle or mode is used	6
Actual journey time	7
Delay and congestion	8
Cost of mode	9
Overall opinion of access	10
Access information	11
Parking cost	12

Source: Ashford 1993.

TABLE 13.8   Ranked Importance of Selected Attributes in Passengers' Choice of Access Mode

In work carried out in the United Kingdom that examined passengers' perceptions of the level of service provided by access to a major airport, it was found that passengers placed the highest importance on such factors as ease of luggage handling, convenience with which the access mode connected to the check-in area, and journey time (Ashford, Ndoh, and Bolland 1993). Delay and congestion, cost of travel, and parking costs were not ranked highly in assessing level of service. Table 13.8 shows the rankings as expressed by travelers at London Heathrow.

Transportation planners have numerous models ranging from the simple to the complex to explain the modal selection procedure. For details of the modeling approach, the reader should consult planning references (Kanafani 1983; Ashford et al. 2011).

13.7   General Conclusions

From the foregoing discussion it can be seen that the access conditions observed at any individual airport are location-specific. Within a large country with widespread access to air transport via many large and small airports, the conditions in many small and medium-sized airports are very similar, and their access provisions consequently are very much alike. In general, however, the access conditions depend on the nature and volume of airport traffic, the location and geographic setting of the airport and the urban areas it serves, and the economic, social, and political structure of the country in which it is situated. The diversity of road- and rail-based access provisions at a number of the world's larger airports (Table 13.9) indicates that with

U.S. Airport	Percentage by Bus	Non-U.S. Airport	Percentage by Bus
Baltimore Washington	6	Oslo	18
Chicago Midway	3	Hong Kong	36
Chicago O'Hare	4	Tokyo Narita	24
		London Heathrow	10
		Paris Orly	17

Source: ACRP.

TABLE 13.9   Examples of Modal Split by Bus for Large U.S. and Non-U.S. Airports

respect to the public transport modes, no generalization is possible. It is possible, however, to discern that road-based access modes are currently vital to the operation of most airports and are likely to remain so in the foreseeable future. Except at a few very large airports in major metropolitan areas where urban traffic congestion is severe, the car and taxi will continue to dominate all other modes of airport access.

References

Ashford, N. J. 1982. *Heathrow Terminal 5 Inquiry* (BA8I). London: British Airways.
Ashford, N. J., and R. Taylor. 1998. "Effect of Variability of Access Conditions on Airport Passenger Terminal Requirements," unpublished report, AAETS, Loughborough University, Loughborough, UK.
Ashford, N. J., S. Mumayiz, and P. H. Wright. 2011. *Airport Engineering: Planning, Design, and Development of 21st Century Airports,* 4th ed. Hoboken, NJ: Wiley.
Ashford, N. J., N. Ndoh, and S. Bolland. 1993. "An Evaluation of Airport Access Level of Service," *Transportation Research Record 1423.* Washington, DC: Transportation Research Board, National Research Council.
Coogan, M. 1995. "Airport Ground Access by Rail at Various International Airports: Report to the FAA," Federal Aviation Administration, Washington, DC.
de Neufville, R. 1976. *Airport Systems Planning.* Boston: MIT Press.
de Neufville, R. and A. Odoni. 2003. *Airport Systems Planning, Design, and Management.* New York: McGraw Hill.
Kanafani, A. 1983. *Transportation Demand Analysis.* New York: McGraw-Hill.
La Magna, F., P. B. Mandle, and E. M. Whitlock. 1980. "Guidelines for Evaluating Characteristics of Airport Landside Vehicular and Pedestrian Traffic," *Transportation Research Record No. 732.* Washington, DC: Transportation Research Board.
Transportation Research Board (TRB). 1987. "Measuring Airport Landside Capacity" (Special Report 215). Washington, DC: TRB, National Research Council.
Transportation Research Board (TRB). 2000. "Improving Public Transportation Approach to Large Airports" (TCRP Report No. 62). Washington, DC: Transit Cooperative Research Program, TRB, National Research Council.
Transportation Research Board (TRB). 2002. "Strategies for Improving Public Transportation Access to Large Airports" (TCRP Report 83). Washington, DC: TRB, National Research Council.
Whitlock, E. M., and E. F. Clearly. 1976. "Planning Ground Transportation Facilities" (Transportation Research Record No. 732). Washington, DC: TRB, National Research Council.

CHAPTER **14**

Operational Administration and Performance

14.1 Strategic Context

Airport operations or *airport logistics* may be defined as the entire series of activities that must take place to process passengers and goods from surface and air transport modes to the aircraft. These activities also may extend to accommodate users and merchandise transiting through the airport to connect to other flights.

Among others, airport operations activities include guiding aircraft for landing, takeoff, and also maneuvering through the runways to parking positions at various sections of an airport; servicing aircraft; clearing international passengers and goods through government inspection services; passenger and luggage check-in; security screening processes; VIP handling; maintenance and upkeep of facilities for safety and convenience; snow removal and deicing (in some parts of the world); provision of ground transportation services; and so on. Operational activities cover the entirety of the physical space of aerodromes as illustrated in Figure 14.1.

Although the integrity of the airport operations "task" is the responsibility of the airport operator, the various processes bring a multitude of players who have mandated jurisdictions to deliver parts of the required services. It is also clear that airports now operate in an increasingly complex business environment with rapid commercialization of airport enterprises, growing capacity constraints, the expanding role of the private sector, new technologies, the consolidation of airlines and the advent of low-fare carriers, the corporate responsibility for promoting a sustainable environment, and in the past decade or so the obvious need to keep safety and security considerations a priority. This context requires greater, broader, and more sophisticated expertise to ensure the successful management of airports.

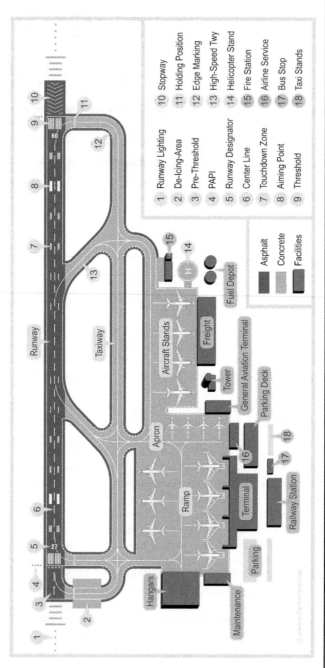

FIGURE 14.1 Schematic airport layout. (*Courtesy of Robert Aehnelt.*)

The backdrop for all this has been a sharp evolution in airport ownership and governance models. An authoritative report published recently (Momberger Airport Information, "Who Owns and Operates Airports," March 2012) has determined the number of companies that own or operate airports worldwide at well over 200. These include long-established entities such as Aéroports de Paris, British Airports Authority (BAA), Fraport AG, Vancouver Airport Services, and others that have entered the field more recently on a relatively large scale, such as GMR from India, Malaysia Airports Holding Berhad, and TAV from Turkey. There is little doubt that full or partial privatization was essentially brought about by the need for large infrastructure investments that governments could no longer afford relative to other pressing societal demands such as health care and education. As a result, the landscape of the governance of aviation-sector enterprises has evolved in a radical manner, especially in the airport sector. Initially, in some cases this change took the form of full privatization, followed by long-term concession agreements such as build-operate-transfer (BOT) schemes and, more recently, a growing spectrum of public-private-partnerships (PPP) arrangements whereby the development and management decisions are made in a cooperative manner with the aim of providing quality infrastructure and services to users while balancing social imperatives and profit motives.

One thing that is sure is that there is no "one size fits all," as evidenced by some state-owned enterprises that run highly successful airports. The interaction between governance models and performance is a complex topic that has been drawing increasing interest and priority for academic and empirical research. Notwithstanding this fact, the adoption of entrepreneurial models clearly has been yielding advanced performance. Although this subject is relevant to the discussion of airport operations as defined earlier, its analysis warrants being treated in another forum.

Irrespective of governance, the operational performance of airports in terms of level of service (LOS) delivered to users is essentially a function of two dimensions: the quality/adequacy of the infrastructure and the effectiveness of the overall logistics management. The matter of optimization of infrastructure falls outside the purposes of this book. Readers interested in that specific topic may find more information in other specialized references, such as *Airport Engineering*, fourth edition, by N. Ashford et al.

This chapter rather focuses on examining the drivers of operational performance that relate or arise from the framework for managing airport logistics. In our view, there is merit in addressing the subject outside airport infrastructure considerations. In fact, there are cases of modern infrastructure suboptimal utilization owing to inefficient operational management, whereas, on the other hand, one also can argue that the negative effects of obsolete or highly congested

infrastructure can be somewhat mitigated by a superior airport logistics management system.

From a strategic perspective, one must recognize that there is a
wide variety of airports in terms of their purpose and mission. For
example, some airports are major hubs with a high proportion of connecting passengers, others exist primarily to provide access to major
tourist destinations, some are located in important political or financial centers, some provide a vital link to remote areas, and others
serve as a base for large global courier companies. Obviously, infra-
structural and operational requirements for each airport will vary
depending on its particularity, although a common goal exists with
respect to the provision of an optimal level of service (i.e., best possible LOS at an appropriate cost).

Recognizing the specificity of each airport in terms of its business
environment and mission is the foundational piece of an *airport strategic business plan* (SBP). The development and implementation of
such a plan are part of the series of best practices and arguably the
most fundamental ones to be implemented as the determining path
to high performance for the airport.

A *strategic business plan* can be defined as

> A comprehensive, action-orienting, top-level corporate plan which
> clearly defines, following a thorough analysis of the business environ-
> ment in which it operates, the specific vision, mission, areas of excel-
> lence and the mission-critical objectives of the enterprise, the means to
> realize them and measure results as well as the financial implications of
> the overall corporate strategy [P. Coutu, "Airport Strategic Business
> Planning Module Course Notes," Aviation MBA, Krems Danube Uni-
> versity, Austria].

As illustrated in Figure 14.2, SBPs drive the formulation and coordination of lower-level functional plans that support the realization
of the overall corporate strategy in the context of predetermined and
airport-specific areas of excellence.

The purpose of each airport would influence its strategic orientation, but when any airport is managed entrepreneurially, the common denominator is the drive to satisfy the customers. Traditionally,
publicly owned and operated airports often were subsidized, and
their focus would be one of providing a "public service," whereas in
a commercialized environment, patterns of influence gradually would
shift from the "owner-sponsor" to the customers who require and
benefit directly from the services.

The scope of the "operations" function of the airport enterprise is
to plan, execute, and monitor the transfer of passengers and goods
through the airport platform in a safe, secure, environmentally
friendly, efficient, cost-effective, and financially sustainable manner for
the benefit of airport users under normal and emergency conditions.

FIGURE 14.2   Airport strategic business plan: Functional-level plans relationship. (*Aviation Strategies International.*)

Achievements are usually measured against key performance indicators (KPIs), as discussed later in this chapter.

Success is achieved through the development of policies, procedures, and processes described in an airport operations higher-level-plan *program* (i.e., an integrated service-delivery plan) and aligned on the overall corporate strategy enunciated in the airport SBP. When driven from a customer perspective, the formulation of an airport operations plan will be inspired by various determinants of service effectiveness as defined, for example, in Figure 14.3.

As stated earlier, one key contemporary challenge that airport operators face in delivering quality service is the presence of many entities that have jurisdiction over specific segments of the airport processing system. In the absence of strong coordination mechanisms and cooperation incentives, this can lead to chaotic and possibly conflicting situations where the duties of airports as landlord could entail significant liability issues that cannot be ignored or assigned to third parties. Another challenge, internal to airport enterprises, is the tendency of functional airport departments to operate somewhat in silos where optimizing each function separately will not necessarily lead to an optimal user experience. This multijurisdiction context, combined with a natural tendency toward narrow/segregated functional management, often increases cost-management inefficiencies, which make it difficult overall to achieve the optimal airport-wide balance

Determinants	Definition	Examples
Reliability	Consistency of performance and dependability	Accuracy of billing Keeping records Performing the service at the designated time
Responsiveness	The willingness or readiness of employees to provide service	Calling the customer back quickly Giving prompt service
Competence	Possession of the required skills and knowledge to perform the service	Knowledge and skill of the contact personnel Knowledge and skill of operational support personnel
Access	Approachability and ease of contact	Reasonable waiting time to receive service Convenient hours of operation
Courtesy	Politeness, respect, consideration, and friendliness of contact personnel	Consideration for the customer's property Clean and neat appearance of the contact personnel
Communication	Keeping customers informed in language that they can understand and listening to them	Explaining the service itself Assuring the customer that a problem will be handled
Credibility	Trustworthiness, believeability, honesty	Company reputation Personal characteristics of the contact personnel
Security	Freedom from danger, risk or doubt	Physical safety Financial security
Understanding the customer	Making the effort to understand	Learning the customer's specific requirements Providing individualized attention
Tangibles	Physical evidence	Physical facilities Appearance of personnel Tools or equipment used to provide the services Physical representations of the service

Figure 14.3   Determinants of service effectiveness. (*D. E. Bowen, R. B. Chase, T. G. Cummings, et al., Service Management Effectiveness. Copyright © 1990. Reproduced with permission of John Wiley & Sons, Inc.*)

between providing a good LOS and the cost of providing it. A vast majority of airports experience problems of this nature. High-performance airports may apply a series of best practices to meet these challenges, some of which are described herein, including leadership frameworks, stakeholder management strategies, and decision-support systems.

14.2   Tactical Approach to Administration of Airport Operations

The implementation of a successful airport operations management program that would yield high-performance results with generally accepted industry benchmarks needs to tackle the challenges described earlier and marshal the efforts of all parties toward an optimal service delivery. From a tactical perspective, the administration of airport operations should be subdivided in two different dimensions handled ideally by two different organizational units:

1. *Dimension one.* Development and monitoring of the airport operations program incorporating LOS policies, procedures, processes, and corresponding allocation of resources.
   → Assigned to airport operations department.

2. *Dimension two.* Ongoing execution of the airport operations program and service-delivery integration, optimization, and reporting.
   → Assigned to airport operations control center (AOCC; see Chapter 16).

The recommended tactical approach advocates separation of the planning/control of the airport operations program, which is an administrative function, from its execution, which is an operational function. These two dimensions require different types of expertise and focus on separate, interconnected, but discrete tasks. They build on different areas of excellence. An analogy for this would be architects with construction managers.

This tactical approach to the delivery of the airport operations program is also predicated on

- The need to call on highly specialized/contemporary expertise to devise policies, programs, and procedures in increasingly complex functional areas (e.g., safety, security, emergency management, retail, operational facilities utilization, maintenance management, and so on)

- The need to deploy airport logistics multidisciplinary duty personnel specialized in the management of operational response to incidents and occurrences under normal and emergency conditions as well as in the coordination/integration of all phases of the processing of passengers and goods through the airport system

- Inefficiencies that would result from operations functional specialists being required to attempt to resolve real-time incidents

- Inefficiencies that would result from having airport operations duty personnel disregard formally promulgated functional policies and procedures other than under special circumstances whereby safety, security, or customer experience otherwise would be compromised beyond preset tolerance levels

Although there is no standardized approach to this, the development of the airport operations program should be the responsibility of the head of operations, usually a vice president or director of operations. At the administrative level, the program should aim at integrating all functional procedures to optimize the airport-user experience while adhering to regulatory requirements (some of which, including those relating to safety, security, and environmental considerations, are discussed in detail in other chapters of this book). Components of the airport operations program might include

- Air traffic services plan
- Airport emergency plan
- Commercial services plan
- Common-use facilities assignment plan
- Environmental management plan
- Ground-handling management plan
- Ground transportation services plan
- Incident/occurrence management and reporting system
- International inspection services plan
- Operational stakeholders engagement plan
- Operational support maintenance management plan
- Public relations and communications plan
- Safety management system
- Security plan
- Terminal operations management plan

It should be noted that some of these plans may not be under the direct control of the airport operator and, as a result, will require consultation with the responsible individual entities. In addition, some of these plans are components of the aerodrome manual required by International Civil Aviation Organization (ICAO) standards and recommended practices (see ICAO Annex 14: *Aerodromes*).

Coordination mechanisms should be implemented for all aspects of operations. Formal means should be deployed in order to engage all airport stakeholders in an integrated effort toward achieving performance. Collaborative frameworks also should be developed and formalized for accountability purposes as well as to take into account the liability of the operator for the overall integrity and performance of the airport with regard to the terms of the airport license delivered by the national civil aviation regulator.

The responsibility for managing day-to-day operations within the framework of the functional guidance provided by the airport operations department through the airport operations program, as well as the plans and procedures that it incorporates, normally rests with the airport operations control center, whose role is discussed in Chapter 16.

14.3   Organizational Considerations

Airport organizational structures have been designed traditionally around key functional areas such as operations, maintenance and engineering, finance and administration, and so on, as illustrated schematically in Figure 14.4.

Because airport entities have been moving toward a more entrepreneurial model, the traditional structure has been increasingly criticized for not being favorable to optimizing commercial results and not being conducive to a customer-centric focus. This has led some airport enterprises to implement a structure based on *strategic business units* (SBUs), which can be defined as "an autonomous division organizational unit, small enough to be flexible and large enough to exercise control over most of the factors affecting its long-term performance" (BusinessDictionary.com; www.businessdictionary.com/definition/strategic-business-unit-SBU.html#ixzz1zV4BjRX9).

In an airport context, SBUs correspond to a series of physical areas of the property such as the airside or terminal sectors for which senior managers are empowered individually to deal with the full scope of the zone of the aerodrome that is assigned to them. This type of structure is illustrated in Figure 14.5.

Arguably, the drive for commercial results associated with this type of structure would indirectly motivate airport management to

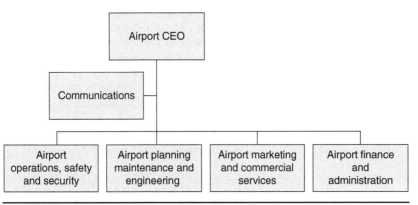

FIGURE 14.4   Traditional airport organizational structure.

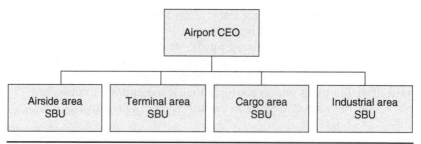

FIGURE **14.5**   SBU-based airport management organizational structure.

be more responsive to its customers by creating a larger series of focal points that are accountable for balancing costs and customer satisfaction. However, it is clear that such structures have limitations relative to their applicability to smaller airports. As a result, a number of airports eventually adopt hybrid organizational designs, that is, a mix between SBU- and functional-type structures.

As we defined earlier, from an airport operations perspective, any structure that is predicated on a predominantly customer-service objective would be welcome (e.g., at Toronto Pearson International Airport, the head of operations carries the title of vice president, operations and *customer experience*). By focusing on customer satisfaction, the business processes normally would be expected to minimize unnecessary bureaucracy. The establishment of effective structures aligning people, processes, and technology with documented customer needs would help to avoid poor coordination, miscommunication, silo mentalities, and low motivation levels. Further research should take place in this field because no model has yet emerged as a commonly recognized best practice.

Many airports have performed organizational impact assessments and implemented stakeholder management plans, but too often there is still a loss of perspective in terms of genuine customer orientation. Innovative organizational structures could be inspired from systematic mapping of site-specific customer experience. According to http:// desonance.wordpress.com/2010/06/16/customer-experience-mapping/, a website that discusses customer service, a *customer experience map* (see example in Figure 14.6) is a "graphical representation of the service journey of customers. It shows the customers' perspective from the beginning, middle, and end as they engage in a service to achieve their goal, showing the range of tangible and quantitative interactions, triggers and touch points, as well as the intangible and qualitative motivations, frustrations and meanings" (http://desonance .wordpress.com/2010/06/16/customer-experience-mapping/; retrieved June 25, 2012).

A number of airports, irrespective of their structure, have innovated in the area of customer service, recognizing it as a strategic lever

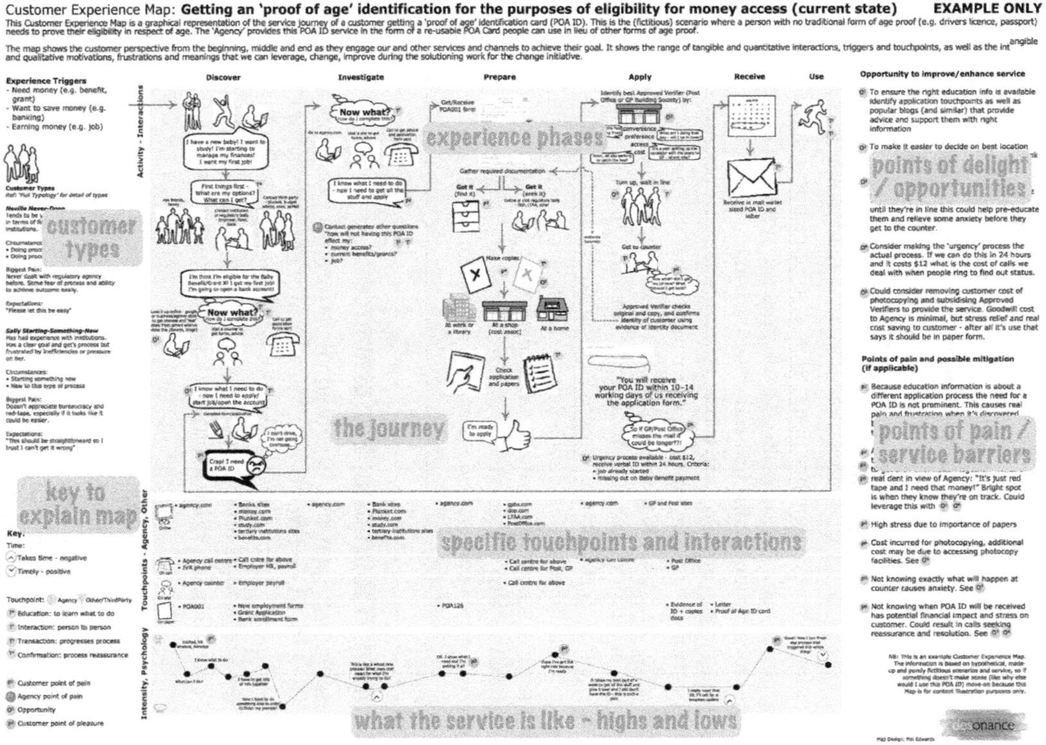

FIGURE 14.6   Sample customer experience map.

in the pursuit of "best in class" operational performance. For example, Auckland Airport has established customer-service representative positions within its airport operations control center (see Chapter 16), and Changi International Airport has implemented a series of customer-centric initiatives such as its "Service Workforce Instant Feedback Transformation (SWIFT)," an interactive system that lets airport patrons tell management what they think of the airport service at various customer touch points so that issues can be identified quickly and corrective action taken in real time.

Another element likely to contribute to improvements in operational performance is the introduction of a performance compensation system for managers and employees. The Greater Toronto Airports Authority (GTAA) has reported that

> A performance-based compensation system also contributed to the improvement of performance results. The compensation system at GTAA takes into consideration individual and collective performance to avoid competition among employees and encourage teamwork. Forty percent of the assessment is based on group performance; therefore, there is an incentive for an individual manager and the whole team to meet the targets and improve organizational performance. Together with improvement in performance, Toronto Pearson experienced an increase in communication efforts among the organizational units when the performance-measurement system was implemented [Transportation Research Board, Airport Cooperative Research Program, *Developing an Airport Performance-Measurement System*, Report 19, April 22, 2011].

14.4   Managing Operational Performance

The successful management of operational performance requires airport operators to act on three fronts. The airport operations program should be *planned*, *executed*, and *controlled*. The main underlying theme is the creation of an enterprise culture that is fundamentally driven by providing a service to airport users as customers.

Planning for Performance

The development of a performance-based operations program must find its foundation in the airport strategic business plan. The enterprise's unique vision, mission, strategic objectives, and areas of excellence must provide the driving force and rationale for the development of operational-level policies, plans, procedures, processes, and performance measures.

Many plans that affect airport users are required under a regulatory framework, for example, those which are part of the aerodrome manual, which is a document that is mandatory for the purposes of airport certification. Others are designed for special purposes, such as the communications plan and the commercial services plan. Even plans that have a technical purpose ultimately exist for the benefit of passengers and other airports users as well as other components of the air transportation system. In essence, what is required is a change of perspective. With a strong strategic impetus to create a positive customer experience, all plans should have a clear statement of their intent and action plan with regard to their contribution to customer satisfaction, the idea being to organize the operational aspects of the airport business *around* the clients or customers.

One effective way to accomplish this goal is to create a *service delivery plan* that is hierarchically positioned immediately below the airport strategic business plan but above all the operational-level plans, focusing squarely on the delivery of an optimal *customer experience* for all users, including passengers, freight forwarders, and also airlines in this context (Figure 14.7).

It is well known that over time airport users view the customer experience in a holistic manner. A bad experience at a security screening point or queuing in a congested food fair may tarnish their perception of an individual airport irrespective of the fact that other steps in the process may have been performed smoothly. This justifies the need for a holistic service-delivery approach of the airport enterprise as the *service provider*.

Typically, a service-delivery plan would include the following elements:

- Statement of purpose
- Description of hierarchical and logical linkage between the airport enterprise vision → mission → strategic objectives → areas of excellence → LOS policy → strategic-level KPIs

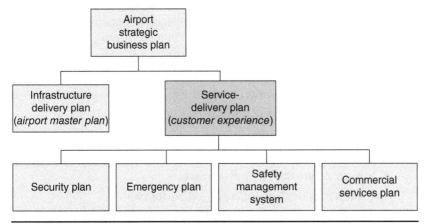

FIGURE 14.7 Service delivery plan—focusing on the customer experience.

- Airport-wide customer experience map
- Real-time service index and KPI dashboards
- Service-delivery, decision-making, and support system
- Customer experience statements for all functional areas (excerpts from functional plans)
- Facilitation committee and service-delivery consultative mechanisms
- Airport/partners LOS agreements
- Passenger charter of rights
- Management and staff customer-service awareness training program
- Customer-service accountability matrix
- Customer experience management enabling technology

Another important lever when planning for high-performance results at an airport is the use of *performance-based contracts* designed for the delivery by third parties of various elements of the airport operations program. This approach typically is very different from traditional contractual arrangements because it

- Emphasizes results related to output, quality, and outcomes rather than how the work is performed
- Has an outcome orientation and clearly defined objectives and time frames
- Uses measurable performance standards and quality-assurance plans
- Provides performance incentives and ties payment to outcomes

It normally also brings about the following benefits (adapted from "Best Practices and Trends in Performance-Based Contracting," FCS Group, study performed for the Office of Financial Management of the State of Washington, December 14, 2005):

- Encourages and promotes contractors to be innovative and find cost-effective ways of delivering services
- Results in better prices and performance
- Maximizes competition and innovation
- Achieves cost savings
- Expects contractors to control costs
- Creates better value and enhanced performance
- Gives the contractor more flexibility in general to achieve the desired results

- Shifts risk to contractors so that they are responsible for achieving the objectives
- Provides incentives to improve contractor performance and ties contractor compensation to achievement
- Allows contractors to have buy-in and shared interests
- Requires less day-to-day monitoring

This method has proven to be very effective in terms of influence on the LOS offered in key airport areas having an impact on customer service.

One example of this is a performance-based contract for mechanical maintenance of escalators, moving sidewalks, elevators, and baggage belts where prequalified companies that meet minimum competency requirements are asked to propose a guaranteed rate of availability of facilities (minimum downtime). The contract then is awarded to the company that proposes the best combination of guaranteed rate of availability and fee. The compensation of the selected contractor is reduced for periods when the set standard is not met, but it is increased if the same standard is exceeded. Performance in this area directly affects passenger convenience.

Another high-impact example is in the field of ground-handling services when the airport operator decides to open the market to third-party service providers. Instead of developing detailed specifications for equipment, operating procedures, and inventories of supplies, bidding contractors, once again following a prequalification phase, are asked to submit a series detailed plans covering safety, equipment, service levels, management, recovery following incidents, and so on in addition to a financial proposal committing to a fixed fee and generally a percentage of gross income. A growing number of airports worldwide can testify to the major impact of performance-based contracts. Results in this area directly affect on-time performance of flights and passenger convenience at various stages of the process.

This being said, notwithstanding all the plans and procedures put in place to favor a customer-oriented approach to airport operations, success will be heavily dependent on the commitment of the airport senior leadership to making this a reality.

Operations Program Execution

The *execution* part of an airport operations program must recognize that airports differ from many other enterprises in a number of aspects that affect their performance:

1. The end product is a service rather than manufactured goods.
2. They deal with a transformation process that is relatively complex and calls on the participation of a large number of stakeholders.

3. They operate in a highly regulated environment (e.g., safety, security, environment, customs, immigration, public health, and so on).

4. They deliver services using technologically sophisticated tools and information systems.

5. They operate in a highly political framework.

6. They operate in an international environment (directly or as a feeder airport).

7. Operation is frequently on a continuous, 24-hour basis.

8. Emergencies can be routinely anticipated at any time.

9. Although they provide ground-based aerodrome-related services for the air traveler or the cargo shipper, the contractual relationship for transport by air lies with the airlines. Also, many terminal services are provided by concessionaires and other third parties.

10. Investment decisions are relatively infrequent and cyclic in nature. The costs involved (e.g., for runways, taxiways, aprons, terminals, and access) are very high, and the results of investment decisions are long lasting.

Still the main purpose of the airport operator is to serve its customers, and in the end, it is the airport users who pass judgment on operational performance, even with respect to regulated aspects of airport activities, which are controlled by outside parties theoretically for their benefit.

It is commonly known that airport user satisfaction correlates highly with matters related to process effectiveness, comfort, freedom from danger, courtesy and helpfulness of staff, and so on. It therefore would follow that "best-in-class" airport operators make real-time management of airport logistics a formally identified *strategic area of excellence*.

The best practice for pursuing excellence in this area is through the implementation of an AOCC concept, as will be described in Chapter 16. In essence, the AOCC is the central nervous system of the airport. Under the supervision of airport duty managers who are fully empowered by senior management and staffed with controllers who are expertly trained in the use of specialized software and telecommunications equipment, the center's primary purpose is to oversee all phases of the airport activities in order to optimize the deployment of resources and to manage operations in real time under normal and emergency conditions. The most important asset of the AOCC is its human capital, comprising specially qualified and trained personnel who are intimately aware of the airport infrastructure and its systems and stakeholders and are especially qualified for anticipating and/or resolving operational incidents competently.

The AOCC personnel must be multidisciplinary and possess a proven ability to work under pressure. Their mandate is to implement the plans and procedures that have been developed by specialized functional departments of the airport administrative team and for them to report on incident and occurrences that will allow for analysis, determination of trends, and taking corrective action by the responsible units or external entities. Much of the work may involve consultations with a number of stakeholders who have jurisdiction over various key elements of the passenger experience.

Operations Program Control

Monitoring the operational performance of an airport can be divided into internal and external assessments. The purpose of the first type of control is to inform the airport management and the board of directors (or its equivalent) of the extent to which the strategic and tactical objectives of the enterprise are being met, and the driver behind the second type of assessment is usually to respond to a regulatory requirement or to benchmark the performance of one airport against another comparable facility for the purpose of competitive analysis or even pure marketing, as in the case of outstanding results.

Most of these assessments call for the use of KPIs or their equivalent, such as key success indicators (KSIs). KPIs can be defined as a series of metrics that an organization uses to measure its achievements against its key strategic objectives in the context of its chosen areas of excellence. Not all metrics should be labeled KPI, and when using this concept, the emphasis should be clearly placed on the notion of what is *key*. Genuine KPIs should be very limited in number because they should be tied directly to the strategic objectives of an airport enterprise as reflected in its strategic business plan. Also, if properly designed, KPIs will in fact result from the integration of many, more detailed factors, considerations, and measures.

W. Wayne Eckerson (*Performance Dashboards: Measuring, Monitoring and Managing Your Business.* Hoboken, NJ: Wiley, 2006, p. 201) developed a list of 12 characteristics of effective KPIs, which are:

"1. *Aligned.* KPIs are always aligned with corporate strategy and objectives.

2. *Owned.* Every KPI is 'owned' by an individual or group on the business side who is accountable for its outcome.

3. *Predictive.* KPIs measure drivers of business value. Thus they are *leading* indicators of performance desired by the organization.

4. *Actionable.* KPIs are populated with timely, actionable data so users can intervene to improve performance before it is too late.

5. *Few in number.* KPIs should focus users on a few high-value tasks, not scatter their attention and energy on too many things.

6. *Easy to understand.* KPIs should be straightforward and easy to understand, not based on complex indexes that users do not know how to influence directly.

7. *Balanced and linked.* KPIs should balance and reinforce each other, not undermine each other and suboptimize processes.

8. *Trigger changes.* The act of measuring a KPI should trigger a chain reaction of positive changes in the organization, especially when it is monitored by the CEO.

9. *Standardized.* KPIs are based on standard definitions, rules, and calculations so they can be integrated across dashboards throughout the organization.

10. *Context-driven.* KPIs put performance in context by applying targets and thresholds to performance so users can gauge their progress over time.

11. *Reinforced with incentives.* Organizations can magnify the impact of KPIs by attaching compensation or incentives to them. However, they should do this cautiously, applying incentives only to well-understood and stable KPIs.

12. *Relevant.* KPIs gradually lose their impact over time, so they must be periodically reviewed and refreshed."

Although all these characteristics provide valuable insight on the matter of performance measurement through KPIs, items 1 ("Aligned"), 2 ("Owned"), 4 ("Actionable") and 5 ("Few in number") are particularly relevant to their use in the airport industry in view of the current observable trends. If anything, we seem to have a plethora of suggested KPIs. One way for airports to deal with this particular issue is to put any KPI being considered through the test of their decision-support systems, namely, "What type of critical decisions (if any) will be better informed by the information provided by this particular KPI?" Obviously, the cost of collecting the data also should be less than the financial or economic gains generated from making the information available.

Internal Assessment

Interesting examples of airport performance metrics may be found in the following five sources:

Doganis, R., *The Airport Business.* London: Routledge, 1992. *Although this textbook has been published more than 20 years ago, many of the performance measures suggested by the author still endure today and have in part inspired the recent publication of the* ACI Guide *on the same topic.*

Airport Cooperative Research Program, "Developing an Airport Performance Measurement System," Report 19, prepared by Infrastructure Management Group et al., Transportation Research Board, Cambridge, MA, 2010. *This report provides step-by-step guidance on how to develop and implement an effective performance-measurement system. It is of interest to airport enterprises desirous of designing an internal performance management system as well as to external stakeholders for benchmarking purposes.*

Airport Cooperative Research Program, "Resource Guide to Airport Performance Indicators," Report 19A, prepared by Jan Blais, Robert Hazel, et al., Transportation Research Board, Cambridge, MA, 2011. *This resource guide provides descriptions of 840 airport performance indicators (APIs) from which airports can select specific APIs to use in benchmarking, an important component of a successful performance-measurement system. The indicators are divided into three categories, namely:*

- *Core APIs: Important for airport overall operation or otherwise important to the airport executive level (CEO and aviation director) and/or the airport's governing board.*

- *Key (Departmental) APIs: Important for the operations of key airport departments or functions (e.g., finance and maintenance).*

- *Other APIs: Not considered as useful for the airport overall operation, to the executive level, nor to key airport departments/functions. However, these APIs may be useful as secondary departmental unit APIs at or below the manager level.*

Airports Council International, "Guide to Airport Performance Measures," prepared by Oliver Wyman, Montreal, 2011. *This guide addresses performance in the context of six key performance areas (KPAs): core, safety and security. service quality, productivity/cost effectiveness, financial/commercial and environmental. Performance indicators (PIs) are listed and described under each of the KPAs. Useful advice and the need for caution are discussed regarding the use of benchmarking.*

The KPIs presented in these references are too numerous to list here. Many of them may be used equally for internal assessments and external industry-type benchmarking. It also should be noted that only a subset of the measures contained in the listed references pertains to airport operations per se, even when incorporating the customer service perspective. Finally, the main challenge for airport enterprises is to use only a limited number of highly relevant, action-orienting performance indicators.

External Assessment

Impact of Airport Industry Commercialization

The need for external assessments of airport operational performance has grown manyfold since the second edition of this book was released in 1996. This is mainly due to a progressive yet important shift in the governance and, in many cases, also in ownership of airports involving the private sector. In many parts of the world, shortages of public funds and pressing needs for social infrastructure such as schools and hospitals have led governments to seek private-sector funding for airports through the outright sale of assets or a variety of public-private-partnership schemes.

This has brought about some concerns regarding airport privatization from stakeholders such as the International Air Transport Association (IATA), which has taken a strong stand regarding the need for robust regulation of airports:

Ten Key Lessons for Successful Airport Privatisation

"1. Customers as key stakeholders should be engaged from the outset and involved on an ongoing and regular basis through agreed processes.
2. A strong focus should be placed on achieving a more efficient management of the airport assets through the transfer to private ownership.
3. Good governance is extremely important if the privatisation is to be in the public interest.
4. Independent, robust economic regulation is essential in order to create incentives for efficiency improvements. Government interference in airport regulation automatically creates an unacceptable conflict of interest.
5. The economic regulator should also be overseen by an independent Competition authority to which airports and their customers have the right to appeal.
6. Economic regulators have, so far, been more effective at extracting efficiencies from existing assets rather than ensuring cost-effectiveness from new investment.
7. Mechanisms to incentivize cost efficiency must be built into the process from the outset. Regulation must avoid preserving monopoly profits or inefficiencies from the start.
8. Service level agreements (or similar systems) must also be put in place to deliver a good-quality as well as a cost-effective service.
9. Controls must be put in place to prevent unjustified asset revaluations or regulatory structural changes that burden airlines and their passengers with substantial charge increases.
10. Customer involvement in new investment is essential to ensure it appropriate, cost-effective and delivered on time and on budget. The 'gold plating' of investment must be avoided."

Airport Privatisation, IATA Economics Briefing No. 1; available at www.iata.org/SiteCollectionDocuments/890600_Airport_Privatisation_Summary_Report.pdf; retrieved June 25, 2012.

On a global scale, safety regulation of airports has been in effect since the signing of the Chicago Convention and the creation of the International Civil Aviation Organization (ICAO) toward the end of World War II (see box "ICAO").

ICAO. In November, 1944, at the Chicago Convention on Civil Aviation attended by 52 nations, a basic framework for civil aviation was agreed upon. The Convention set out this framework in the form of 96 articles that provided for the establishment of international recommended practices. Twenty-six national states ratified the convention (today ICAO has 191 Member States), and on April 4, 1947, the International Civil Aviation Organization, ICAO, came into being, with headquarters in Montreal. ICAO functions with a sovereign body, the Assembly, and a governing body, the Council. One of the main duties of the Council is to adopt international standards and recommended practices for safety, environment and infrastructure (ATC and airports). Once adopted, these are incorporated into the Annexes to the Convention on International Civil Aviation. There are 18 Annexes. Airport administrators will find that Annex 14 "Aerodromes", Annex 9 "Facilitation", Annex 16 "Environmental Protection" and Annex 17 "Security" are of prime importance in the operation of their facilities.

There is no doubt that the commercialization of airports has created some discomfort and doubt about the motivation of the private sector to invest for the purposes of meeting safety, security, and environmental regulations unless it is kept under close scrutiny. There has been a noticeable global increase in audits based on ICAO standards and recommended practices pertaining to safety, security, and environmental matters. ICAO is now mandated to conduct periodical safety and security audits of its Member States.

In addition, ICAO has published economic guidance on matters such as the need for transparency in rate making for the establishment of aeronautical charges. ICAO also has provided guidance on airport performance management by requesting states to ensure that airports have performance management systems in place in key performance areas. This guidance can be found in ICAO's *Policies on Charges for Airports and Air Navigation Services* (Document 9082) and ICAO's *Airports Economics Manual* (Document 9562). Most ICAO Member States have adopted ICAO guidance for safety, security, and environmental considerations that are translated into their national regulatory framework.

Airport Economic Regulatory Oversight

There is no common approach to the way states regulate or monitor the LOS delivered by corporatized airports. In most cases, the service levels are defined in the public-private-partnership agreements, which are not public documents. However, there are cases where it is the competition (or antimonopoly) regulator that acts on service considerations. This has been the case recently in the United Kingdom, where the Competition Commission decided to break up the monopoly of BAA (see box "BAA").

BAA's breakup – timeline and summary of regulatory issues

In 1987, the world's second largest airport operator (after AENA of Spain), the British Airports Authority, was privatized in a stock offering to become BAA plc. This watershed transaction stands in the history of airport ownership and governance as the first real acknowledgment by private markets that airports could be a solid investment. This move consequently led in 2006 to BAA's purchase by a foreign company, subsequent de-listing from the London Stock Exchange and more recently, a partial break-up of the company, forced on it by the UK Competition Commission because of a perceived *lack of competition* (monopoly) and reduction of standards, especially those of customer service.

Prior to the Commission's ruling BAA plc had owned and managed seven airports. The privatization of BAA can be considered an initial success as BAA plc went well beyond its earlier brief and began diversifying revenue streams. BAA was considered one of the most successful airport operators in terms of generating income from non-aeronautical sources, particularly airport retail. The company opened a subsidiary which managed retail operations at several airports in North America. Operating profits steadily increased.

BAA's good financial results ultimately contributed to its unraveling as a company, when a successful hostile takeover occurred in 2006, led by the Spanish property and construction conglomerate Group Ferrovial (56 percent equity), who mounted a successful challenge and bought BAA for GBP£10.3 billion. BAA was de-listed from the Stock Exchange and became BAA Limited, a private limited company with majority foreign ownership.

BAA plc was a unique player in the world of airports. It benefitted from strength in the world's largest aviation market, the greater London area where its three airports served nearly 120 million passengers in 2009. But a number of criticisms had been directed both at BAA plc for its business model and at the British Government for the way it handled the initial privatization.

In brief, the chief comments about BAA plc were that it took on too many ventures outside its core business, eventually spreading itself too thin. Customer service levels were poor after privatization, only becoming respectable when a full scale passenger interviewing regime to obtain feedback was put in place. For the UK government there are two overriding criticisms, namely, the privatization process neglected to give the UK Government a "golden share" in BAA plc. That "share" would essentially be a veto power over a hostile takeover like the one from Ferrovial. Regulators, including the Competition Commission and the CAA, "over-regulated" BAA, creating inefficiencies and high costs for the company, which had to maintain a large department solely for the purposes of responding to Government requests for information.

There is also the case of India, where airport performance regulatory monitoring falls under two categories: (1) airports with annual traffic of 1.5 million passengers or above and (2) airports with fewer than 1.5 million passengers. The former falls under the jurisdiction of the Airport Economic Regulatory Authority (AERA), and the second one falls under the Airport Authority of India (AAI). In both cases, the service standards are laid down by the Ministry of Civil Aviation and monitored by AERA and AAI. The performance of the airports falling under the jurisdiction of AERA is monitored via ACI airport service quality (ASQ) surveys. The airports governed by AAI are monitored using a system that is similar to ASQ, but administered by another provider. Noncompliance to preagreed service levels leads to the imposition of fines.

Industry Benchmarking

With respect to industry benchmarking, the purpose is generally different. One of the motivations of airports is to reach "best-in-class" recognition that enhances the image of award winners, gives them an additional tool to market themselves as a destination, and in the case of global airport operators, provides them with an interesting selling point when pursuing new markets. Benchmarking may be achieved in at least three different ways: (1) An airport can enter a competition where results are determined by passenger surveys [there are two major global and somewhat competing providers of this service, ACI-ASQ (www.airportservicequality.aero/) and SKYTRAX (www. worldairportawards.com/index.htm)], (2) an airport can hire a specialized firm to conduct a benchmarking study on some aspects of its activity, or (3) an airport can participate in confidential surveys where results are made known only to survey participants (ACI Europe runs a confidential benchmarking service with 39 different performance measures; www.aci-europe.org/key-performance-indicators.html).

Various aspects of current industry benchmarking systems are criticized from time to time for weaknesses in their design, their policy, their cultural biases, and as if they truly measure the passenger experience from a holistic perspective. It appears that more research and user consultation are warranted in this field.

14.5 Key Success Factors for High-Performance Airport Operations

The following list provides some of the key success factors required for achieving high performance in airport operations:

- Recognize the strategic implications of the changes in the airport business environment, including the increasing role of the private sector, in the approach to the management of airport operations and the assessment of the related performance.

- Closely align the operations program on the airport strategic business plan, more specifically the vision, mission, strategic objectives, and areas of excellence of the airport enterprise.

- Design and implement the operations policies, plans, procedures, processes, and organizational framework around the needs of airport customers and users.

- Implement service-delivery plans to tackle the management of customer service in a holistic/integrated manner (develop airport-specific customer experience maps).

- Clearly delineate the tasks of airport operations planning/monitoring from those of executing the operations program in the field. Optimize both aspects and their interface.

- Apply proven best practices in the selection key performance indicators (i.e., "Aligned," "Owned," "Actionable," and "Few in number"). Prioritize internal assessments through time over industry benchmarking schemes.

- Implement effective leadership, coordination, and consultation mechanisms for interfacing with entities involved in various phases of service delivery to airport users.

- Create an operational work environment that welcomes continuous-improvement processes and innovation.

- Study the practices of top-ranking airports regarding customer service delivery and how they nurture their "areas of excellence."

CHAPTER **15**

Airport Safety Management Systems

15.1 Safety Management System Framework

Safety in aviation activities has always been one of the overriding considerations of the International Civil Aviation Organization (ICAO). Article 44 of the Chicago Convention of 1944 charges ICAO with ensuring the safe and orderly growth of international civil aviation throughout the world.

But why is safety so important in an aviation context? What sets it apart from other industries? Is it just the mode of transport and the fact that if something goes wrong, it can go very wrong? One could respond that it is more seemingly a matter of confidence if we look at the significant downturn in public travel for some time after the tragic September 11, 2001, terrorist attacks on the United States. Since then, security measures have increased at an exponential rate—all for the sake of and in the name of safety—and confidence eventually was largely restored.

If we took a sample population of the traveling public, we would note that the direct and indirect implications of a safety incident can be extreme, whether the mode of transport is by road, by rail, or by air. Any single loss of life is one too many, but the intensity of that loss (or consequence) can be more extreme depending on the mode of transport. Certain rules apply in all these specific means, with set standards to ensure safety that one would hope are followed by a range of stakeholders.

There is also a strong interrelationship among such factors as cost, benefit, risk, and opportunity, particularly in relation to aerodrome certification. Some 44 percent of the world's airports still do not have certification. Why is this? Is it because airports in some areas of the world have evolved more quickly over time in terms of specific planning and location criteria? It is important to note, however, that an airport could be certified despite noted deviations from the standards with the view that it would comply at some point—although, in reality, in some cases the airport may never comply because of its physical location and environment (e.g., mountainous terrain). [*Note: The Manual on Aerodrome Certification* (ICAO Document 9774) covers matters of this nature.]

In simpler terms, such an airport may make a decision (together with its state) not to proceed with aerodrome certification based on costs versus benefits as they would apply to the identified residual risk to be managed. Airlines, of course, may choose not to operate from such airports, but generally, they still agree to do so based on their calculation and acceptance of the risk together with their own business rationale. The International Civil Aviation ICAO is well aware of the issue of risk management in its attempt to support the Member States and airports throughout the aviation community. In particular, ICAO recognizes the potentially catastrophic nature of aviation accidents (in flight and from runway incursions and excursions) that has led Member States to focus on flight operations and on ground operations in the maneuvering area.

Aircraft are easy to damage but expensive to repair and any delay could result in heavy indirect costs, whether they may be schedule disruptions or passenger inconvenience. Even slight damage to an aircraft, if gone unreported, could be the cause of a subsequent in-flight emergency.

Regulatory Framework

From a regulatory perspective, the role of ICAO is to provide procedures and guidance for the safe conduct of international aircraft operations and to foster the planning and development of air transport. This is achieved largely by developing standards and recommended practices (SARPs), which are contained in Annexes to the Chicago Convention and reflect the best operational experience of Member States.

Airport operations consist of the activities necessary to expedite air traffic, including the movement of aircraft, passengers, baggage, cargo, and mail. For airport operations to be successful, these activities need to be *safe, secure, efficient,* and *environmentally sustainable.*

Importantly, Member States have the prerogative of determining the structure of civil aviation within their borders. They also can establish different degrees of Member State control in airport ownership and management arrangements. Those arrangements invariably affect the scope of services assigned to the airport operator.

ICAO Standards and Recommended Practices

ICAO defines *standard* in this way:

> Any specification for physical characteristics, material, performance, personnel or procedure, the uniform application of which is recognized as necessary for the safety or regularity of international air navigation and to which Contracting States *will conform* in accordance with the Convention; in the event of impossibility of compliance, notification to the Council is compulsory under Article 38.

Standards therefore are specifications that are *necessary* for international air navigation, and in the case of deviations, it is *compulsory* to notify other contracting states.

ICAO defines *recommended practices* in this way:

> Any specification for physical characteristics, configuration, material, performance, personnel or procedure, the uniform application of which is recognized as desirable in the interest of safety, regularity or efficiency of international air navigation, and to which Contracting States will endeavor to conform in accordance with the Convention.

Recommended practices, therefore, are considered *desirable* specifications to which Member States will *endeavor* to conform.

ICAO Annex 14 Aerodromes (Volume 1: *Aerodrome Design and Operations*)

This document sets forth the minimum specifications for aerodromes based on the characteristics of the aircraft that operate there, current or future (aircraft are categorized by codes based on wing span, main gearwheel span, and length). Thus the SARPs applicable to an aerodrome with smaller aircraft may differ to some degree from those for an aerodrome with larger aircraft, but both aerodromes will be covered in Annex 14.

At the outset in Chapter 1, Annex 14 establishes the requirements for Member States to certify aerodromes for operation against the applicable SARPs. An aerodrome certificate must be issued by the appropriate authority under applicable regulation for the operation of an aerodrome. Annex 14 also establishes that aerodromes must have a safety management system (SMS), which must be documented in the airport operations manual (see Annex 14, Chapter 15: "The Airport Operations Manual").

In Chapter 2, Annex 14 addresses the standards for *reporting aeronautical and aerodrome data*, ranging from the geographic location to the service levels for emergency services. The reporting of these data is carefully standardized and presented in the *Aeronautical Information Publication* (AIP) so that even aircraft operators unfamiliar with a particular aerodrome will have the essential information for safe aerodrome operations there.

Chapters 3 through 8 of Annex 14 present *technical details* for aerodrome layouts, restriction and elimination of obstacles, visual navigation aids, obstacle marking, visual aids, and electrical systems.

Chapter 9 addresses *airport services,* including emergency services, disabled-aircraft removal, reduction of bird risk, apron administration, refueling operations, vehicle operation, surface guidance systems for air traffic, and others. Chapter 10 offers *general guidance on aerodrome maintenance programs.*

ICAO's Stance on the Implementation of a Member State's Safety Program

The ICAO safety management SARPs are included in Annex 1: *Personnel Licensing*; Annex 6: *Operation of Aircraft*; Annex 8: *Airworthiness of Aircraft*; Annex 11: *Air Traffic Services*; Annex 13: *Aircraft Accident and Incident Investigation*; and Annex 14: *Aerodromes.*

Annexes 1, 6, 8, 11, 13, and 14 include the requirement for Member States to establish a *state safety program* (SSP) in order to achieve an acceptable level of safety in civil aviation. An SSP is a management system for the management of safety by the Member State.

A *State Safety Program (SSP)* is defined as an integrated set of regulations and activities aimed at improving safety. It includes specific safety activities that must be performed by the Member State as well as regulations and directives promulgated by the Member State to support the fulfillment of its responsibilities concerning safe and efficient delivery of aviation activities in the Member State.

Clearly, there is a need to understand the relationship between an SSP and the service provider's SMS. First, the SSP.

The introduction of requirements regarding an SSP is a consequence of the growing awareness that safety management principles affect most activities of a civil aviation authority, including safety rule making, policy development, and oversight. Under an SSP, safety rule making is based on comprehensive analyses of the Member State's aviation system; safety policies are developed based on hazard identification and safety risk management; and safety oversight is focused toward the areas of significant safety concerns or higher safety risks. An SSP thus provides the means to combine prescriptive and per States.

It is further envisioned that Member States developing internal resources will cooperate to assist other Member States in the implementation of their SSPs and the development of safety data management capabilities, thus achieving the synergistic partnership recognized as necessary for the global implementation of safety management practices.

15.2   Safety Management Systems and Aerodromes

Introduction of SMSs to Aerodromes

SMSs are not new and are found in numerous industries that manage risk as an integral part of their operations, such as the chemical, nuclear, manufacturing, and construction industries. Historically, SMSs were geared mainly toward occupational health and safety (OH&S) rather than the specific application of aviation safety as they exist at aerodromes. At many airports, a more holistic approach is now taken to cover not only the mandatory aviation safety matters but also OH&S. The application of SMSs to aerodromes is more complex than for other industries and environments given the fact that airports host a large community of employers. Traditional SMS models are based on one employer who essentially has direct control over all activities within the one workplace. They also may extend to include an employer's contractors, but do not cut across other employers.

Indeed, *all aerodrome service providers* need to have an organized approach to managing safety, including the necessary organizational structures, accountabilities, policies, and procedures. ICAO refers to the approach and support structure identified above as a *Safety Management System* (SMS), the implementation of which has been a mandatory requirement for aerodrome certification since November 2005.

At a minimum, according to ICAO, a SMS should achieve the following:

- Define lines of safety accountability, including direct accountability on the part of senior management
- Implement and extend the Member State's safety program at the aerodrome, including definition of appropriate safety objectives and procedures for incident reporting, safety investigations, safety audits, and safety promotion
- Identify safety hazards
- Ensure that remedial actions necessary to mitigate the risks/hazards are implemented
- Provide for continuous monitoring and regular assessment of the safety level achieved

The safety management SARPs are aimed at two audience groups: Member States and service providers. In this context, the term *service provider* refers to any organization providing aviation services. The term thus encompasses approved training organizations that are exposed to safety risks during the provision of their services, aircraft operators, approved maintenance organizations, organizations responsible for type design and/or manufacture of aircraft, air traffic service providers, and certified aerodromes, as applicable.

The ICAO safety management SARPs address three distinct requirements:

1. Requirements regarding the SSP, including the Acceptable Level of Safety (ALoS) of an SSP

2. Requirements regarding safety management systems (SMSs), including the safety performance of an SMS

3. Requirements regarding management accountability vis-à-vis the management of safety during the provision of services

The ICAO safety management SARPs introduce the notion of an ALoS as the way of expressing the minimum degree of safety that has been established by the Member State that must be assured by an SSP, as well as the notion of safety performance as the way of measuring the safety performance of a service provider and its SMS.

In establishing a Member State's requirements for the management of safety, ICAO differentiates between safety programs and SMSs. Specifically:

1. A *safety program* is an integrated set of regulations and activities aimed at improving safety.

2. A *safety management system* is an organized approach to managing safety, including the necessary organizational structures, accountabilities, policies, and procedures.

The ICAO's SARPs require that Member States establish a *safety program* to achieve an acceptable level of safety in aviation operations. The acceptable level of safety shall be established by the Member State(s) concerned. In effect, Member States within regions could differ on the specifics of an ALoS. The airport operator, aircraft operator, maintenance organization, and air traffic service provider also must be consulted in relation to the establishment of an ALoS, and in some cases, airports within Member States may have differing views.

A safety program will be broad in scope, including many safety activities aimed at fulfilling the program's objectives. A Member State's safety program embraces those regulations and directives for the services, aerodromes, and aircraft maintenance. The safety program may include provisions for such diverse activities as incident reporting, safety investigations, safety audits, safety promotion, and so on. To implement such safety activities in an integrated manner requires a coherent SMS.

A clear understanding of the relationship between an SSP and an SMS is essential for concerted safety management action within Member States. This relationship can be expressed in the simplest terms as follows: States are responsible for developing and establishing an SSP; service providers are responsible for developing and establishing an SMS. Therefore, in accordance with Annex 14 provisions, Member States shall require that individual aircraft operators, maintenance organizations, air traffic service providers, and certified aerodrome operators implement *safety management systems* approved by the Member State. Figure 15.1 illustrates some of these interdependencies.

The concept of an SMS may be broken up into the following components:

- The term *safety* is used to mean the condition where risks are managed at acceptable levels.

- The term *management* may be defined as the allocation of resources.

- The term *system* refers to an organized set of things that interact to form a whole that is required for the delivery of goods or services.

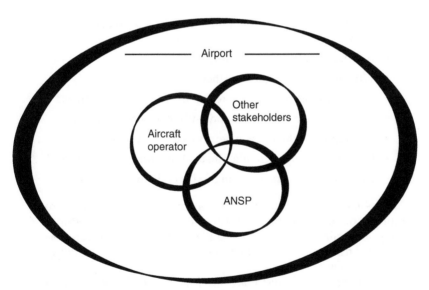

[Note: Whilst Aircraft Operators, ANSP's, and other stakeholders may have their own SMS's, the ultimate aim is for them to operate independently as required but 'dovetail' into, and be truly integrated with the Airport SMS].

Figure 15.1 Airport safety management system model.

One could say that an SMS is an organized set of interrelated processes to allocate resources in order to achieve the condition where risks are managed at acceptable levels. At a minimum, such SMSs shall

- Identify actual and potential safety hazards
- Ensure that remedial action necessary to maintain an acceptable level of safety is implemented
- Provide for continuous monitoring and regular assessment of the safety level achieved

Assessment of the Current Safety Level (Where Are We At Now?)

In order to assess the current level of safety, there is a need to undertake a thorough gap-analysis exercise. Obviously, some systems and processes are already in existence, as is the recording of incidents and other related data. Some simple steps to take include the following:

- What systems and processes are in existence now?
- Have all the hazards relative to operations been identified?
- Of the hazards identified, have the risks been assessed for each?
- Are there controls in place for the risks?
- What residual risk gaps are there?
- How does this relate to the rest of the airport's operations?
- Is there some common ground with other stakeholders?
- What impacts do others have on us and us on them?
- How do we help each other instead of working in isolation?
- How do we integrate existing SMSs and any new ones?

Assessments of the current position may prompt upgrades to equipment or infrastructure or improvements to policies, procedures, staffing, and training or even the overall organizational structure. *The SMS should reflect the current situation and then be adjusted accordingly when things change.*

An SMS goes beyond just being a tool with which to manage safety compliance. It also should reflect what is being done proactively to manage safety and reduce risks.

An organization's SMS, approved by the Member State, also should clearly define lines of safety accountability, including a direct accountability for safety on the part of senior management. ICAO has produced a manual on safety management (ICAO Document 9859:

Safety Management Manual). This manual includes a conceptual framework for managing safety and establishing an SMS as well as some of the systemic processes and activities used to meet the objectives of the Member State's safety program.

- *Oversight* should include *audits* of *each* certified service provider's SMS. Who should do this? The Civil Aviation Authority (CAA), internal, independent, or all three?

- The effectiveness of Member State safety programs, in turn, is audited periodically by ICAO through the *Universal Safety Oversight Audit Program* (USOAP) which is soon to be complemented by the implementation of a formal Continuous Monitoring Approach (CMA). This also extends to ICAO auditing of an airport within the Member State.

- ICAO Annex 14 establishes that Member States require that the aerodrome manual submitted for aerodrome certification contains details of the SMS. To reinforce the link between certification and the SMS, ICAO Document 9774 states that "suspension of an aerodrome certificate may be considered if an aerodrome operator's SMS is found to be inadequate."

Note: At the time of writing, ICAO was in the process of developing a new edition of the *Safety Management Manual* and a new Annex (Annex 19). The new Annex will collect in one document all the safety management requirements now spread across various guidance reference publications. ICAO is also undertaking the transformation of its safety oversight inspection system into a CMA practice.

Acceptable Level of Safety

In any system, it is necessary to set and measure performance outcomes in order to determine whether the system is operating in accordance with expectations and to identify where action may be required to enhance performance levels to meet those expectations. The concept of an ALoS is, in effect, an agreement on the safety performance that service providers should achieve while conducting their core business. In determining an ALoS, it is necessary to consider such factors as the level of risk that applies, the cost/benefits of improvements to the system, and public expectations on the safety of the aviation industry. The ALoS therefore becomes the reference against which the oversight authority (e.g., CAA), the aviation industry, and the public can determine the safety performance of the aviation system.

In practice, the concept of an ALoS is expressed in two measures, or metrics—safety performance indicators and safety performance targets—and implemented through various safety requirements. The following explains the use of these terms:

Safety performance indicators are a measure of the safety performance of an aviation organization, components of a Member State safety program, or an operator/service provider's SMS. Safety indicators therefore will differ between different segments of the aviation industry, such as aircraft operators, aerodrome operators, and air traffic service providers.

Safety performance targets (sometimes referred to as *goals* or *objectives*) are determined by considering what safety performance levels are desirable and realistic for individual operators/service providers. Safety targets should be measurable, acceptable to stakeholders, and consistent with the Member State safety program.

Safety requirements are needed to achieve the safety performance targets and safety performance indicators. They include the operational procedures, technology, systems, or program to which measures of reliability, availability, performance, and/or accuracy can be applied.

A range of different safety performance indicators and targets will provide a better insight into the ALoS of an organization or sector of the industry than the use of a single indicator or target.

The relationship between an ALoS, safety performance indicators, safety performance targets, and safety requirements is as follows: The ALoS is the overarching concept. Safety performance indicators and safety performance targets are the measures, or metrics, used to determine if the ALoS has been achieved. Safety requirements are the means to achieve the safety targets and safety indicators.

There is seldom a national ALoS because each agreed-on ALoS should be commensurate with the complexity of the individual operator/service provider's operational context. Establishing an ALoS for its safety program does not relieve a Member State from compliance with ICAO standards and recommended practices. Likewise, establishing an ALoS for its SMS does not relieve an operator or service provider from compliance with applicable standards and recommended practices and/or national regulations and requirements.

Typical examples of safety indicators in the aviation system include, among others:

1. Fatal airline accidents
2. Serious incidents
3. Runway-excursion events

4. Ground-collision events

5. Development/absence of primary aviation legislation

6. Development/absence of operating regulations

7. Level of regulatory compliance

Typical examples of safety targets in the aviation system include, among others:

1. Reduction in fatal airline accidents

2. Reduction in serious incidents

3. Reduction in runway-excursion events

4. Reduction in ground-collision events

5. The number of inspections completed quarterly

Safe and efficient aviation requires significant infrastructure and aeronautical services, including airports, navigation aids, air traffic management, meteorologic services, flight information services, and so on. Some Member States own and operate their own air navigation services and major airports; others own and operate their own national airline. However, many Member States have corporatized these operations, operating under the oversight of the State. Regardless of the approach taken, Member States must ensure that the infrastructure and services in support of aviation are maintained to meet international obligations and the needs of the Member States.

Where the regulatory function and the provision of particular services are both under the direct control of the one Member State body (such as the CAA), a clear distinction must be maintained between these two functions, namely, the service provider and the regulator.

15.3 SMS Manual

Overview

Guidelines for the development of an SMS Manual itself are provided in ICAO Document 9859. This provides the corresponding theory behind successful safety management programs and systems, as well as a wealth of support material, including useful checklists for risk management.

First, there should be a *systematic approach to safety*. Second, safety should be managed and controlled with *proactive management*. Third, there should be a *structured organization with defined responsibilities*. Fourth, there should be *procedures*. Fifth, there should be a *safety policy*. Sixth, the SMS's ultimate *objective should be the safe operation of the aerodrome*.

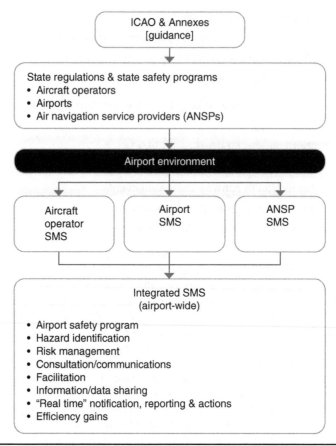

FIGURE 15.2 Airport SMS regulatory context.

The aerodrome operator has the obligation of ensuring that the aerodrome organization, facilities, equipment, and systems are designed and operated to control hazards and keep risk at an acceptable level. For example, in apron operations, most day-to-day services generally are not provided by the aerodrome operator. An effective SMS should ensure that the level of safety of the aerodrome is not degraded by the *activities, equipment, and supplies* provided by external organizations.

ICAO Document 9774 requires that all parties operating on the aerodrome should comply with the aerodrome safety requirements and that they should participate in the SMS (see Figure 15.2). In other words, the SMS is applicable to all levels and fields of aviation, including to a subcontractor at the aerodrome, to third parties operating at the aerodrome, to the aerodrome operator itself, and to other service providers.

The Key Elements of an SMS Manual

Policy, Organization, Strategy, and Planning

- Declare and promulgate a safety policy and a safety management process.
- Organize the SMS structure, including staffing and individual/group responsibilities.
- Conduct SMS strategy and planning, including setting performance targets and allocated resources.

Risk Management

- Develop a framework for risk analysis and control, including establishing an ALoS.
- Identify all safety-related procedures, facilities, and critical safety areas, including identifying hazards and determining risks.
- Develop mitigation measures to be implemented for risks that are higher than acceptable levels.

Safety Assurance

- Enforce safety requirements.
- Provide for continuous safety monitoring.
- Inspect safety-related facilities and document results.
- Process accidents, incidents, complaints, defects, faults, discrepancies, and failures.
- Conduct internal safety audits of the SMS itself.

Safety Promotion

- Create a positive safety culture.
- Communicate safety messages effectively, including reporting.
- Ensure adequate staff training and competency.

The effectiveness of an SMS is best achieved through "proper" acceptance, coordination, and implementation throughout the aerodrome community and the continual promotion and training on the SMS required to sustain it.

At this point, one would turn to each of these major areas and examine best practices as they are currently found in the field and the relevant literature.

Policy, Organization, Strategy, and Planning

- Implementing an SMS on top of the numerous individual safety-related programs that are already likely to be functioning at the airport is not easy and requires very strong commitment by management.
- Management needs to take an active interest in safety and *assume accountability* for the safe conduct of operations within its control. This commitment should be expressed in the organization's safety policy.
- Specific safety objectives (i.e., concrete goals with measurable indicators) should be proposed to "operationalize"' the safety policy at the task level. Management has the responsibility to review the indicators, evaluate the performance of the system, and decide on ways to improve.

Safety Policy

A safety policy is the highest expression of an organization's commitment to safety. The safety policy should be developed by *management and staff* and *signed by the CEO*. It is a core value of the organization, equal in importance to other policies. This policy

- States the safety goal, which, in turn, is consistent with the objectives and operating efficiency of the organization
- Is relevant to the industry—compliance with required safety standards is included
- Is applicable to all employees in the workplace, including those of other organizations where applicable
- States the responsibilities and accountabilities for directors, managers, and employees
- Provides direction for implementing the policy

Once developed, the safety policy must be communicated effectively to all staff and then reviewed periodically to ensure that it maintains its currency.

Safety Organization

- Safety is the responsibility of all employees. It is critical that job descriptions include specific safety responsibilities because the main responsibilities for safety will and must always remain in the line organization.
- The SMS is a tool with which to manage safety and nothing more. The SMS requires an explicit organization in addition to the traditional line organization to function effectively

as a system. This includes a management representative (the safety manager), an effective committee structure, and an audit/analysis capability.

- The organizational structure facilitates (1) lines of communication between the safety manager and the CEO and with the line managers, (2) a clear definition of authorities, accountabilities, and responsibilities, thereby avoiding misunderstandings, overlap, and conflict (e.g., between the safety manager and line management), and (3) hazard identification and safety oversight.

- The line organization must be supported by a top-level safety committee.

- The aerodrome also needs an operational safety committee that includes all key operators at the aerodrome. A well-functioning external safety committee greatly expands the vision of senior managers and generates a safety commitment on the part of third parties.

Safety Planning

Safety planning involves several interrelated activities, including resource allocation, adoption of operational safety standards, setting organizational safety goals, agreeing on indicators to measure performance against those goals, and establishing procedures for the control of safety information and documentation. Some examples include (1) access to all relevant safety publications, including ICAO annexes and manuals, national legislation, and CAA norms and regulations, (2) technical training to support responsibilities related to risk assessment, mitigation measures, and accident investigation, and (3) administrative support to manage safety information and safety documentation.

Safety Standards

Throughout the process of establishing an SMS, management always should keep in view the requirements of the ICAO standards and recommended practices and the *national regulations, standards, rules, or orders*. As part of the aerodrome certification process, compliance with these standards must be verified and any deviations reported to the CAA. Under some circumstances, the Member State can waive compliance of some standards if risk levels are found to be acceptable following a formal aeronautical study.

Many of these standards, such as declared distances, pavement condition, firefighting capability, and category of navigational aids, must be published in the *Aeronautical Information Publication* (AIP). Standards and conditions also can be modified temporarily through *Notices to Airmen* (NOTAMs) when construction activities or other potential hazards affect flight operations.

The Member State's safety program also should set overall safety objectives and may include direction on specific activities, such as incident reporting, safety investigations, safety audits, and safety promotion. These directives also become standards and should be incorporated into the SMS.

Goals and Indicators

Given the set standards, it is important to measure the degree of compliance in some cases where it is not a clearly prescriptive requirement. For example, with a performance-based safety system, this may include occurrences (accident and incidents), bird strikes, foreign-object debris (FOD) events, runway friction levels, and so on, where compliance can be measured according to indicators.

One of the best approaches to accident and incident reporting at an aerodrome is offered by the Airports Council International (ACI). Accidents and incidents are reported in the following categories:

1. Damage to stationary aircraft by apron equipment

2. Damage to moving aircraft

3. Property/equipment damage from jet blast

4. Equipment/equipment damage

5. Equipment/facility damage

6. Spillages (fuel and others)

7. Injuries to personnel or passengers relating to reported incidents

The ACI-defined categories of an apron accident/incident differ from the ICAO definition in Annex 13, which applies to aviation accidents. Annex 13 defines an *accident* as an occurrence during the operation of an aircraft that entails (1) a fatality or serious injury, (2) substantial damage to the aircraft involving structural failure or requiring major repair, or (3) the aircraft is missing or is completely inaccessible.

James Reason's *model of accident causation* also has played a key role in reactively investigating aviation safety accidents but also has been used proactively by many safety practitioners in helping to better understand how accidents can occur when all things (even seemingly subtle at the time) line up (Figure 15.3).

Safety Information and Documentation

It is critical to devote thinking into how safety information should be managed and protected. The SMS itself should be well documented in an updated and readily available safety management manual. Other existing documentation requirements are extensive, especially those related to training records, accident/incident investigations and follow-up actions, risk analysis and mitigation proposals, and tracking of safety indicators (especially occurrences

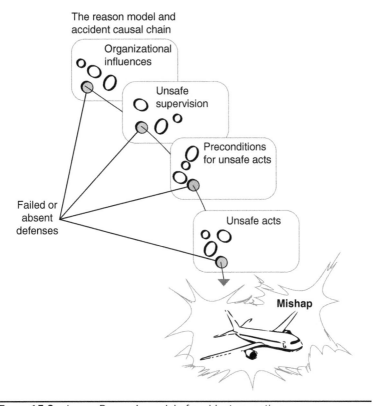

The reason model and accident causal chain

Organizational influences

Unsafe supervision

Preconditions for unsafe acts

Failed or absent defenses

Unsafe acts

Mishap

FIGURE 15.3 James Reason's model of accident causation.

and infractions). *Database management becomes increasingly important as the system matures in order to adequately support safety analysis and performance monitoring.*

Risk Management

Definition

- What is a hazard?
- What is a risk?
- How do they relate to each other?
- When does one apply risk management?

According to *best practices*, risk is managed to a level "as low as is reasonably practicable." The combination of the *likelihood* of an event and its potential *consequences* defines the risk associated with the event. An example of risk analysis, targets, and indicators will help to explain the process.

In the case of bird strikes, for instance, the wildlife officer should have on file an updated study of the bird species that present a hazard at the airport, including size and approximate weight, flying and flocking patterns, density of population, nesting locations relative to the approaches, and so on. From this information, the probability of a bird strike happening (during a season or at a certain time of the day) and the degree of damage to an aircraft can be estimated.

Basically, safety is defined in terms of risk. *There is no such thing as absolute safety.* The question to ask is whether a system has an acceptable level of risk or not. Evaluation of the acceptability of a given risk associated with a hazard must take into account the likelihood of its occurrence and the severity of the potential consequences. Some hazards already have risk controls in place, but are they adequate? Can the risk be reduced further?

The combination of *consequence* and *likelihood* produces three levels of risk (normally color-coded red, yellow, and green). The ICAO *Safety Management Manual* defines these levels of risk as (1) risks so high that they are unacceptable, (2) risks so low that they are acceptable, and (3) risks that will require some measure of mitigation to reduce the risk to a level that is acceptable, bearing in mind the ratio between costs and benefits.

The risk level is regarded as tolerable if the following three criteria are met:

- The risk is less than the predetermined unacceptable limit.

- The risk has been reduced to a level that is as low as reasonably practicable.

- The benefits of the proposed system or changes are sufficient to justify accepting the risk.

Hazard Identification—Take a Step Back!

A partial list of known *hazards* in the apron area, adapted from a list compiled from the ICAO *Safety Management Manual*, follows:

- Traffic volume and mixture (including high-density periods)

- Vulnerability of aircraft on the ground (fragile, etc.)

- Abundance of high-energy sources (including jet blast, propellers, fuels, etc.)

- Extremes of weather (i.e., temperatures, winds, precipitation, and poor visibility)

- Wildlife hazards (i.e., birds and animals)

- Aerodrome layout (especially taxiway routings, congested apron areas, known hot spots, and building and structures design limiting line of sight, possibly leading to runway incursions)

- Inadequacy of visual aids (i.e., signs, markings, and lighting)

Hazard identification is site-specific and not only may involve something *physical* but also could relate to a particular *process*. It does not have to be seen.

- For instance, the aerodrome layout may feature an apron with minimum space between parking positions and an active taxi lane. This clearly would complicate aircraft circulation and make positive control of pushback operations critical for safe operations.

- A physical and process example could be the operation of stop bars and advanced surface movement guidance and control systems (A-SMGCS) for controlling and monitoring the movement of aircraft and vehicles on the maneuvering area.

- At dual-use airports, a military presence may create challenges owing to military activities that are potentially dangerous to civil aviation, such as military flight training. The mere fact that another agency could be managing part of the aerodrome movement area could create issues relating to access, wildlife control, traffic mix, and emergency response, among others.

- Traffic patterns, operating hours, culture and language differences, governance issues, and interfaces all may intensify the incidence of hazards and complicate mitigating efforts.

Risk Mitigation
The ICAO *Safety Management Manual* describes three levels of risk mitigation:

- *Level 1 (engineering actions).* The safety action eliminates the risk. This involves equipment, tools, or infrastructure.

- *Level 2 (control actions).* The safety action accepts the risk, but adjusts the system to mitigate the risk by reducing it to a manageable level. For example, by increasing operating restrictions, posting adequate signage, or increasing the preventative maintenance frequency of older equipment.

- *Level 3 (personnel actions).* The safety action taken accepts that the risk can neither be eliminated (level 1) nor controlled (level 2), so personnel must be taught how to cope with it, for example, through procedures and work instructions.

Factors that increase risk include the introduction of untrained personnel, use of temporary access through construction projects, and the risks of unauthorized access and hence runway incursions. Controlling these risks requires a detailed work plan with detailed movement routes, communications, evacuation procedures, scheduled briefings, inspections, turnover procedures, and control measures.

When Does One Apply Risk Management?

There are no specific rules on when risk management should be applied, but consideration within the airport operations context should be made at various stages:

- The introduction of anything new or a change to a process, infrastructure, or plant/equipment
- A modification to something that already exists
- The conception (to look at both risks and opportunities) and then planning
- The design (to get it right) followed by construction
- Commissioning followed by handover (for use and maintenance)
- Asset-life reevaluation

There is a need to consider the overlay between *operations* and *maintenance*, as well as *who* is to be affected.

Safety Assurance

Safety assurance is the set of interrelated activities that ensures the operational controls designed to mitigate risks are functioning properly. Safety-assurance activities range from routine enforcement of aerodrome regulations to external audits of the entire SMS.

While no accident usually will result from only one cause, it is possible to make some general statements as follows:

1. Regulations and procedures at times are not followed.
2. Poor discipline can exist.
3. Loss of situational awareness can occur.
4. Weather limitations exist at times.
5. Unskilled and inadequately trained workers exist.
6. General human factors can contribute to accidents.

Nearly all these factors can be mitigated to a great extent by adequate supervision and safety leadership.

15.4   Implementation

Issues

Although all stakeholders concur with the need for airport-wide SMSs in line with the ICAO requirement referred to in the early part of this chapter, there are a number of factors that affect the effectiveness of their implementation, as discussed herein.

Complexity

One major consideration in the implementation of integrated airport SMSs is the complexity brought about by the large number of entities that conduct activities on airport property in support of passengers, freight, and aircraft processing, as well as the various authorities that have related regulatory or enforcement jurisdiction. Table 15.1 illustrates the range of stakeholders involved at airports and their jurisdictions.

Moreover, the ICAO standard requires that certain aviation service providers such as airports, airlines, and air navigation services implement SMSs. This creates a de facto overlap situation between various entities notwithstanding the fact that some organizations have relatively structured systems in place that must be modified to interface with the airport SMSs that are required in the context of aerodrome certification. Currently, many countries and a large number of aviation service providers report widespread implementation problems. A number of Member States have informed ICAO that delays are expected regarding compliance. It is therefore not completely surprising that deficiencies are still observed regularly in the field and reported in many safety assessments worldwide such as

- Unclear responsibilities and accountabilities of stakeholders
- "Silo" perspectives of stakeholders
- Lack of coordination for both planning and operational response
- Inadequate management training

Many experts are of the opinion that it is not possible to implement an SMS from following a rigid "one size fits all" model. Local institutional environments and scope of activities, as well as the number of agencies involved, will influence the deployment approach to be adopted. The focus must be on effective implementation of all SMS elements, not just some form of simple administrative compliance.

It is in this context that in the United States, for example, the Federal Aviation Administration (FAA) launched a pilot program in 2007 to have airports gather initial data, complete a gap analysis, and write draft SMS manuals. Twenty-two airports participated in the initial program, mostly larger air carrier airports. Recognizing the importance of

Entity	Mandate	Airport Operations	Systems and Facilities Maintenance	Safety	Security	Air Traffic Control	Customer Experience	International Inspection Services	Commercial Activities
Civil Aviation Administration	Aviation safety and security enforcement; Issuer of Aerodrome Certificate; Issuer of other aviation regulatory compliance licenses and permits	◆	◆	◆	◆	◆		◆	
Airport Enterprise	Holder of Aerodrome License; Facilities maintenance	◆	◆	◆	◆	◆	◆	◆	◆
Air Traffic Services	Positive control of aircraft and vehicles on maneuvering areas	◆	◆	◆		◆			
Airlines	Aircraft operation; passenger and baggage and cargo processing	◆	◆	◆	◆		◆	◆	◆
Security Services	Airport physical security and passenger/cargo screening	◆		◆	◆		◆		

Area of Jurisdiction								
Police — Prevention, handling and investigation of criminal activities	•		•	◆			•	
Government Inspection Services — Inspection of international passenger and goods at arrival (Immigration, Customs, Health and Agriculture)	•			•		◆	◆	
Ground Handlers — Aircraft, passenger and cargo handling on contract basis	◆	•	◆	•	•	◆	•	•
Ground Transportation Companies — Transportation of passengers to and from airports	•		•			◆		◆
Employee Unions — Representations on labor conditions	•	•	◆	•	•	•	•	•
Occupational Health Regulator — Occupational safety regulation/inspection	•	•	◆	•	•			•

Legend: Primary Role ◆
Secondary/Support Role •

TABLE 15.1 Areas of Jurisdiction

program scalability, the FAA initiated a second pilot program comprised of mostly smaller airports. Nine airports participated, and the same general requirements for a gap analysis and development of an SMS manual applied. With more than 30 U.S. airports that had SMS manuals following the FAA's pilot-program guidance, a third pilot study was initiated, with the intent to identify challenges and lessons learned when actually implementing SMS programs. Fourteen airports participated in the third pilot program, and the findings were submitted to the FAA in late 2011.

Some airports, such as San Antonio International Airport (SAT), a medium hub air carrier airport participated in the first and third pilot-program studies, recognizing the value of SMS early on, with a desire to have a role in how it would be implemented in the United States. As such, SAT has implemented SMS practices campus-wide (including airside and landside) ahead of rule making and has integrated additional programs such as foreign-object debris and wildlife-hazard management into its SMS program.

As of June 2012, the FAA was in the rule-making process. It was anticipated that some form of SMS guidance would be determined and regulation issued in late 2012. In the meantime, the findings of the pilot programs indicated that in order to achieve a smooth integration and adoption of SMS processes, existing procedures and processes in place should be used as much as possible; moreover, it was determined that the adoption of SMS processes does not and should not entail a complete retooling of existing airport practices.

Safety Promotion and Culture

Another key factor in the implementation of airport SMSs relates to the culture of the organizational environment. Even the best-designed system cannot function properly unless it is enabled by a workplace culture that is supportive of and consistent with SMS goals. Culture is of paramount importance because the SMS, regardless of how detailed its checklists and procedures are, relies to a large extent on the voluntary reporting of safety information. A culture that supports the SMS is termed *generative*, a term that captures the proactive positive environment that is conducive to safe operations.

James Reason (creator of the *swiss cheese model of accident causation;* see Figure 15.3) has identified some of the major elements of a *positive safety culture:*

- Informed culture (i.e., knowledge of the safety factors in the workplace)
- Just culture (i.e., atmosphere of trust)
- Reporting culture (i.e., a willingness to report errors and near misses)
- Flexible culture (i.e., safety is the responsibility of all, not just leaders)

- Learning culture (i.e., an ability to draw the right conclusions from safety data and take corrective action)

The development of such a culture does not happen overnight. It normally results from conscious decisions on the part of management.

Communication

Communication plays a critical role in the effective implementation of any safety program, and in the context of an airport SMS, it is very much multifaceted. Interestingly, although in another area of aviation safety (crew management), Kanki and Palmer (1993, p. 112) developed a classification of communication based on purpose that could be useful when planning and managing safety management systems:

- Communication provides information.
- Communication establishes interpersonal relationships.
- Communication establishes predictable behavior patterns.
- Communication maintains attention to task and monitoring.
- Communication is a management tool.

Effective communication is a "two-way street." It requires constant encouragement and follow-up to reach all players. Involving all operators creates ownership in the SMS, an important step toward creating a generative culture.

Effective communication is essential for the dissemination of safety lessons learned at an aerodrome. The reporting systems must encourage timely reporting, and information should reach the top levels of the organization quickly and without filter.

Training and Competency of Personnel

There are two broad categories of training to consider in the context of implementing an SMS: (1) training on the SMS itself and (2) competency training on safety-relevant tasks. Indeed, to facilitate the airport SMS, training is not the only aspect to consider in terms of human resources, but it is important to assess the competency of personnel relative to tasks and duties performed, particularly those with a core operational focus.

Operational needs should be analyzed carefully against the organizational chart and job descriptions in a matrix format to ensure that specific training requirements are identified properly for each position. In most instances, individuals who will be recruited into roles already possess the relevant knowledge, skills, qualifications, and personal attributes required of the position. However, the aforementioned may not necessarily be specific to airports (i.e., it may be vocational-based) and capture localized "gap training" such as the SMS and safety-relevant tasks to airport operations.

SMS training should be conducted at all levels, with presentations tailored to the hierarchical level addressed. This training needs to be documented and subjected to audit. Refresher training also should be conducted on (1) airport familiarization, (2) specific procedures, and (3) airport communications, including procedures for reporting unsafe conditions.

Guidance and Resources

The basic guidance material for airport SMSs can be found in ICAO Annex 14, Document 9774: *Manual on Certification of Aerodromes*, and Document 9859: *Safety Management Manual*. Some practical implementation advice, although somewhat geared to a U.S. context, has been published by the Transportation Research Board as part of the Airport Cooperative Research Program as *Safety Management System for Airports*, Vol. 1: *Overview*, and Vol. 2: *Guidebook*. In addition, Airports Council International (ACI) has published, *ACI Best Industry Practice Safety Management System (SMS) Gap Analysis and Audit Tool*.

Airports have started to cooperate by sharing documents and lessons learned regarding the implementation of airport SMSs through industry committees and by disseminating their own SMSs over the Internet. For example, Bangaluru International Airport, India (BLR), and SAT have placed full versions of their SMS manuals on the Web.

As a reference point, one can consider the table of contents in Table 15.2 as relatively typical for an airport SMS.

Airport Safety Management System: Typical Table of Contents
Section 1: General
1.1 Acronyms, Abbreviations, and Definitions
1.2 Introduction
1.3 ICAO Statement
1.4 Civil Aviation Authority Statement
1.5 Manual Control
1.6 SMS Organizational Structure
Section 2: Safety Policy and Objectives
2.1 Safety Policy Statement
2.2 Safety Objectives
• Management Responsibility
• Consultation
• Achievement of Objectives
• Company Operations
• Airport Organizational Chart

TABLE 15.2 Airport Safety Management System Manual: Typical Table of Contents

Airport Safety Management System: Typical Table of Contents

2.3 Structure and Responsibility
- General
- Chief Executive Officer (CEO)
- Board Safety Subcommittee
- General Managers
- Safety Services Office
- Manager Airport Safety
- Other Managers and Supervisors
- Aerodrome Reporting Officer
- Works Safety Officer
- All Employees
- Contractors
- Visitors and Other Nonemployees

2.4 Safety Action Groups (SAGs)
- Safety Committee
- Other Safety Groups/Committees
- Stakeholder Meetings
- Ramp or Apron Operations Committee
- Terminal Operations Committee
- Risk Management Workshops
- Airport Authority Contractor Meetings

2.5 Document and Data Control
- General
- Core Elements of the SMS
- OH&S Manual
- Airport Environmental Strategy
- Airport Authority Contractors
- Contractor Safety Manual
- Retail Tenancy Manual
- Terminal Users Guide

2.6 Emergency Preparedness and Response
- General
- Emergency/Safety Equipment
- Airside
- Terminal
- Landside
- Spills

2.7 Legal and Other Requirements
- Updating Legal and Other Requirements

Section 3:   Safety Risk Management

3.1 Overview of Safety Risk Management
- Risk Register
- Risk Assessments
- Corporate Risk Profile
- Board Safety Subcommittee

TABLE 15.2 Airport Safety Management System Manual: Typical Table of Contents (*Continued*)

Airport Safety Management System: Typical Table of Contents

3.2 Risk Management Tools
- Risk Assessment Tool
- Risk Assessment Methods
- Safe Work Procedures (SWPs) Tool
- Airport Operations Manual (AOM)
- Risk Management Workshops

3.3 Managing Operational Risk
- General
- Safety Risk Management: Airport Tasks/Activity-Based Risks
- Safe Work Procedures (SWPs)
- Risk Workshops for Projects
- Risk Categories
- Risk Matrix and Ratings for Likelihood and Consequence

3.4 Effective Safety Reporting
- Safety Performance Reporting
- Reporting of Incidents and System Failures
- Reporting on Hazard Identifications
- Hazard/Risk Assessment Reporting
- Board Safety Subcommittee
- Ramp or Apron Data
- Public Liability
- Airport Emergency Plan
- Security Occurrences
- Spills
- Airports Council International

3.5 General Reports
- Statutory Reporting
- Reporting to the Civil Aviation Authority
- Reporting to the OH&S Authority

Section 4:   Safety Assurance

4.1 Objectives, Targets, and Plans
- Acceptable Level of Safety
- Safety Performance Indicators
- Safety Performance Targets
- Safety Requirements

4.2 Monitoring and Measurement
- Incident Investigation, Corrective and Preventative Actions

4.3 General Investigations
- Investigation Process
- External Investigations
- Record Keeping

4.4 Audits
- General Audits
- Follow-up Actions
- Hazardous Substances
- Environmental Audits

4.5 Management Review
- General

TABLE 15.2   Airport Safety Management System Manual: Typical Table of
Contents (*Continued*)

Airport Safety Management System: Typical Table of Contents

Section 5: Safety Promotion
5.1 Training and Education
  • Commitment to Training
  • Provision of Safety Training
  • Induction Training
  • Awareness Training
  • Specialist Training
  • Task/Hazard-Identification-Specific Training
  • Training to Meet Compliance Obligations
  • On-the-Job Training
  • Training Record Keeping
5.2 Safety Communication
  • Safety Committee
  • Internet and Intranet
  • Publication Booklets
  • Signage and Posters
  • Quick Reference Cards
  • Alert Messages
  • Safety Alerts/Bulletins
  • Capital Works Review
  • Apron Operational Procedures

Source: Courtesy: Aviation Strategies International.

TABLE 15.2 *(Continued)*

15.5 Key Success Factors in Airport SMS Implementation

As inferred in various parts of this chapter, several variables can affect the degree of success in implementing an airport-wide SMS in line with the ICAO standards and recommended practices. However, when looking ahead, two elements stand out among others as having a determining role: integration and communication technology.

Integration

The paramount success factor for any airport SMS implementation is *integration*. While other factors such as leadership, commitment, responsibility, and accountability obviously are important, without true integration, both internally (e.g., airport operator) and externally (e.g., other service providers), the airport SMS may be doomed to failure.

In many ways, the problem is that SMSs have been developed and implemented in isolation by various on-airport service providers.

The timing of these also has complicated matters because no one SMS has a significant or controlling influence over the others. As stated previously, ICAO Document 9774 requires that all parties operating on the aerodrome should comply with the aerodrome safety requirements and that they should participate in the SMS. The question remains, "With which SMS?" As such, if the airport created its own SMS, as mandated by ICAO, then it would be participating *in the SMS*.

The effectiveness of an SMS is best achieved through proper acceptance, coordination, and implementation throughout the *aerodrome community* and the continual promotion and training on the SMS needed to sustain it. The best method to achieve this would be through appropriate and *measured* consultation. It is presumed that each SMS existing on-airport will have common principles. The difficulty would be for the airport operator to gain the commitment of other entities to share information and nonsensitive data for the purpose of establishing a holistic position on airport safety.

The airport safety "system" is composed of an integrated set of safety-related processes that reach across many boundaries, including the airlines and other third-parties certified to provide services. Because of this complexity, the "system" must be managed in a systematic way if its risks are to be controlled or mitigated, and operations are to be safe.

In the final analysis, the airport operator should be required to obtain aerodrome certification by meeting the related requirements that include the obligation to submit an aerodrome manual as per ICAO Document 9774: *Manual on Certification of Aerodromes*. Since 2005, SMSs have been an integral part of aerodrome manual submissions that airport operators must file with the civil aviation regulator for securing or maintaining their aerodrome certificate.

Communication Technology

Following on the need for effective communication, it has become increasingly important to recognize the value in technological advances in terms of SMS implementation. For example, many SMS software solutions exist to support SMS implementation, but they are frequently created in isolation from other on-airport SMSs, and they are not truly automated, still relying on selective inputs.

The trend toward developing mobile solutions for many operational requirements, with a versatility to share "real time" data, could create significant efficiencies for airports—not only improved efficiency but also producing the positive benefit of safer outcomes. The value of such approaches cannot be understated, and this is a field where expected and advanced change will occur in the future.

Emerging solutions will make it compelling for all service providers at an airport to use perhaps one common communications platform, and this likely would create true opportunities for the advent of e-SMSs.

References

ACI Operational Safety Subcommittee. 2010. *ACI Best Industry Practice Safety Management System (SMS) Gap Analysis and Audit Tool.* Montreal, Canada: ACI.

International Civil Aviation Organization (ICAO). 2009. *Chicago Convention, Annex 14, Aerodromes, Volume I-Aerodrome Design and Operations.* 5th edition, incorporating Amendments 1–10-A. Montreal: ICAO.

International Civil Aviation Organization (ICAO). 2001. *Manual on Certification of Aerodromes* (Document 9774). 1st edition. Montreal: ICAO.

International Civil Aviation Organization (ICAO). 2006. *Safety Management Manual* (SMM) (Document 9859). 1st edition. Montreal: ICAO.

Kanki, B. G., and M. T. Palmer. 1993. "Communication and Crew Resource Management." In E. Wiener, B. Kanki, and R. Helmreich (eds.), *Cockpit Resource Management* (pp. 99–136). San Diego: Academic Press.

Raman, R. C. 2010. "Problems and Solutions in the Implementation of Safety Management Systems," submission for the ACI Asia-Pacific Young Executive Award 2010, November 30, 2010. Available: http://aciasiapac.aero/upload/page/817/photo/4f2fa5b9b8ed5.pdf; retrieved June 17, 2012.

CHAPTER 16

Airport Operations Control Centers

16.1 The Concept of Airport Operations Control Centers

Introduction

When discussing the issues pertaining to airports and their business environment, it is important to recall the role that they occupy in the overall transportation network. Airports are, as other transportation terminals, "portions of the total transportation system which are concerned with the transfer of passengers and their baggage between vehicles and between modes" (Peat, Marwick, Livingston & Co., "Terminal Interface System, Northeast Corridor Transportation Project," report prepared for USDOT, December 1969). In this respect, the main purpose of an *air terminal* is to serve as an interface between ground and air transportation means, where different airport actors are engaged in the processing of users and goods from one mode to another.

The size of the facilities required to accommodate exponential traffic growth has increased tremendously over the years and reached surprising proportions since the early 1970s, as illustrated by the development of several new large airports (e.g., Athens, Bangkok, Buenos Aires, Dallas–Fort Worth, Denver, Dubai, Durban, Hong Kong, Incheon, Kuala Lumpur, Mexico City, Munich, Oslo, and Shanghai), as well as the undertaking of major expansion programs at existing sites (e.g., Atlanta, Beijing, Delhi, Frankfurt, Johannesburg, London Heathrow, Moscow Domededovo, Paris Charles de Gaulle, Sao Paulo Guarulhos, and Vancouver).

Following these gains in size and physical complexity, new problems mainly related to congestion, asset utilization, environmental considerations, and operational logistics surfaced, and ensuing concerns over ground access, internal transportation, and processing efficiency led to important changes in the philosophy of airport planning. For example, many newer airport designs are based on a multiterminal configuration, where individual terminal units form comprehensive operational and administrative modules with their own aircraft apron and car parking facilities.

Most airports became more difficult to manage efficiently in an attempt to establish and maintain an optimal balance between acceptable (or more than acceptable) service and cost considerations. At the same time, airlines were under pressure to revisit their own operating processes to implement enhanced efficiencies in view of intensifying competition and rapidly growing equipment, labor, and fuel costs, which resulted in demands for more cost-effective airports. To this effect, management practices were in many cases modified to facilitate common use by airlines of such facilities as check-in counters, aircraft gates, and baggage-claim devices, thereby reducing the extent of the overall capital expenditures required to provide an acceptable level of service to the users. All these changes resulted in additional responsibilities for the airport operator in its role as coordinator and integrator of airside and groundside activities. The need for a more systematic and proactive management of increasingly complex processes, procedures, facilities, and equipment involving a multitude of entities with a variety of functional jurisdictions in real time grew very rapidly and gave birth to what is generally referred to as *airport operations control centers* (AOCCs).

An *AOCC* can be defined as the command and control center tasked with overseeing the integrity of day-to-day airport operations for the purpose of optimizing efficient aircraft parking, asset utilization, safety, security, levels of service, and overall performance for the benefit of airport patrons and partners.

The American Association of Airport Executives believes that airport operators should implement an efficient "communications coordination mechanism" in order to oversee overall airport activities. It also notes that such airport "central nervous systems" are referred to under many names such as "Communications Center, Emergency Operations Center, Airport Operations Center, Dispatch Center, and Airport Communications Centers." However, it is clear that these names are not synonymous because the centers they refer to often focus on some of the functions that represent only a subset of the more comprehensive scope of activities of an AOCC.

IBM, a company that has been involved in the development of a number of command and control centers in different industries, gives an interesting perspective on what AOCCs are:

> The purpose of an Airport Operations Control Center (AOCC) is to oversee and align all airport processes from a single, trusted source, creating a common focus on punctuality, process quality and continuous improvement.

AOCCs feature modular, flexible airport operating systems and an information architecture that can receive information from anywhere in the airport and route it to where it needs to be to support all airport operations processes. For example, airport staff and enterprise resource planning systems can receive predicted passenger departure traffic volume so [that] they can match resource levels with demand. Such alignment can help reduce the high cost of overstaffing and correct poor service levels. The AOCC can also use communication channels such as secure Web portals to share information that can help integrate financial processes by creating a unified situational and analytical view for airport management [ftp://public.dhe.ibm.com/common/ssi/ecm/en/ttw03003usen/TTW03003USEN.PDF; retrieved June 13, 2012].

Origins to the Present

The need for command and control centers to optimize the performance of complex multifaceted systems is not new. They have been in place in military contexts for as long as one can remember and grew in popularity for the management of large service enterprises with physically distributed assets. Even before the first such centers appeared at airports in the late 1970s, a number of organizations already had relatively sophisticated logistics coordination centers in place and operating around the clock, for example, Bell, for telephone system networks across North America and British Columbia Hydro, in Canada, for its extensive power generation and distribution facilities across all kinds of terrains, including the Rocky Mountains. In the aviation sector, many of the larger airlines also had sophisticated command centers in place. United Airlines for one had a "network control center" in its Chicago headquarters, where senior executives of the company, including its CEO, received short daily briefings on operational performance prepared by a staff of a dozen analysts who focused on the causes of disruptions as well as the projection of various trends in key areas.

As information and telecommunication technologies evolved and more efficient and affordable software, hardware, and solutions for the monitoring and decision-support activities of complex systems became more readily available, control centers mushroomed in many industries. Today some centers constitute worldwide benchmarks for such facilities. Two superb examples of these are the AT&T Global Network Operations Center (GNOC) located in Bedminster, New Jersey, and the SITA Command Centers (SCCs), which are more specifically related to aviation, located in Montreal, Canada, and Singapore (Figures 16.1 and 16.2). These AT&T and SITA centers aim primarily at anticipating, preventing, offering priority alternative solutions to, and/or troubleshooting disruptions in the services that they supply to their customers globally, ideally before these customers suffer any significant inconveniences.

Figure 16.1 AT&T Global Network Operations Center. (*Courtesy of AT&T Archives and History Center.*)

Figure 16.2 SITA Command Center (Montreal). (*Courtesy of SITA.*)

AT&T describes the role of its center as follows:

The AT&T GNOC is the largest and most sophisticated command-and-control center of its kind in the world. The GNOC staff monitors and proactively manages the data and voice traffic flowing across AT&T's domestic and global networks twenty-four hours a day, seven days a week. From their workstations on the GNOC floor, they can quickly survey a sweeping wall of 141 giant screens showing different aspects of network activity, network topography and news events. At their consoles, each team member monitors a different segment or technology in the network using the most advanced diagnostic and management tools available. The condition of AT&T's global network is continually monitored in the GNOC. When an anomaly occurs that threatens or actually impacts the performance of our network, the response is managed by the GNOC staff through a practiced and proven incident command process called 3CP (Command, Control, and Communications). The incident command team is led by a Network Duty Officer in the GNOC, a role that is staffed around the clock, every day of the year. The GNOC coordinates the network incident response across AT&T organizations, assessing the impact of the event in near-real time and prioritizing the restoration efforts. In response to a catastrophic event, the GNOC would activate AT&T's Network Disaster Recovery Team and would coordinate its response [http://www.corp.att.com/gnoc/; retrieved May 11, 2011].

SITA explains the role of its command centers based through the following rationale:

As customer operational needs continue to evolve, customers increasingly expect more consistent, responsive and proactive service support with little tolerance for disruptions or downtime. Customers need to have their incidents resolved quickly and efficiently no matter where the fault lies—whether at the infrastructure level or the application level. The Air Transport Industry (ATI) needs an end-to-end service management approach that can ensure maximum uptime of operational systems and services. To match these expectations, and provide enhanced global customer support for its products and services, SITA has two SITA Command Centers (SCCs). The SCCs deliver operational excellence through proactive monitoring of applications and network services and full end-to-end service management—all handled by a single unified central operations team. Located in Montreal, North America, and Singapore, Asia, the SCCs operate in a follow-the-sun mode of operations to ensure full business continuity.

SITA is the only provider of air transport communications and IT solutions that is focused solely on the ATI, so the SCCs are dedicated 100 percent to ATI customers. And they are staffed with people who combine their specific technical expertise with extensive ATI knowledge. Both an incident resolver and a service provider, the SCCs proactively identifies performance issues and resolves them before they impact service. This means SITA customers benefit from an enhanced service experience, improved business continuity, increased responsiveness and faster resolution times. At the same time, SITA also moves one step closer to its ultimate goal of delivering zero downtime for the ATI [SITA correspondence, May 9, 2012].

Management Philosophy

The main purpose of an AOCC is to facilitate the achievement of high levels of operational performance. Under proper conditions, as discussed later in this chapter, the effective implementation of an AOCC will enhance a smooth transfer of passengers and goods through an airport, resulting in increased user satisfaction and cost-effective service delivery, measurable in terms of consistency of performance, reliability, minimization of congestion and delays, quality customer interface, efficient communications, safety, security, and relative comfort.

Part of the critical importance of a control center in an airport environment arises from the fact that from a "systemic" perspective, the airport enterprise or operator that is the entity that runs the airport is never in a position of directly controlling all aspects of the facilitation of passengers and goods on its own, contrary to what is usually the case in a military aerodrome setting. Government inspection services such as customs and immigration, police services, security services, airlines, ground handlers, and air traffic control (ATC) agencies, to name a few, that have legal and functional jurisdictions on elements of airport operations all have to be involved in achieving "efficient" airports. In practice, airport operators must leverage their position as landlords as well as various elements of the overall safety and security responsibilities conferred to them by regulatory authorities in order to marshal the efforts of all concerned stakeholders toward the optimization of airport logistics.

Command-and-control centers present a logical tactical solution when the achievement of performance targets is a direct function of the quality of the coordination of complex systems that operate under a multitude of managerial jurisdictions. Additionally, for an airport to excel in service delivery, it usually would require an exceptional set of physical infrastructures, but if the management of the day-to-day logistics is deficient, it will not be able to meet the ultimate performance expectations. It could, however, also be argued that, to a certain extent, inadequacies in physical infrastructure could be compensated for by exceptional operational management. This is the context within which AOCCs find their justification.

Strategic Significance

From the strategic management perspective of airport enterprises, their operational performance is a critical indicator of success. The experience worldwide to date has shown that the implementation of an AOCC has a determining influence on the level of service delivered to users, as measured by industry-accepted key performance indicators (KPIs) related to an airport's operations. In addition, "best practice" AOCCs not only capture and feed high-significance data into the KPI reporting requirements, but they also are designed to contribute to innovation based on the outcomes of built-in trend analyses.

Additionally, assuming that one of the major goals of an airport operator is to maximize the service offered to the users while minimizing the cost of providing it, management must know

- What level of service is being offered
- How much it costs to provide it

Effective management of all organizations, including service enterprises such as transportation terminals, logically should be based on, among other key factors, the availability of meaningful feedback on costs and outputs, the output being, in the case of an airport, the service of transferring passengers and goods between modes. Leading airports now also aim at providing users with a total-journey experience involving ancillary services such as a wide range of retail and entertainment offerings.

Still, all airport operators are concerned with making ongoing decisions aimed at achieving an optimal balance between service and cost. The current best practice for addressing this is the development and implementation of comprehensive management information systems (MIS) that deal with the operational performance and costs as well as the interrelationship between each of these dimensions. These decision-support systems come under many generic and commercial brand names that in essence constitute an airport operations control system (AOCS) that actually constitute the MIS that integrates real-time information on all elements of the overall airport activities needed for effective and timely decisions to be made within the AOCC. Conceptually, the AOCC is the highest component of an AOCS hierarchy with other elements of the system focusing on different physical sectors of the airport.

The AOCS in its optimal form should interface with the enterprise's financial administration system to allow for invoicing of operational services the airport operator provides to users but, more important, to allow the ongoing analysis of operating costs.

The AOCS and its main component, the AOCC, fit under the concept of *control* as one of the five major managerial functions along with planning, organizing, resource gathering, and supervision. The cyclic relationship between these activities is of an ongoing, albeit linear, nature, where feedback information is continuously sent to the initial planning step to allow for readjustment and corrective action once the control system has identified departures from expected results.

In this context, four major considerations usually influence the decision of an airport operator to implement an AOCC:

- The necessity to administer the proper functioning of a complex airport system involving multiple service providers in order to provide an integrated service to airport users

- The need to optimize on the common use of critical airport facilities such as gates, check-in counters and kiosks, baggage-handling systems, flight information display systems, aircraft parking areas, groundside transportation services, etc.

- The requirement for management to be systematically informed on the level of service offered to airport users

- The cost efficiencies associated with centralizing the management of several traditionally distributed coordination activities such as the control of operations, emergencies, critical maintenance work, security, and access control as well as airside safety in a single physical location

Regulatory Requirement for AOCCs

A review of International Civil Aviation Organization (ICAO) guidance as it relates to the operation of aerodromes indicates no direct reference to AOCCs, as found at many airports and as defined in this chapter. However, there are two relevant requirements that cover part of the functions normally attributed to AOCCs.

ICAO Document 9137: *Airport Services Manual*, Part 8: *Airport Operational Services*, 1st ed. (1983), Chapter 2 (Section 2.4, "Operations Room"), states

2.4.1 A co-coordinating centre should be established where information relating to the operation of the airport can be received and distributed. This may combine the functions of the Apron Management Unit as well as the Movement Area Safety Unit.

2.4.2 The room should be provided with direct telephone lines to ATC and any other operational control rooms as well as Meteorological Services (MET) and Airport Information Services (AIS). Radio communications should be provided so that operational staff can be contacted whether on foot or in vehicles. Arrangements should be made for the preparation and issue of Notice to Airmen (NOTAMs).

2.4.3 Communications should be established with any management duty control room which is provided to cover the overall operation of the airport.

And ICAO Document 9137: *Airport Services Manual*, Part 7: *Airport Emergency Planning*, 2nd ed. (1991), Chapter 5 (Section 5.2, "Emergency Operations Centre"), states

5.2.1 The main features of this unit are:
- its fixed location;
- it acts in support of the on-scene commander in the mobile command post for aircraft accidents/incidents;
- it is the command, co-ordination and communication centre for unlawful seizure of aircraft and bomb threats; and
- it is operationally available 24 hours a day.

5.2.2 The location of the emergency operations centre should provide a clear view of the movement area and isolated aircraft parking position, wherever possible.

5.2.3 The mobile command post will usually be adequate to coordinate all command and communication functions. The emergency operations centre is a designated area on the airport which is usually used in supporting and co-coordinating operations in accidents/incidents, unlawful seizure of aircraft, and bomb threat incidents. The unit should have the necessary equipment and personnel to communicate with the appropriate agencies involved in the emergency, including the mobile command post, when this is deployed. The communication and electronic devices should be checked daily.

16.2　Airport Operations Control System

AOCS Dynamics

An airport is a physical system that is dedicated to the transfer of people and goods by means of air and ground transportation. The demand exerted on its *physical processing system* depends on a specific variable that is the airline flight schedule because it translates into a number of people and tons of freight going through the airport per unit of time. Close and continuous monitoring of this schedule is required to make the necessary adjustments to the airport facilities assignment plan. It should be remembered that the optimal utilization of all its existing resources will allow an airport system to attain its ultimate capacity.

The ongoing use of the airport facilities is a dynamic situation that needs to be monitored and optimized to ensure high performance. At leading airports, this is achieved through the implementation of what may be referred to as an *airport operations control system* (AOCS; Figure 16.3). The AOCS is in fact a decision-support system that allows airport management to be informed on the status of airport operations in real time. It anticipates and detects related problems and provides the means to allow timely resolution of situations through preventive or corrective actions.

The AOCC, in its role as the AOCS central unit, performs the task of adjusting the "supply" to the level of "demand" by fulfilling its

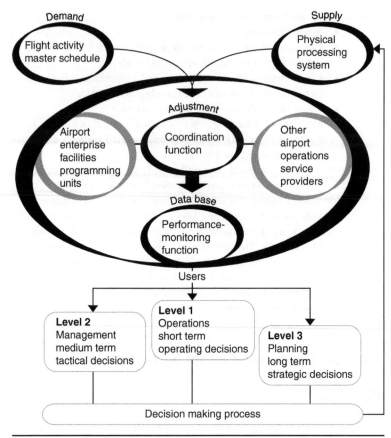

Figure 16.3   AOCS dynamics.

coordination function with the help of airport enterprise facilities programming units (i.e., the units, subordinate to the AOCC, that deal with assignment of critical airport facilities such as aircraft parking gates and remote stands, boarding lounges, check-in counters, baggage-delivery devices, ground transportation, etc.) and other airport operations service providers (i.e., customs and immigration, ground handlers, ATC, etc.). In practice, individual seasonal airlines' schedules are provided to the AOCS computer, which generates an overall master schedule; this schedule is distributed daily to the different units responsible for the assignment of specific facilities such as aircraft parking gates, airside buses serving remote gate positions, baggage carrousels, buses and taxicabs, public information desks, and so on. Last-minute editing related to flight delays, cancellations, or additions is also communicated by the AOCC to other parties involved in the delivery of services critical to airport operations, following updating of the master-schedule data by the airlines. Many other airport services, namely, government inspection services, ground handlers, retailers, and public information services engaged in passenger processing or related activities, also maintain continuous contact with the center, which enables them to improve the use of their personnel and provide a better service to the public.

At most airports, information related to flight activity and characteristics such as flight number, origin, destination, aircraft type, arrival/departure times, passenger load, and so on is stored in what is commonly referred to as an *airport operations database* (AODB). The flight-activity data are used to plan the assignments of airport facilities such as check-in counters, gates, and baggage-delivery devices to flights on a daily basis. The AODB serves as the mechanism to capture revisions to flight information from the airlines as they occur (through an interface between computerized airline operational systems and the AODB) to integrate it and via the AOCS to display the assignment of airport facilities to flights and disseminate the information to all interested parties, including passengers and well-wishers, through the public flight information display system (FIDS) and baggage information display system (BIDS) in terminal buildings.

The airport facilitation programming units in charge of gates, counters, baggage-claim devices, and so on are required to keep the AOCC informed at all times of the up-to-date facilities-assignment details and processing problems. Other major incidents or emergency situations also must be reported to the center, which then would switch from its activity-monitoring role to positive and direct control of the airport facilities/services until conditions return to normal. Such procedures greatly enhance task coordination and improve troubleshooting response times significantly.

Interestingly, because the AOCC coordinates and channels a large quantity of data that it gathers, filters, and analyzes, it facilitates a practical evaluation of the airport's performance, a matter that is discussed later in this chapter.

An example of this would be the computation of *processing delay indicators* (i.e., delay in minutes at a facilitation step of the process multiplied by the number of passengers affected). Considering that efficiency may be expressed in terms of *service* related to user satisfaction and that passengers are very sensitive to processing time, potential *zones* of delay in the transfer process can be identified (e.g., check-in, security screening, boarding, immigration, customs, aircraft taxiing, groundside access to terminals, etc.). Procedures implemented will ensure the reporting of any significant disruptions in the processing flows that would cause passenger inconvenience. The AOCC can log all delays exceeding known tolerance levels standards and analyze them in terms of trends over time or refer its findings to the relevant functional units of the airport enterprise for further study.

AOCS Users

The AOCS is designed to support the action of different levels of decision makers within the airport operator's organization. Typically, there are three categories of AOCS output users, namely, operations, management, and planning, because the airport operator decision-making processes are assumed to be distributed among three categories of intervention levels illustrated in Figure 16.3 as operating, tactical, and strategic.

In order to clarify the characteristics of this distribution, we will use the example of congestion occurring in the main automobile parking lot of an airport. Let us assume that on a sunny Saturday afternoon, the main lot reaches maximum capacity so that traffic even backs up in long waiting lines to the airport access road. Information on the problem is stored by the AOCS as being of interest to the three levels of users. The airport duty manager (level 1 user category) will be advised of the problem by the AOCC as soon as it occurs and will require information on flight activity, number of employees on duty in the car parks, availability of other parking lots, and so on. Based on these data, he or she will take immediate action to minimize the effects of the congestion. The concerned functional department heads of the airport management team (level 2 user category) will be informed of the occurrence of the problem, its extent, and the attempts that were made by operations staff to solve it; they will draw the appropriate conclusions and take the necessary steps to avoid a recurrence of a similar situation mainly in the form of changes to standard operating procedures and staffing levels. Finally, airport planners (level 3 category users) will require to be informed of such parking-lot capacity problems if they tend to reoccur under the same conditions despite changes in operating procedures and staff deployment; using, among other sources, the information retrieved from the AOCS database, they will determine what policy or physical modifications to the parking system might be needed. Cases such as these need to be handled efficiently in terms of decision making in view of the potential impact on the rest of the terminal and passenger processing.

This example shows that the same information can be used and filtered to suit the specific needs of the system users. The AOCS aims at facilitating this process by pinpointing the problem areas. It assesses the efficiency of airport operations through the ongoing monitoring of the level of service offered to users.

16.3   The Airport Operations Coordination Function

Purpose

The AOCC coordination function (Figure 16.4) rests on the centralization of operational decision power in one physical location where pertinent data on airport activities are continuously sent to

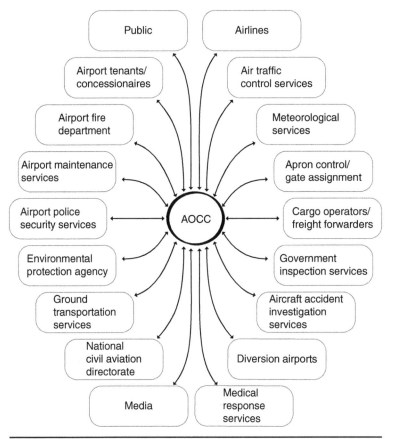

Figure 16.4   AOCC coordination function.

inform managers, who issue directions to the different airport services.

During emergency situations, the AOCC would normally assume the role of emergency operations center (EOC), as discussed earlier under the ICAO guidance section of this chapter. A colocated *situation room* also would be activated, where senior officials from airport management and other concerned entities would assemble to oversee the situation. The AOCC would ensure that a mobile command post is established in the field to manage the response at the scene of the incident. It also should be noted that an *alternate AOCC* should be established and procedures for its activation clearly spelled out in the airport emergency manual in case of a situation that would render operation of the AOCC impossible.

The determination of this AOCC coordination role during normal and emergency conditions was greatly inspired by the network management philosophy of the telecommunications industry, whereby a central control point staffed by duty officers oversees the operation of landline and cellular transmissions and switching networks by coordinating the work of regional centers on a continuous basis and providing technical direction when major or unusual problems occur and ensures rapid system recovery following emergencies. The hierarchical organization and operational environment of telecommunication systems are indeed quite similar to those of an airport system with its multiple terminal facilities.

The coordination function objectives are as follows:

- To allow for highly efficient coordination of all airport services by reducing response times and increasing the quality of intervention under normal and emergency conditions

- To provide management with real-time information as to aircraft, people, and vehicular movements within the airport in order to support the decision-making process

- To optimize utilization of airport facilities

- To supply the general public and other stakeholders with adequate flight and baggage information

These objectives are achieved through

- Supervision/coordination of all services engaged in the processing of passengers, freight, and aircraft

- Coordination of maintenance activities as they relate to the operation of critical systems and facilities

- Coordination of the action of police, security, and emergency forces in accordance with operational needs

- Supervision of the control of access to the airside and other restricted areas

- Enforcement of all operational procedures such as the airport emergency plan, snow control program (when applicable), and so on

- Relay of information between terminal services units, dissemination of action directives from management to operations personnel

- Update of information status displays, collection of airline seasonal flight schedules for preplanning the assignment of airport facilities, supervision of the real-time assignment of common-use facilities, monitoring of the flight and baggage-delivery display systems

- Ongoing analysis of the impact of current and anticipated flight and passenger activity levels on operational systems

Applications

Coordination Function Example: Anticipating Passenger-Processing Problems

The following example explains how passenger-processing incidents related to the capacity of some components of the airport system can be predicted and avoided. Through the AODB flight master schedule and corresponding amendments provided by the airlines, the AOCC is kept informed of the estimated arrival and departure times. The system's data also include the corresponding number of enplaning and deplaning passengers. This information can be either displayed on digital screens or printed.

On request, an AOCC computer software program calculates the peak hour for passenger (PHP) activity for any given day in the future based on predictive forecast data. If necessary, figures may be subdivided into enplaning and deplaning PHPs. Typically, the program computes the number of anticipated deplaning passengers for each 60-minute period that follows the arrival of a flight and indicates, for example, which three hours of the day will be the busiest. This information must be updated and based on revisions to the estimated arrival times and passenger loads.

This type of analysis generates valuable information and provides management with a general idea of what level of demand/pressure will be exerted on airport arrival facilities during a specific period of time. In line with its coordination role, the AOCC will inform all relevant operational units, including stakeholders, of expected peak periods. This, for example, should enable customs and immigration, ground-transportation dispatchers, parking-lot operators, and police services, among others, to react proactively to any special circumstances, thus providing the traveling public with the best service available.

The data on deplaning peak periods may be coupled with a standard distribution of passenger flows expressed in percentages and validated through periodical surveys (i.e., how passengers "statistically" typically deploy through the terminal and exit the airport on deplaning). For example, knowing that 985 passengers will deplane between 19:45 and 20:44 (according to computer-projected deplaning peak hour passenger data) and that approximately 19.5 percent of them probably will use bus transportation to the city or that 69.9 percent will be met by greeters whose vehicles are parked in the short-term lot gives an interesting insight into potential problem areas, assuming that the throughput capacity of related airport subsystems has been predetermined.

The same analytical approach may be equally applicable to departing flows using information on enplaning peak hour for passenger (EPHP) figures. In some cases, it could even be desirable to consider the total peak-hour passenger (TPHP) figures to forecast processing problems linked to the simultaneous use of the airport facilities by arriving and departing passengers.

Understandably, one must exercise caution in the use of the preceding analytical technique, although its predictive value is considered acceptable, as evidenced by some airports assessing the level of demand/pressure that will be exerted on its facilities for charter (i.e., irregular) flight operations.

Coordination Function Example: Snow-Removal Operations Control

One of the functions assigned to the AOCC is the control of the snow-removal operations at many airports operating in colder climatic conditions. Airports that fall into this category normally develop detailed snow-removal plans aimed at maintaining operations at an optimal level under storm conditions. Such plans cover the assignment of responsibilities to various units, removal priorities by sector, and communication protocols. Responsibilities of the AOCC and the field maintenance section are, respectively, tied to the enforcement of the priority plan ("WHAT") and to the actual removal operations ("HOW"). The priority plan is normally agreed on by the concerned parties before the beginning of the snow season and ensures that critical facilities such as priority runways, taxiways, and groundside access roads are looked after first and foremost. The control of snow-removal operations involves a complex coordination task. Figure 16.5 provides a general idea of the different services and organizations taking part in or impacted by these special operational conditions.

The AOCC is kept informed of the weather forecasts by the airport's meteorological office and passes the relevant details on to the field maintenance duty supervisor. From the beginning of the precipitation, the AOCC snow desk is activated and the priority plan put into effect. ATC advises the center of any changes in the use of the runways and transmits any pilot comments on the condition of the maneuvering area. The AOCC controllers maintain radio contact with the maintenance crews and follow the progression of the work. Appropriate *Notices to Air Men* (NOTAMs) on runway surface condition as well as breaking action reports are issued. Field-condition reports are recorded on a dedicated telephone line and updated regularly. Weather and field conditions of "neighboring" airports serving as alternate landing sites are also verified periodically to give an early warning of potential traffic surges owing to diverted traffic. Ongoing information on the operational status of the airport is provided to the traveling public and the media public through the public relations department.

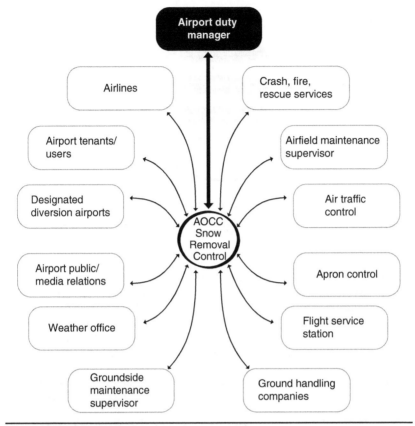

FIGURE 16.5   Communication chart: snow removal.

16.4   Airport Performance-Monitoring Function

Purpose

The second AOCC role relates to its performance-monitoring function (Figure 16.6), which pertains to the collection and recording of all information required to perform analysis of the level of service offered to airport users.

It is important to put in perspective the role that management should play in establishing adequate levels of service at airports:

> The management role, aside from reviewing the validity of service standards and local applicability of these standards, is to balance the stated needs of all users against the manager's perception of these needs. The manager is the balancer, by definition as well as by function, among competing factions whose pecuniary interests may have a way of tipping the

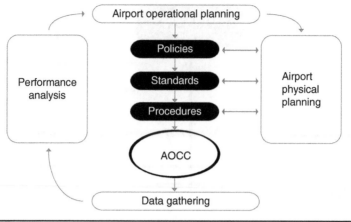

Figure 16.6   AOCC performance-monitoring function.

scales away from the common good. Although it may be natural enough and in some cases acceptable, it simply will not do for airport management to forego or downplay the importance of the service measurement role [Robert S. Michael, *Airport Landside Capacity: Role of Management* (Special Report No. 159). Washington, DC: Transportation Research Board, National Research Council, National Academy of Sciences, 1975].

The following objectives underline the importance of positive monitoring of airport performance as part of AOCC responsibilities:

- To monitor facilities performance and level of service delivery
- To maintain a reliable and justifiable cost-benefit activity database containing detailed and comprehensive airport operations–related information such as aircraft and passenger statistics
- To serve as a planning tool by offering adequate information display capabilities

These objectives are achieved by

- Logging all important operational requests made by users and subsequent action
- Recording aircraft, vehicular, and passenger movements at the airport
- Producing standard statistical reports on operational activity (i.e., aircraft mix, peak activity periods, etc.)
- Producing analytical reports allowing the
  - Evaluation of the level of service given to the users in light of cost-benefit criteria

- Assessment of the level of facility utilization
- Appraisal of the adequacy of planned infrastructure improvements against observed trends in the performance of airport systems and facilities
- Generation of data supporting the long-range planning of airport financial viability

Application

Performance-Monitoring Function Example: Automated Data Search, Operations Log

As mentioned in the preceding section, an AOCC normally maintains a log of all operations-related events occurring at the airport for incident management purposes. The information is fed in chronologic order into a computerized database by AOCC controllers on duty. Figure 16.7 presents the typical structure of an incident management system database (IMSD) structure.

The information contained in the database encompasses relatively standard categories, such as

- Date

- Time

- Originator of request/report

- Event coding (nature of event)
  - Code 1: Emergency
  - Code 2: Collision on airside
  - Code 3 Processing incident (aircraft, passenger, cargo)
  - Code 4: Unusable facility, related to aircraft operations
  - Code 5: Unusable facility; airport operations equipment
  - Code 6: Unusable facility: other equipment
  - Code 7: Adverse weather conditions

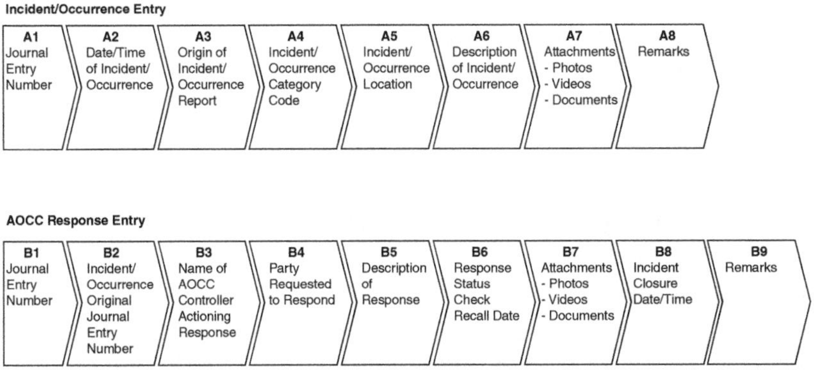

Incident/Occurrence Entry

A1 Journal Entry Number	A2 Date/Time of Incident/ Occurrence	A3 Origin of Incident/ Occurrence Report	A4 Incident/ Occurrence Category Code	A5 Incident/ Occurrence Location	A6 Description of Incident/ Occurrence	A7 Attachments - Photos - Videos - Documents	A8 Remarks

AOCC Response Entry

B1 Journal Entry Number	B2 Incident/ Occurrence Original Journal Entry Number	B3 Name of AOCC Controller Actioning Response	B4 Party Requested to Respond	B5 Description of Response	B6 Response Status Check Recall Date	B7 Attachments - Photos - Videos - Documents	B8 Incident Closure Date/Time	B9 Remarks

FIGURE 16.7   AOCC incident management system: Database structure.

- Occurrence/incident description
- Photo, video, or documentation attachment
- Party responsible for taking action
- Response description
- Response status check recall date

Data searches may be tailored to fit all types of requirements (i.e., events and time periods) by selecting any one field or a combination of the fields just listed. For example, with regard to performance appraisal, a qualitative indication of the level of service offered to pilots in terms of safety and availability of facilities with respect to aircraft maneuvering operations can be obtained by searching for "code 4" in the "Event Code" field. Search results also could reveal, on a given date, events such as "the guidance light system at gate 121 was out of service from 16:00 to 17:02 hours," "the taxiway lights on the west side of Bravo 3 were reported unserviceable at 18:12 hours" and are not likely to be fixed before a specific date in the future "due to the nonavailability of spare parts," and finally, "at 20:40 hours, field maintenance and emergency services were dispatched to gate 6 in the cargo area to take care of a fuel spill incident."

Searches of all types may be made over any desired time period; in the preceding case, if the "code 4" search had been requested a few hours later, further details would have been given on the fuel spill at gate 6.

It is also possible to search the log on the basis of a combination of fields. An example of this feature would be a search for a situation whereby events pertaining to "processing incidents—code 3" (first search criterion) involving, say, Air France (second search criterion) and requiring subsequent action by the "Terminal 2 Services Unit" (third search criterion) over a given year (fourth search criterion) in anticipation of the annual meeting between the airport operator and the airline.

Finally, another interesting characteristic of the system relates to an incident/occurrence status check "recall date." For instance, every time that an operational facility becomes unserviceable and cannot be repaired immediately, a recall date can be inserted in the appropriate database field, and every 24 hours, around 08:00 in the morning, a search is made for the current day date in that field, allowing for all previously faulty conditions earmarked for status verification to be listed.

16.5   Design and Equipment Considerations

Physical Layout

Typically, an AOCC physical layout (Figure 16.8) includes three rooms: the control room where the staff positions dedicated to different

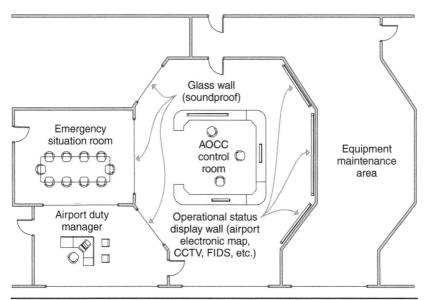

FIGURE 16.8   Typical AOCC layout.

functions are located, the airport duty manager's office (i.e., AOCC shift supervisor's office), and the emergency situation room. Ideally, the control room has a large display wall that contains summary status information that can be seen from any staff position as well as from the duty manager's office and the situation room. If direct line of sight of wall displays is a problem because of room height restrictions, control positions might be recessed into the floor of the control room by 2 to 3 feet (approximately 0.6 to 1.0 meter).

The number of positions in the control room is a function of the size of airport. The duty manager's office is normally equipped with telecommunications systems that duplicate the AOCC positions from where situations also could be handled, when necessary. Finally, the situation room is intended to serve primarily as a meeting point for senior airport management personnel and other entities involved in the handling of emergencies and when a segment of the AOCC is required to assume the EOC role. All three rooms are usually divided by see-through soundproofed walls. Communications between rooms is carried out via intercom.

Among other facilities, the control room normally contains relatively sophisticated telecommunication and information technology (IT) equipment. Typical requirements are described in the next section.

AOCC Systems and Equipment

A typical list of AOCC equipment would include (without being limited to)

- Access control system
- Airport electronic map
- Airport facilities database including digitized floor plans of buildings
- Airport operations control system with dashboard status displays
- Airport operations database
- Audio and video recording system
- Baggage information display system
- Closed-circuit television system
- Computerized airport operations log
- Critical equipment-monitoring system
- Digitized aerial photographs of the airport and environs
- Electronic area topographic map
- Flight information display system (operational and public channels)—arrivals and departures
- Intercom and hotlines to strategic units and locations
- Master clock
- Operational, maintenance, security, and emergency telecommunication systems
- Public-address system
- Television screens with cable
- Uninterrupted power supply
- VHF air-ground transceiver

Remark: The degree of sophistication of the equipment just listed will vary according to the size of the airport and also will depend on the policies, resources, and structures in place. These factors also influence whether systems actually serve to either manage or monitor activities.

It should be noted that the most crucial system is the *computerized airport operations log,* which is, in essence, at the heart of the airport incident/occurrence management system and constitutes an essential tool for the effective performance management of airport operations. Clearly, in addition, maximum payoffs are derived if this system is of an intelligent nature.

Finally, as demonstrated in a number of smaller airport facilities handling even fewer than 500,000 passengers, an AOCC concept can be implemented at a minimal cost and can be limited to basic telecommunications equipment dedicated to the handling of operations and maintenance (under normal and emergency conditions), maps and drawings for reference purposes, and possibly a limited closed-circuit television (CCTV) system.

Ergonomics

One key consideration when establishing an AOCC is the issue of *ergonomics*, which "is concerned with the 'fit' between the user, equipment and their environments . . . and takes account of the user's capabilities and limitations in seeking to ensure that tasks, functions, information and the environment suit each user" ("Ergonomics," Wikipedia.com; retrieved May 11, 2012) (Figure 16.9).

The matter is important enough for the International Organization for Standardization (ISO) to have developed a seven-part standard entitled, "Ergonomic Design of Control Centers" (ISO 11064), which covers in great detail issues such as layout of control rooms and workstations, displays, and controls, as well as environmental requirements.

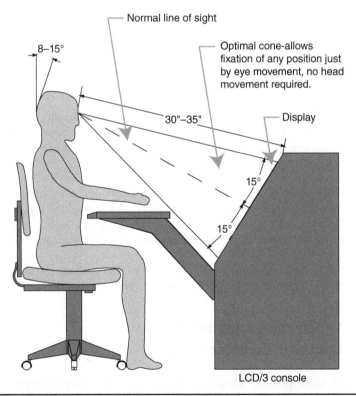

FIGURE 16.9 Ergonomics—Sample console design criteria. (*Courtesy of the Windsted Corporation.*)

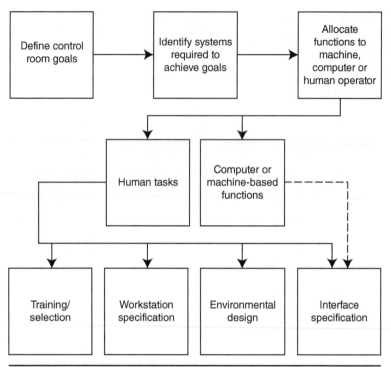

Figure 16.10 Ergonomics—Control center design process flowchart. (*Courtesy of the Windsted Corporation.*)

It should be remembered that AOCCs usually operate on a 24/7 basis, where personnel work long hours and sit at communications positions performing tasks requiring sustained periods of mental concentration and visual awareness.

As mentioned previously, most AOCCs also assume the function of EOCs during crises, when optimized layouts and equipment positioning are of even greater importance. In all cases, the layouts of workstations should be based on functional adjacency as well as level of utilization priority for communications equipment and visual displays.

Figure 16.10 presents a flowchart of the ideal path for the systematic introduction of ergonomics in the design of control centers.

16.6 Organizational and Human Resources Considerations

AOCC Management Structure and Reporting Relationships

With introduction of the AOCC concept, one noteworthy problem that may arise relates to all operational decision making being vested into the airport duty manager. This centralized process could be met with some resistance from technical supervisors in the field. In addition, middle and lower functional managers may feel that they are losing control over the preestablished work priorities they set for their subordinates. Moreover, to a certain extent, the center's approach can be perceived as going against the sacred rule of the *unity of command* by instituting a formal matrix management system (i.e., operational versus administrative/functional expertise management).

There are several proven methods to overcome these problems, including an education process where it is explained that the AOCC is there to support those engaged in operational activities and to foster the coordination of work that is essential to properly serve the airport's patrons. With the buy-in of the entire organization for the center and its mandate, that the airport as a whole is likely to gain in terms of performance and reputation.

It is recommended that the AOCC unit should report to the CEO of the airport enterprise (i.e., the airport general manager); otherwise, the approach will stand a good chance of falling short of achieving most of its intended results. For example, experience has shown that having the center report to the head of airport operations, which would seem logical, might create serious communication and authority problems with the maintenance department unless maintenance comes under operations, as is the case at a few airports. Also, by its very nature, the AOCC is engaged in performance evaluation of the overall airport operations and, therefore, should be associated with "neutral ground." It also should be realized that a direct AOCC reporting relationship might create an additional workload for the CEO and the senior management team that should, however, be counterbalanced by the benefits of being more rapidly and better informed of problems through the close monitoring of key performance indicators.

Second, a comprehensive set of operating rules based on the differentiation between the operational and administrative chains of command should be established. Normally, the AOCC would be assigned the responsibility of enforcing the procedures (*operational line*) developed by the functional managers (*administrative line*). In practice, this means that in coordinating the actions of airport services, the AOCC is bound to follow established procedures and, in this respect, is subject to an "after the fact" formal appraisal by functional managers in their respective fields of competence. In short, all airport employees should be made aware that the center is there primarily to ensure effective and efficient operations and that the purpose of senior management in implementing an AOCC is not one of centralization but of performance optimization.

Third, the quality-control objective should be transparent, but lower and middle management should be informed of the results and encouraged to propose corrective measures concerning performance problems. This can be done, for example, by distributing relevant data from the AOCC log to the different managers and allowing them to verify the accuracy of the information.

In short, the basic ground rules should be as clear as possible for everyone; AOCC personnel should be trained in the art of diplomacy, and most of all, every effort should be made to implement a teamwork atmosphere.

Staffing and Key Competencies

From an organizational point of view, the AOCC should come under the supervision of the airport senior duty manager, who ideally should report to the airport CEO (Figure 16.11).

This structure is warranted by two basic essential factors:

- The type of service that the center provides to airport users and that, from time to time, requires the pooling of resources from different branches

- The center's quality-control mandate derived from its performance-analysis role

In line with the foregoing considerations and management's objectives, the AOCC personnel resources should be tailored to support the coordination and performance-analysis roles described earlier. Analysts should develop and maintain performance-reporting systems, whereas duty managers and AOCC agents, working in shifts, should ensure the monitoring, supervision, and troubleshooting related to airport operations. A brief summary of the duties of typical AOCC personnel is given below.

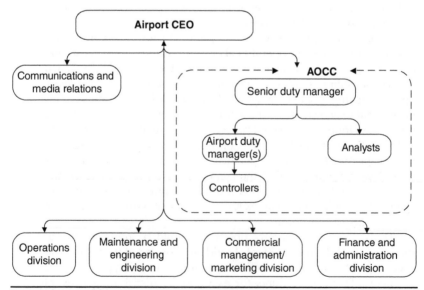

Figure 16.11   AOCC organizational position/structure.

Senior Duty Manager

Under the general direction of the CEO, the senior duty manager is the executive member of the senior management team responsible for the day-to-day operational management of the airport; the incumbent focuses on liaising with colleagues who manage functional areas, such as maintenance, safety, security, planning, emergency services, and commercial services, to ensure that the AOCC is operating within the set functional guidelines and to report as well as make recommendations on issues that might prompt modifications to these guidelines. The senior duty manager is also responsible for the selection and training of airport duty managers, AOCC controllers, and analysts, and this role is of paramount importance in view of the mission-critical nature of the AOCC positions.

Airport Duty Manager

Under the supervision of the senior airport duty manager and on a rotating-shift basis, the airport duty manager directs the operation of the airport on behalf of the CEO and the airport senior management team; oversees and programs the activities of the AOCC in order to provide airport management with real-time information as to aircraft, people, and vehicular movements within the airport; maintains an effective public relations program in order to handle complaints and provide a VIP reception service; coordinates all airport services for the effective and efficient processing of aircraft, cargo, passengers, and their baggage; oversees the day-to-day implementation of airport security and fire-protection programs for the safety of passengers, the public, and airport property; and performs other related duties as required.

AOCC Controller

Under the direct supervision of the airport duty manager, the AOCC controller coordinates on a rotating-shift basis all phases of passengers, baggage, freight, and aircraft processing at the airport through monitoring of the AOCC facilities and equipment; acts as an assistant to the duty manager; and performs other related duties as required.

Operations Analyst

Under the supervision of the senior airport duty manager, the operations analyst reviews the efficiency of airport operations and prepares related reports for managerial consideration; directs and coordinates the developments and activities required for the effective performance of the operational and administrative computer system; recommends improvements to the airport operations database; supports and advises the user community (i.e., operations, management) in the optimal use of the systems features; ensures that the AOCC documentation is maintained and updated when required; remains current regarding the technological developments in the field of database management systems; and creates and maintains historical files on airport activity.

The unique role of AOCC controllers requires further mention. First, overall success of the concept rests on them because they are the ones physically handling all first-line communications. Second, they more or less fulfill a resource role in support of other airport employees, and the very nature of their work places them at the heart of all airport activities. Moreover, they are able to discern short-term operational trends with a surprising degree of speed and accuracy and therefore can predict their impact or even prevent incidents from occurring. They develop a working knowledge of problem situations and the methods to resolve them. Quite often they become invaluable to the organization.

On the other hand, the nature of their duties dictates that they possess some very special skills and aptitudes in order to react positively to the support that is sought from them. A closer look at their daily work reveals that the tasks they perform are quite similar to those of an industrial dispatcher. Usually, people assuming dispatching responsibilities are highly experienced in their field of competence, having worked their "way up the line"; this is not always possible to achieve in exactly the same manner for AOCC agents owing to the great scope of knowledge they must possess. All efforts should be made to fill the AOCC agent positions by hiring people with different technical backgrounds encompassing security, electrical and field maintenance, meteorology, ATC, computer programming, and airline operations. Even when experienced, AOCC agents should undergo a formal, competency-based training program requiring successful completion and covering topics in airport familiarization and operations and AOCC procedures and equipment.

16.7   Leading AOCCs

Although no formal widespread survey was conducted regarding the extent of the implementation of AOCCs at airports around the world, a number of such facilities have been visited or examined through documentation and telephone interviews. From this, one can reasonably conclude that the implementation of these centers in the comprehensive manner described earlier is still in an early stage. This is due to a number of factors that complicate the deployment of full-scope AOCCs, such as the growing complexity of the airport business and the multitude of entities with jurisdiction over key elements of the airport processing system (resulting in some confusion as to how to operationalize the notion of collaborative decision making in the context of the accountabilities and liabilities of stakeholders). Moreover, it is clear that the development of software solutions for the integrated/intelligent management of airport operations is still largely fragmented and in its infancy, which is compounded by the uncertainty regarding the usability of mobile telecommunication technologies to deliver cost-effective operational solutions in an airport environment.

The question of the multitude of players, for example, ATC, police, customs, immigration, health services, airlines, ground handlers, facilities maintenance services, safety and rescue services, the accident investigation body, civil aviation regulator, and so on, with legitimate mandates that relate to various crucial aspects of airport operations, presents a particularly serious challenge. This is due to the fact that all these entities tend to focus on optimizing the segment of activities for which they are responsible, but as explained in the literature on "system theory," optimization of the individual parts of a system rarely leads to optimization of the system as a whole. In this context, the airport enterprise, as landlord of the premises and the holder of the aerodrome license (i.e., certification), must exert decisive leadership to ensure that all activities are managed in an integrated manner to deliver performance, safety, and security for all users. In so doing, it must secure the active support of all stakeholders. The AOCC is the privileged mechanism for pursuing this goal and for reconciling, in real time, a diversity of requirements.

Despite these challenges, a number of airports and their AOCCs currently stand out in terms of some of their characteristics and/or special achievements.

Auckland Airport: A Focus on Customer Service

Auckland International Airport Limited (Auckland Airport; AKL) was formed in 1988, when the New Zealand government corporatized the management of Auckland International Airport. As New Zealand's major transport hub, the airport is investing in traveler experience in support of the growing popularity of the country as one of the world's leading tourism destinations. Interestingly, this corporate driving force is even reflected in the nature of AOCC operations, whereby some positions are occupied by a team of customer-service representatives that takes external calls from the public—including complaints, etc. This team also fields phone calls from courtesy telephones around the terminals. In addition, management is evolving its AOCC into a more proactive stance, allowing it to anticipate adverse operational situations and prevent them from occurring (Figure 16.12).

Beijing Capital International Airport: Tightly Aligned on Best Practices

Beijing Airport is owned and operated by the Beijing Capital International Airport Company, Limited, a state-controlled company that falls under the Civil Aviation Administration of China (CAAC). In 2011, it handled over 78 million passengers and was the second busiest

FIGURE 16.12   Auckland Airport. (*Courtesy of Auckland International Airport Limited.*)

airport in the world behind Atlanta Hartsfield-Jackson International Airport. It is a complex facility with three terminals. Terminal 3 officially opened on a full scale on March 26, 2008, a few months prior to the 2008 Beijing Olympics, which took place in August of that year. In terms of size, with its 10.6 million square feet (approximately 986,000 m^2), it is bigger than the five terminals of London Heathrow Airport combined. Smooth handling of the opening of a complex facility of this nature and of a logistically demanding event such as the Olympics requires top-level coordination mechanisms that can only be guaranteed with a strong and efficient command and control system delivered by a state-of-the-art AOCC. The opening of the new terminal and the airport response to the Olympics also called for the development and implementation of operational readiness plans as discussed in Chapter 5. Such plans are almost impossible to execute properly without an effective AOCC in place, especially in the case where new facilities are being built at an existing airport that has to be maintained in operation during the construction phase (Figure 16.13).

Of the airports surveyed in the preparation for this segment of the book , it appears that the Beijing Airport AOCC is the closest to meeting the best-practices criteria described in this chapter. It is interesting to note that the Beijing Airport AOCC functions are not regrouped primarily according to airport physical locations (e.g., airside, terminal, groundside, etc.) but rather on a purpose basis, which seems to

FIGURE 16.13 Beijing Capital City Airport AOCC. (*Courtesy of BCIA.*)

facilitate a systemic and, by design, more integrated approach to the management of airport operations. This is illustrated in Figure 16.14.

The information management position is responsible for obtaining and disseminating all data required to coordinate airport logistics; the operational control position is responsible for decision making and the troubleshooting of incidents; the emergency management position is responsible for coordinating all emergency-response activities; the resources management position handles the deployment of

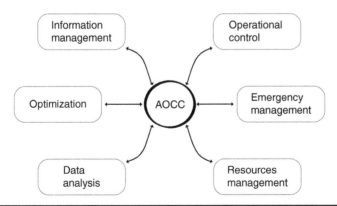

FIGURE 16.14 Beijing Capital City Airport AOCC functional positions. (*Courtesy of BCIA.*)

airport of staff and other assets according to real-time requirements; the data-analysis position focuses on study of trends, risk management, and scenario planning; and finally, the system optimization position focuses on performance management by monitoring deviations from standards and reporting against KPIs (key performance indicators).

Dublin Airport: Real-Time Automated Level-of-Service Measurement

Dublin Airport is owned by the government of Ireland and is operated by the Dublin Airport Authority (DAA). It handled more than 18.5 million passengers in 2011, with the new Terminal 2, in its first year of operation, processing 8 million of the total.

Similar to Beijing, the logistics of having to build and launch the operation of a new terminal played an important role in the decision of DAA to implement a more comprehensive AOCC. This was done in the context of four specific objectives:

(1) Streamline day-to-day management of the operation; (2) enable better recovery after disruption; (3) improve decision making and communications by having all functions co-located and working side-by-side; (4) improve public perception of DAA and illustrate a high level of professionalism [Hughes, John, Dublin Airport Authority, personal communication, April 30, 2012, and Murphy, Gráinne, Dublin Airport Authority, personal communication, June 22, 2012].

The fourth stated objective is particularly interesting because it implies a direct connection between the AOCC concept, the impact on the level of service actually delivered, and the perception of stakeholders. In Dublin, one of the important features of the operations management system linked to the AOCC is the ability to monitor passenger queues for security screening in real time and to create alerts. This is achieved through the positioning of sensors in the terminals capturing Bluetooth activity in the queuing areas and by using dashboard software to display the relevant data in the AOCC as illustrated in Figure 16.15.

Fort Lauderdale–Hollywood International Airport: Self-Audit and Improvement Plan

The Fort Lauderdale–Hollywood International Airport, which is owned and operated by Broward County in southern Florida, processed nearly 23.5 million passengers in 2011. The airport AOCC oversees operations that span over four terminals and an airfield comprising three runways.

Airport management recently performed a thorough assessment of AOCC future requirements in consultation with all concerned parties.

Figure 16.15 Dublin Airport AOCC queue management dashboard. (*Courtesy of Dublin Airport Authority.*)

The findings brought about recommendations for the implementation of new equipment and technologies that actually are somewhat typical of what many airports around the world are contemplating in the area of operations management improvements, although sometimes on an ad-hoc basis. Some of the features being considered include

- Automatic staff notification system
- Co-sharing of information between all entities involved in airport operations processes
- Employee tracking system
- Full integration of databases
- Full integration of telecommunications equipment
- Integration of the customer-service function
- Interface with the airport maintenance management system
- Joint location of the AOCC and the emergency operations center
- Voice-recognition system
- Voice recording

Figure 16.16 shows the existing FLL AOCC.

Figure 16.16   Fort Lauderdale–Hollywood International Airport. (*Courtesy of Broward County, Florida*.)

Kuala Lumpur International Airport: Monitoring a Network of Airports

Kuala Lumpur International Airport (KLIA) is owned by the government of Malaysia and operated by Malaysia Airports Holdings Berhad (MAHB). The company also operates 38 other airports in Malaysia comprising international, domestic, and short-takeoff-and-landing (STOL) ports.

Since its opening in 1998, KLIA has pioneered the use of state-of-the-art technology in airport management known as a *total airport management system* (TAMS), which consists of more than 40 systems and airport functions that are monitored by the AOCC, which has recently been modernized (Figure 16.17).

One interesting feature of the KLIA AOCC is that it is now being used to monitor incidents and performance at the other airports in the MAHB network, including docking at aerobridges, queue lengths at check-in, security, and immigration.

Los Angeles International Airport: Most Recent and Comprehensive

In 2011, Los Angeles International Airport (LAX) was the sixth busiest airport in the world, with over 62 million passengers (Figure 16.18). The airport is owned by the City of Los Angeles and operated by

FIGURE 16.17   Kuala Lumpur International Airport AOCC. (*Courtesy of Malaysia Airports Holdings Berhad.*)

FIGURE 16.18   Los Angeles International Airport AOCC. (*Courtesy of Los Angeles World Airports.*)

Los Angeles World Airports (LAWA), which also manages LA/Ontario International (ONT) and Van Nuys (VNY). Van Nuys is currently the world's busiest general aviation airport.

The LAX AOCC, locally referred to as the Airport Response Coordination Center (ARCC), has recently been completely modernized and transformed into a high-tech facility that probably will serve as one of the leading benchmark centers for years to come. Its role is defined as follows:

> The ARCC was created to enhance LAX's operational efficiency and crisis management capabilities by centralizing communications and streamlining management of all the airport's many operations, while improving service to passengers, airlines, concessionaires, tenant service providers, governmental agencies, and the surrounding community. The Center provides day-to-day, round-the-clock operational support, facility management, flight information, security coordination and ensures compliance with all federal aviation regulations [Yaft, J., Los Angeles World Airports, telephone interview, April 11, 2012].

Through the ARCC, LAX has been able to implement a "unified command" approach with the following functional positions present in the center:

- Airport duty manager
- Common-use facilities management (e.g., gates, check-in counters, and baggage carousels)
- Operations base (communications and airport operations logs)
- Maintenance work request desk (liaison with facilities management units)
- Airfield bus operations
- Transportation Security Administration (TSA)
- Airport police (safety and security enforcement)
- Communication/information (mass notification system for messages related to incidents/emergencies)

At LAX, the role of the airport duty manager is in line with the various notions discussed earlier in this chapter:

> The Duty Managers supervise a group of employees engaged in directing the airside and landside operations at Los Angeles International Airport; they direct employees in the Airport Response Coordination Center (ARCC) to provide a coordinated and timely response to daily safety, security, and operational issues; they assume the role of Incident Director in an emergency incident to form a Unified Command; and may act for Airport Management in their absence [Tenelle, Regina M., Los Angeles International Airport, personal communication, April 16, 2012].

Munich Airport: Direct Impact on Minimum Connecting Times

In recent years, Munich Airport (MUC) has earned global recognition as a benchmark for high performance in the category of airport connecting hubs (in 2011, approximately 40 percent of the total MUC passengers were in transit). Flughafen München GmbH (FMG) is the limited-liability company co-owned by the City of Munich, the State of Bavaria and the German government that operates the airport.

At Munich Airport, the AOCC located in Terminal 1 works in constant liaison with the Terminal 2 Hub Control Center to ensure exceptional minimum connecting times (MCTs): 35 minutes for Terminal 1, 30 minutes for Terminal 2, and 45 minutes between Terminals 1 and 2 including the transfer of checked baggage.

As indicated on the Munich Airport website:

> The HCC is a highly successful team effort between Munich Airport, Lufthansa and Terminal 2's "command center" for maximizing connectivity, full of activity day and night where a busy staff of 35 specialists is responsible for coordinating the entire range of handling processes. This ensures that all interactions are direct, face-to-face and immediate. A key element of the HCC is the Connex Center, where a team of experts is in charge of ensuring that passengers and their baggage reach their connecting flights. They maintain constant contact with air traffic control, can request priority landing clearance and reassign gate positions to minimize the distance that connecting passengers have to cover. Even when incoming flights are delayed, leaving less than the required 30 minutes to make connections, the HCC team pulls out all the stops, dispatching the special "ramp direct service" to pick up passengers and their luggage right at the gate and drive them directly to their connecting flight [http://www.munich-irport.de/en/consumer/aufenthalt_trans/airportstop/minconntime/hcc/index.jsp; retrieved June 11, 2012].

It is clear that achieving such an outstanding operational performance cannot be realized without a very effective AOCC and ancillary systems.

Zagreb Airport: Proving the Concept for Small and Medium Airports

Zagreb Airport (ZAG) is the main international airport of Croatia, and it provides air access to the capital of the country. In 2011, it handled over 2.3 million passengers and is the primary hub for the national flag carrier, Croatia Airlines. The airport is owned by the government of Croatia, but is due to embark on a major redevelopment phase starting in 2012 facilitated by a major foreign investment and will evolve into a commercial management model.

Despite its lower traffic levels relative to the other airports discussed herein and its public-ownership model, Zagreb Airport, like

Figure 16.19   Zagreb Airport AOCC. (*Courtesy of the Government of Croatia.*)

other international airports in the country, such as Split Airport and Dubrovnik Airport, has a long history of making use of AOCCs to manage airport logistics in an integrated manner, resulting in enhanced operational performance. These airports have been successful at implementing effective AOCCs and automated operational systems adapted to the small/medium-sized category of airports (Figure 16.19).

16.8   Best Practices in Airport Operations Control Center Implementation (Key Success Factors)

In summary, the successful implementation of airport operations control centers (AOCCs) to meet the purposes defined in this chapter depends on a number of minimum factors that can be described as follows:

- The role of the AOCC has to be defined and recognized as the command-and-control focal point charged with marshalling all airport resources in a coordinated manner for the purpose of optimizing the handling of aircraft, passengers, and goods through the airport processes and services in the most efficient, safe, and secure manner achievable under normal and emergency conditions.

- The AOCC has to be empowered by the CEO and the senior executives of the airport enterprise to fulfill its role with the full delegated authority to implement all operational procedures and contingency plans and, in the absence of applicable plans and procedures, use best judgment to make decisions aimed at preventing or troubleshooting operational incidents in an optimal manner; ideally, it should be organizationally positioned as a department reporting to the CEO and executing its mandate on behalf of the entire management team.

- Airport operational decision-making functions, including those falling under a jurisdiction other than that of the airport enterprise (e.g., security and local air traffic control at many airports), should be integrated physically or virtually linked with the AOCC to ensure that the overall airport operations process is fully optimized.

- The AOCC should be equipped to a minimum level, even at smaller airports. This should include a reliable airport radio or cellular communication network making it possible to reach out to key staff, access to mobile and landline telephone networks, detailed airport layout and area maps, as well as drawings of all buildings within the airport property (ideally in a digital format that can be projected on a wall for several people to see), and a computer/database for keeping a log of all airport incidents/occurrences, allowing for trend analyses.

- There should be an AODB in place at the airport to allow the AOCC to monitor the status of flights in real time to facilitate the effective assignment of facilities such as gates and baggage-delivery devices and the dissemination of relevant information to all interested units involved in airport operations as well as to passengers, well-wishers, and other airport users.

- The AOCC should be staffed with competent airport duty managers, controllers, and analysts possessing a mix of functional expertise/competencies and intimately familiar with all airport physical characteristics, systems, and equipment; these personnel should be rigorously trained in airport operations management under normal and emergency conditions.

In addition to the preceding, AOCC best practices are evolving to include

- Sophisticated communication systems progressively taking advantage of mobile communication technology that allows improved incident/occurrence management

- Features based on advanced ergonomics for all command-and-control positions and sophisticated architectural designs allowing status-display walls, contiguous rooms for operational supervision (e.g., duty manager), and emergency management with ready access to critical information in visual and audio formats

- Intelligent and interactive software as well as information-capturing/dissemination systems integrated into a high-performance decision-support system (DSS) and allowing for real-time risk management

- Computerized incident/occurrence management historical database allowing for trend analysis and feeding into the airport enterprise key performance indicators reporting system

- Leveraging the airport operators committee into supporting the work of the AOCC for the benefit of all stakeholders and in working toward the continuous improvement of safety, security, and customer-service levels

CHAPTER 17

The Airport Operations Manual

17.1   The Function of the Airport Operations Manual

The complexities of today's airports cover a great variety of disparate elements from special-purpose buildings to complex equipment needed to meet the needs of aircraft and passenger and cargo handling. Procedures to enable airports to function are equally complex and involve many different organizations, all of which need to have the fullest possible information if the airport is to operate efficiently. It is for this reason that an operations manual can provide such a valuable tool for management. Since one of the primary considerations for the air transport industry is safety, both in the air and on the ground, a substantial part of any operations manual will be devoted to this subject. In most countries, the primacy of safety is recognized by a requirement that airports must be licensed before they can be used for air transport. As part of the licensing process, airports may be required to produce and maintain an airport manual. An airport manual required by national regulations is usually limited in its contents to the elements of the airport related to the safe conduct of aviation. A particular example is described in Section 17.4.

In Section 17.2, a document of much broader scope is discussed. The airport operations manual there is conceived as a document that describes the operation of the airport across its entirety. In scope, it will contain all the aspects required as part of the licensing process and in addition all the additional aspects that contribute to the working of the airport system (e.g., the facilities and procedures necessary for the safe and efficient movement of passengers, baggage, and cargo). The operations manual, therefore, can serve as a basic reference document in which the facilities and procedures of the various operational areas are recorded. It also can serve as a useful introductory training aid for newly hired or transferred personnel.

17.2   A Format for the Airport Operations Manual

Ideally, the operations manual will embrace the airport in its entirety, outlining all procedures to be adopted under the various conditions of operation. It also will contain details of the principal infrastructure and equipment for reference purposes. The manual should be seen as the updated standard reference that describes the condition of the airport and how it is operated in all its aspects. A suggested format for a comprehensive operations manual for an airport follows:

1. Introduction
   a. Purpose and scope of manual
   b. Format and table of contents
   c. Licensing bodies
   d. Type of license, terms, and conditions of operation
2. Technical administration
   a. Status
      (1) Legal identification, addresses of airport, and licensee
      (2) Operating hours
      (3) Latitude and longitude of airport, grid reference, location of airport reference point, elevations of aprons and runway thresholds, airport reference temperature
      (4) Runway lengths, orientation
      (5) Description of aprons and taxiways
      (6) Level of rescue and firefighting service protection
      (7) Level of air traffic control
      (8) Level of airport radio and navaids
      (9) System of aeronautical information service available
      (10) Organizational structure of airport administration: names and addresses of key personnel
      (11) Airport movement statistics: passengers, freight, operations, airlines served, based aircraft, aircraft types served
      (12) Airport statistics: staff numbers, visitors, physical areas of facilities by type
   b. Administrative procedures
      (1) Procedures for complying with requirements relating to accidents and incidents
      (2) Procedures for promulgation of operational status of facilities
      (3) Procedures for recording aircraft movements
      (4) Procedures for control of construction affecting safety of aircraft
      (5) Procedures for control of access to airport and for control of vehicles while on airport operational areas
      (6) Procedures for control of apron activities

(7) Procedures for reception, storage, quality, control, and delivery of aviation fuel

(8) Procedures for the removal of disabled aircraft

3. Airport characteristics

   a. Plans showing the general layout of the airport, preferably at some scale in the region of 1:2,500

   b. Elements to be shown include the airport reference point; layout of runways, taxiways, and apron; airport lighting, including Visual Approach Slope Indicator Systems (VASIs) and obstruction lighting; location of navigational aids within the runway strips; location of terminal facilities and fire-fighting rescue facilities

   c. Data necessary for Inertial Navigation Systems (INS)-equipped aircraft, that is, INS coordinates for each gate position

   d. Description, height, and location of obstacles that infringe on the protected surfaces

   e. Survey data necessary for the production of the International Civil Aviation Organization (ICAO) Type A chart, which provides an aircraft operator with the necessary data to permit compliance with performance limitations within the runway, runway strip, clearway, stopway, and takeoff areas

   f. Data for and method of calculation of each of the airport runway cleared distances (take off run available [TORA], emergency distance available [EDA], take off distance available [TODA], and landing distance available [LDA]) with elevations at the beginning and end of each distance (where reduced declared distances are accepted owing to temporary objects infringing on the runway strip or transitional approach or takeoff surfaces, the method of calculating the distances should be indicated)

   g. Operational limitations owing to obstructions and so on

   h. Details of the bearing strengths and surface conditions of runways, taxiways, and aprons

   i. Temporary markings of displaced thresholds

4. Operational procedures

   a. Ground movement control procedures: rules of access, right-of-way, and vehicle obstruction marking; structure of control and duties

   b. Responsibility and procedures for routine operations on apron, including marshaling, docking stand allocation, aircraft servicing

   c. Routine inspection procedures of movement areas, including runway strips, runways, aprons, taxiways, grassed areas, and drainage

   d. Procedures for measurement and notification of runway surface condition: responsibility for reporting, measurement of runway braking action, depth and density of snow and slush, frequency of reporting
   e. Details of the airport snow plan, including categories of snow warnings: preliminary and final with duration and intensity forecasts; organization and chain of responsibility; standing instructions for procedures in snow conditions; equipment including maintenance and care; training of personnel; procedure or clearing runways, taxiways, aprons, airport roads, and domestic areas; hiring additional equipment
   f. Procedures for the general cleaning and sweeping of runways, taxiways, and aprons
   g. Details of bird-hazard control plan including methods of control: technical measures, killing, and environment modification; treatment of runways, taxiways, and aprons, grassland, built-up areas, and special areas such as garbage dumps; control of hazard outside airport boundaries
   h. Procedures for determining the availability of grass runways after heavy rains or flooding

5. Rescue and firefighting services (or emergency orders)
   a. Purpose of emergency orders
   b. Categories of emergency: aircraft accident, aircraft ground incident, full emergency, local standby, weather standby, aircraft bomb warning, building bomb warning, hijack, act of aggression on ground, building fire, approach alert within 10 miles (15 km) of airport
   c. ICAO category of the airport
   d. Range of on-airport and off-airport rescue and firefighting services
   e. Details of appliances and extinguishing media necessarily available on airport to meet the ICAO category requirements
   f. Structure of airport emergency services: organization, qualified staffing. and chain of command
   g. The airport emergency procedures, including in-shore rescue where necessary; rendezvous points, grid map, designation, and maps of remote search areas
   h. Degree of response of service
   i. Training program for rescue and firefighting services (RFFS)
   j. Procedures and limitation of aircraft operations during temporary depletion of RFFS
   k. Other responsibilities: fire prevention, storage of dangerous goods

  l. Procedures for removal of disabled aircraft: general procedures alternative runways, emergency removal, designation of recovery team, and responsibilities; formulation of recovery plan; phasing of recovery

  m. Medical facilities: quantities of equipment available, list of qualified personnel, structure of medical organization and control, arrangements for obtaining external medical assistance, medical procedures in case of accident, allocation of casualties to hospital

  n. Responsibilities in case of emergency: air traffic control, airport fire service, telephone switchboard, airport security, ground services, motor transport pool, airport engineering, airlines and handling agents, airport authority staff, and off-airport services (e.g., fire services, police, medical services)

  o. Location of ministers of religion

  p. Emergency communications systems; emergency direct lines, radio channels, direct ex-directory lines, radio-equipped vehicles, intercom channels, posted instructions

6. Airport security plan

  a. General description and rationale of security plan

  b. Responsibilities for airport security: airport administration and airport director, airlines, and other authorities

  c. Structure of airport security organization, police, and definition of diversion responsibilities

  d. Degrees of security: (1) normal, (2) period of heightened tension, (3) alert

  e. Persons authorized to prevent a flight or order an inspection or aircraft search

  f. Responsibilities for action in case of security alert at airport: general responsibilities, airport switchboard, airport fire service, police, air traffic control, ground services, airlines, and handling services

  g. Search procedure for aircraft: positioning, conduct of search, baggage identification, and cancellation of alert

7. Airport lighting

  a. The lighting scale of each runway, including approach and threshold lights

  b. The airport lighting system: method of operation, general layout diagram and individual circuit loops, ancillary lighting control, control console, constant current regulators, ac distribution switchboard

  c. Maintenance and routine inspection procedures

  d. Procedures for operation, including various brilliancy settings

  e. Failure maintenance and fault-finding procedures
  f. Standby power arrangements
  g. Technical diagrams of all lighting units
  h. Location of obstacle lighting on and off airport; responsibility for maintenance

8. Meteorological services
   a. Organization and structure of meteorological services; staff structure and responsibilities
   b. Class of service and information provided: form of messages
   c. Equipment: uses and maintenance
   d. Supply and use of meteorological services

9. Air traffic control services
   a. Description of the system for managing air traffic in airspace in the vicinity of the airport, including organization and responsibility of air traffic services, aeronautical information service
   b. Rules of the air and air traffic control; general flight rules
   c. Rules governing the selection of the runway in use and the circuit direction
   d. Standard procedures
   e. Noise-abatement procedures
   f. Search and rescue alerting
   g. Method of obtaining and disseminating meteorological information, including runway visual range (RVR), visibility, and local area forecasts

10. Communications and navaids
    a. Description and procedures for the use of general communications channels; Aeronautical Fixed Telecommunication Network (AFTN), Société Internationale de Telecommunications Aéronautiques (SITA), Aeronautical Radio Incorporated (ARINC)
    b. Description and procedures for use of air/ground and operational ground radio where these are not covered by air traffic control procedures
    c. Description of and operating procedures for radio and radar navigation aids; installation procedures; inspection and maintenance

11. Signals and marking
    a. Location of the signals area; procedures for display of temporary and permanent signals
    b. Location of wind socks

12. Access provision
    a. Airport location plan showing regional road network and public transport routes
    b. Access routes in immediate vicinity of airport

  c. Provision of parking: number and location of spaces; contract and obligations of parking operators

  d. Taxis and car hire: names and contractual obligations of operators

  e. Passenger access information system

13. Passenger terminal

  a. Organizational structure of terminal administrative staff: description and responsibilities

  b. Passenger information system: scale, equipment, maintenance procedure, standard information formats

  c. General information: airlines operating from airport, passenger and aircraft handling companies, airline interline handling agreements

  d. Layout of passenger terminal

  e. Arrivals: method of processing

  f. Departures: method of processing, definition of permissible cabin baggage, security arrangements, and definition of dangerous goods

  g. Special handling of passengers: procedures and equipment for handling disabled persons, unaccompanied minors, and VIPs

  h. Airport services and concessions: procedures for operating airline ticket desks; mishandled baggage procedures; listing of, location of, and contractual obligations of concessionaires such as duty-free shops, catering, banks, car hire, insurance desks, valet service, shops, hotel booking facilities, postal and telephone services

  i. Description and operational arrangements of facilities for meeters, senders, and spectators

  j. Description and operational arrangement of support facilities for airport airline and other staff working in the passenger terminal area

14. Cargo-handling facilities

  a. Organizational structure of airport authority, airlines, and other cargo handlers in cargo area

  b. Location of facilities and layout

  c. Cargo-handling procedures: export and import, including customs clearance requirements; special procedures for integrated carriers

  d. Description of airport and customs computer facilitation system

  e. Procedures for handling mail

  f. Procedures for handling airlines' company freight

15. General aviation

  a. Description and layout of general aviation facilities

  b. General regulations relating to general aviation operations

    c. Inbound procedures
    d. Outbound procedures
    e. Scale of fees, method of collection
    f. Aeronautical information services, meteorological services, and lounge arrangements for general aviation

16. Typical appendices
    a. Amendment procedures
    b. Distribution list of operations manual
    c. Necessary telephone numbers, airlines, concessionaires, agents airport staff, other airports
    d. Bylaws and regulations of airport
    e. Two-letter airline codes
    f. Three-letter airport codes
    g. Standard information symbols used at airport

17.3   Distribution of the Manual

The airport operations manual should be regarded by those administering the airport as the definitive document containing a statement of standard operating procedures and a description of airport equipment and facilities. An updated version of the manual should be held by the chief administrator of the principal operations areas of the airport. At a very small airport, the manual would be a rather simple document, and probably only one copy would exist, which would be kept by the director. In a large, complex operation, a more likely distribution list for the manual and its subsequent amendments would be

- Airport director
- Deputy director
- Assistant director, operations
- Assistant director, engineering
- Assistant director, administration
- Chief security officer
- Chief fire officer
- Chief air traffic control officer
- Airline station managers of based airlines

Whether simple or complex, the updating and amendment of such a manual require close attention. Sources of information should be carefully checked. It is possible, as in the case of one major U.S. airport, for an accident investigation to discover uncertainty regarding the origin of information on the precise length of a runway.

Updated sections of the manual that are pertinent to their own sections should be available to line managers of the operations of the passenger terminal, the cargo terminal, the passenger and cargo aprons, air traffic control, engineering meteorology, firefighting and rescue, security, and maintenance. It is also useful to use appropriate sections of the manual as instructional aids in the training of the various grades of manual workers.

17.4 U.S. Example: Federal Aviation Administration Recommendations on the Airport Certification Manual (FAA 2004)[1]

In a number of countries, the airport operations manual or the aerodrome manual, in some form or another, is linked to certification of the facility as an airport. In the United States, *Federal Aviation Regulations (FAR)*, Part 139, which deals with the certification of land airports serving scheduled carriers, requires that each applicant for an airport operating certificate must submit an airport certification manual (ACM) to the FAA (CFR 2011).[2] The required contents of the ACM are set out in checklists in Appendix 2 (Classes I, II, and III) and Appendix 3 (Class IV) of AC 150/5210-22. The classification system is shown in Table 17.1. The FAA provides guidelines on how a manual is to be prepared, recognizing that the suggested format is not all-inclusive but rather is intended to provide broad guidance so that each operator will have flexibility in developing a manual to fit the particular airport. The recommendations are largely related to a description of the data that must be supplied for eligibility for certification and other data to indicate compliance with operational rules. The suggested format for the FAA manual is substantially less broad than that indicated in Section 17.2 and includes the following.[3]

	Class I	Class II	Class III	Class IV
Scheduled large air carrier aircraft (30+ seats)	X			
Unscheduled large air carrier aircraft (30+ seats)	X	X		X
Scheduled small air carrier aircraft (10–30 seats)	X	X	X	

Source: FAA.

TABLE 17.1 Classification of Airports by Air Carrier Operations

[1]Appendix 4 of AC 150/5210-22 contains a list of the numerous applicable advisory circulars.
[2]For detailed guidance, reference should be made to the *Code of Federal Regulations*, Title 139. Subpart D.
[3]Sample airport certification manuals are available online at http://www.faa.gov/certification/part139/.

Suggested Airport Certification Manual: FAA Format

1. Introduction
   a. Table of contents
   b. Name, class, and location of airport
   c. Mailing address of airport manager or operator
   d. Person/title responsible for ACM maintenance
   e. FAA inspection authority
   f. Location of official ACM and distribution list
   g. Date of FAA approval

2. Personnel
   a. List of key personnel and job titles
   b. Brief description of functions
   c. Organizational chart showing operational lines of succession
   d. Description of personnel training

3. Paved areas
   a. Movement areas available for air carriers with their safety areas (use of a map or diagram is recommended)
   b. Describe the procedures for maintaining the paved and safety areas

4. Unpaved areas: If the airport has maintenance procedures for these areas, they should be shown here

5. Safety areas
   a. Location and dimensions
   b. Procedures for maintenance
   c. Inspection and maintenance procedures program if the airport has an engineered materials arresting system EMAS

6. Marking signs and lighting
   a. A legible color plan showing the runway and taxiway identification system, including the location and inscriptions of signs, runway markings, and the holding position markings
   b. Descriptions of and procedures for maintaining marking, sign and lighting systems
   c. Contact information for approach lighting maintenance
   d. Procedures for the shielding of airport lighting
   e. Instructions on determining whether the system should be declared inoperative

7. Snow and ice control
   a. If ice and snow rarely occur at the airport, the ACM should state this
   b. Provide details of the snow and ice control plan, developed according to AC 150/5200-30

    c. Provide specific procedures for notifying air carrier users of airport movement area conditions

    d. Provide instructions and arrangements for snow removal to prevent interference with navaids by snow accumulation

    e. Specify who has the authority to initiate snow removal operations, especially when procedures involve calling in outside assistance

8. Aircraft rescue and firefighting: index determination

    a. The airport's Aircraft Rescue Firefighting Facility (ARFF) Index is calculated according to the requirements of *FAR*, Part 139.315(a) and (b). These requirements are shown in Table 17.2 and override ICAO manuals.

    b. The ACM shall specify the airport's firefighting ARFF Index and designate the largest applicable aircraft that the index can serve

9. Aircraft rescue and firefighting: equipment and agents

    a. The ACM must have a description of the equipment necessary to meet the airport's aircraft rescue and firefighting requirements, including the ARFF equipment and the type and quantities of agent provided/maintained on each vehicle

    b. Specification of the number and type of portable extinguishers the vehicles carry to permit the airport to compute the index the airport can maintain in an equipment outage

10. Aircraft rescue and firefighting: operational requirements

    a. Description of the facilities, personnel, and procedures necessary to meet the airport's aircraft rescue and firefighting requirements

    b. Describe methods of alerting rescue and firefighting crews to an emergency

    c. Show on a grid map the location of the fire station(s) on the airport and primary traffic routes for the fastest response to all air operations areas; also show the designated emergency access roads

    d. Specify the operational requirements relating to on-airport and off-airport responses to emergencies

    e. Name personnel authorized to dispatch (off-airport), reduce, and recall ARFF resources

    f. Procedures for notifying air carriers of any changes to the normal complement of the ARFF unit

    g. List the channels of communication available to the ARFF unit

    h. List and describe rescue and firefighting training programs

    i. Define the role of the air traffic control tower in emergency operations

Index	Aircraft Length	Vehicles	Extinguishing Agents
A	<90 ft (<27 m)	1	Either 500 pounds of sodium-based dry chemical, halon 1211, or clean agent or 450 pounds of potassium-based dry chemical and water with a commensurate quantity of aqueous film-forming foam (AFFF) to total 100 gallons for simultaneous dry chemical and AFFF application
B	90 to <126 ft (27 to <38 m)	1	500 pounds of sodium-based dry chemical, halon 1211, or clean agent and 1,500 gallons of water and the commensurate quantity of AFFF for foam production
		2	One vehicle carrying the extinguishing agents as specified for Index A and one vehicle carrying an amount of water and the commensurate quantity of AFFF so that the total quantity of water for foam production carried by both vehicles is at least 1,500 gallons
C	126 to <159 ft (38 to <48 m)	2	One vehicle carrying the extinguishing agents as specified for Index B and one vehicle carrying water and the commensurate quantity of AFFF so that the total quantity of water for foam production carried by both vehicles is at least 3,000 gallons
		3	One vehicle carrying the extinguishing agents as specified for Index A and two vehicles carrying an amount of water and the commensurate quantity of AFFF so that the total quantity of water for foam production carried by all three vehicles is at least 3,000 gallons
D	159 to <200 ft (48 to <61 m)	3	One vehicle carrying the extinguishing agents as specified for Index A and two vehicles carrying an amount of water and the commensurate quantity of AFFF so that the total quantity of water for foam production carried by all three vehicles is at least 4,000 gallons
E	200 ft and longer (61 m)	3	One vehicle carrying the extinguishing agents as specified for Index A and two vehicles carrying an amount of water and the commensurate quantity of AFFF so that the total quantity of water for foam production carried by all three vehicles is at least 6,000 gallons

TABLE 17.2   ARFF Index Classification (United States)

11. Handling and storing of hazardous substances and materials
    a. This section of the ACM covers the handling of hazardous materials (HAZMAT) such as aircraft cargo and the fuel for operation of aircraft. The airport does not have to have to include HAZMAT procedures if it is not the HAZMAT agent. In the uncommon cases where it acts as the cargo operator, it does have to include
        (1) Special areas for the storage of hazardous materials while on the airport
        (2) Shipper assurance that the cargo can be handled safely, including special procedures
        (3) Designated personnel to receive and handle hazardous materials
    b. If aviation fuel is available at the airport, regardless of who the fueling agents are, the airport operator must
        (1) Indicate whether it is the HAZMAT agent
        (2) Describe the standards set up and maintained for protecting against fire and explosions in storing and handling fuel, etc., including
            (a) Grounding and bonding
            (b) Public protection
            (c) Control of access to storage areas
            (d) Fire safety in fuel farm and storage areas
            (e) Fire safety in mobile fuelers, fueling pits, and fueling cabinets
            (f) Training of fueling personnel

12. Traffic and wind-direction indicators
    a. Indicate the location(s) of wind-direction indicators on the airport
    b. Make arrangements for displaying and maintaining signals and markings, both permanent and temporary

13. Airport emergency plan: The airport certification manual must include a comprehensive emergency plan. *Note:* In view of the extensive technical information to be included, detailed guidance is provided by a separate advisory circular (AC 150/5200-31), which was issued in January 1989.

14. Self-inspection program
    a. Details of self inspection program, including frequency; also, personnel training, information dissemination, and corrective action procedures for unsafe conditions
    b. Appropriate checklists for continuous surveillance, periodic condition evaluation, special inspection
    c. Schedule of self-inspections and identification of who is responsible for the inspections

15. Pedestrians and ground vehicles: The ACM of Class I, II, and III airports must address procedures for controlling access to movement and safety areas
    a. Indicate on a map of the airport the limits of access to movement and safety areas
    b. Designate ground vehicles approved to have access to the movement and safety areas and the procedures necessary for their safe and orderly operation
    c. Detail the arrangements made for communications with and control of ground vehicles operating in the movement and safety areas, including involvement of the air traffic control tower
    d. Consequences of noncompliance

16. Obstructions
    a. Identify each object within the area of authority that is identified in *FAR*, Part 77
    b. Describe the marking and lighting of each of the obstructions
    c. Provide an airport layout plan locating all lighted obstructions that fall within the airport authority, keyed to a narrative description that also identifies any parties, in addition to the airport authority, responsible for their maintenance
    d. Procedures for removing, marking, and lighting obstructions
    e. Maintenance procedures and responsibilities for lighted obstructions
    f. Airspace evaluation procedures for any proposed construction or alteration of the airport
    g. Identification of objects within the airport operator's authority stated to be "no hazard" by the FAA
    h. Responsibility for monitoring obstructions

17. Protection of navaids: The ACMs of Class I, II, and III airports must provide procedures for the protection of navaids, including
    a. Procedures and assignments, depending on the placement of the navaid for security patrols, fence maintenance, etc.
    b. Designate the airport department that should be alerted to all proposed activities that may interfere with the signal of a navaid
    c. Provisions for preventing interference from accumulation of snow

18. Public protection
    a. Detail the procedures, devices, or obstacles used to prevent inadvertent entry of persons or vehicles into any area containing hazards for the unwary trespasser

    b. Provide an airport layout plan indicating the type and location of fencing and fence gates and also showing areas restricted from use by the general public. *Note:* The prevention of intentional infiltration of airport security areas is within the preview of the regulation on airport security, *FAR*, Part 107.

19. Wildlife hazard management
    - *a.* Indicate the nature of the existing conditions on the airport and the control techniques to be employed if a wildlife hazard exists, or show why there is no wildlife-hazard problem
    - *b.* If a problem exists, include a wildlife-hazard-management plan

20. Airport condition reporting
    - *a.* Describe the procedures used for identifying, assessing, and disseminating information to air carrier users of the airport
    - *b.* Describe the various internal means of communication that may be available for urgent dissemination of information
    - *c.* Document any system of information flow agreed on with airline tenant(s)
    - *d.* Describe the conditions and procedures for issuing NOTAMs

21. Identifying, marking, and reporting construction and other unserviceable areas
    - *a.* Describe how construction areas and unserviceable pavement and safety areas are marked and lighted
    - *b.* Describe the provisions made for identifying and marking any areas on the airport adjacent to navaids that, if traversed, could cause emission of false signals or failure of the navaid
    - *c.* Describe how construction equipment and construction roadways on the airport are to be marked and lighted when on or adjacent to aircraft maneuvering areas
    - *d.* Describe procedures for the routing and control of equipment, personnel, and vehicular traffic during periods of construction on the aircraft maneuvering areas of the airport

22. Noncomplying conditions
    - *a.* Conditions under which air carrier operations will be halted
    - *b.* Designate the airport department to be informed if someone discovers an uncorrected, unsafe condition on the airport

References

Civil Aviation Authority (CAA). 2011. *Licensing of Aerodromes* (CAP 168). London: Her Majesty's Stationery Office, CAA.

Code of Federal Regulations (CFR). 2011. *U.S. Code of Federal Regulations*, Part 139: "Certification and Operations: Land Airports Serving Certain Air Carriers," May 2011, Washington, DC.

Federal Aviation Administration (FAA). 2004. *Airport Certification Manual* (AC 150/5210-22). Washington, DC: FAA.

CHAPTER 18

Sustainable Development and Environmental Capacity of Airports[1]

18.1 Introduction

Over the past half century the air transport industry has undergone remarkable growth, resulting in significant economic and social benefits (ACI 2004; ATAG 2008; ICAO 2002; OEF 2006). However, the adverse environmental and social costs associated with that growth are also significant and, as they increase, can constrain the operation of airports and their potential to respond to demand and so support regional development (Thomas et al. 2004). They can have substantial financial implications for airport operators, where investment is required to mitigate these impacts (e.g., emissions) or secure adequate resources (e.g., energy) to ensure efficient operations and meet passenger and service-partner expectations. A number of airports have failed to gain approval for new infrastructure development as a result of either the environmental implications of the construction itself or the additional traffic that would result from it (Upham et al. 2003). Finally, where an airport is offered for sale or where an airport operator seeks funds for further infrastructure development, its environmental impacts, environmental legacy, and environmental capacity can have significant implications for the market value of the asset (Thomas 2005).

---

[1]This chapter was written by Callum Thomas and Paul Hooper.

In broad terms, environmental issues can have a direct impact on the operating capacity of airports and their potential for future growth when

- Noise or emissions exceed regulatory limits, the provisions of planning agreements, or tolerance within surrounding communities (Thomas et al. 2010a; Bennett and Raper 2010).
- The climate-change implications of planning approval for new airport infrastructure (e.g., a new runway) run counter to government objectives (e.g., carbon dioxide reduction targets) (CCC 2009).
- The implications of a changing climate (e.g., extreme weather) have an impact on airport operations (Eurocontrol 2010).
- Airports cannot secure sufficient resources to guarantee normal operations and growth.
- Further infrastructure growth is restricted by ecologically important habitats around the boundary of the airport (Eurocontrol 2003).

A survey of medium-sized and larger airports across Europe found that almost two-thirds were subject to actual or potential *environmental capacity constraints*, with about 80 percent anticipating that it will get increasingly difficult to secure planning approval for growth as a direct result of environmental issues. Such challenges are now spreading to smaller airports in all parts of the globe. This is so because of

- Traffic growth
- Competition with other sectors for increasingly limited resources
- Growing affluence, democratization, and changing public attitudes
- Tightening regulation [e.g., relating to local air quality (Bennett and Raper 2010)]
- Emerging science [e.g., relating to the health effects of noise (WHO 2011) or climate change (Lee et al. 2009a, 2009b)]
- The consequences of a changing climate (Eurocontrol 2010, 2011)

This chapter introduces the key environmental issues associated with the operation and sustainable development of airports, explains how they can constrain operating capacity and growth, and describes the drivers for action and potential responses to minimizing or mitigating those impacts. The principles underlying the development of environmental management systems suitable for airports also will be explained.

The Sustainable-Development Challenge

The term *sustainable development* (SD) is receiving increasing attention within all sectors of the economy, including air transport, but its precise definition is unclear. Despite being popularized by the Brundtland Commission (WCED 1987) as "development which meets the needs of the present without compromising the ability of future generations to meet their own needs," the term has been subject to many different interpretations [see International Institute for Sustainable Development (IISD), www.iisd.org], and there is often difficulty in determining exactly what this means for an individual sector such as air transport or, within that, a single stakeholder such as an airport operator.

In broad terms, *sustainability* is the maintenance of important environmental functions (Ekins and Simon 1998) for present and future generations, but this does not take into account the need to take millions of people out of poverty, which in itself can result in environmental degradation. The Natural Step, a framework by which the practical consequences of a commitment to SD can be determined, defines a sustainable society as one in which nature is not subject to increasing concentrations of substances extracted from the earth's crust or produced by society, where degradation by physical means is minimized, and where human needs are met worldwide (Natrass and Altomare 1999). This perspective addresses the concepts of limited resources, environmental impact, and social equity at a global level and thereby establishes environmental criteria as the main limits to growth.

The implications for airports is that in striving for sustainable development, there is a need to compensate for growth through the introduction of more eco-efficient infrastructure, technologies, and operating systems (Upham 2001). It is only through such actions that airports will be able to continue to operate profitably and avoid or alleviate environmental capacity constraints.

18.2   The Issues

The operation of airports gives rise to a wide variety of impacts on local communities and the natural environment that have the potential to restrict airport development and need to be strategically and systematically assessed and managed to maximize the opportunities for growth.

Noise Impacts

The benefits of airport growth are spread across regions, but there are adverse impacts that are borne by residents of communities living nearby and along approach and departure routes (Thomas et al. 2010). These can generate significant opposition, leading to operational constraints, failure to secure planning approval for further development, and in the extreme, potentially closure of airports and their movement to more remote sites (e.g., Athens, Munich, and Hong Kong). In consequence, noise disturbance has become the single most important local environmental constraint to air transport growth.

As indicated in Chapter 3, noise exposure is related to the number and timing of aircraft movements, the noise generated by each, and the proximity of the aircraft to built-up areas (and people), all of which can be measured and modeled with a degree of accuracy (Ashford et al. 2011; ICAO 2008). The impact of that noise on people's lives, the perceived disturbance and resulting response of communities, is, however, unique to each airport and influenced by personal perceptions (Hume et al. 2001). This creates a challenge for airports in terms of what they should measure and manage.

Noise impacts cannot be measured in decibels because they can affect a wide variety of activities influenced by factors such as lifestyle [e.g., time spent at home and the nature of activities pursued (Hume et al. 2003)]. Perceived disturbance is influenced by a variety of non-acoustic social and economic factors such as expectation of quality of life and levels of home ownership (which are often influenced by income) (DfT 2007; Eurocontrol 2009). It is also influenced by other impacts of airports on local communities that can heighten sensitivity and further exacerbate relations with the airport operator (Table 18.1).

The response to disturbance ranges from complaints, through legal action (e.g., Chicago, Heathrow), to public protest (e.g., Frankfurt, Sydney, Tokyo-Narita). The result can be disruption to airport operations, external intervention, adverse publicity, and increased operating costs.

Whatever the causes of disturbance and fear, airports clearly need to engage with external stakeholders to address these issues and

---

- Disturbance to sleep, leisure, watching TV, reading, etc.
- The smell of unburned aviation fuel and associated health concerns
- Aircraft flying "off track"—where they are not expected to be
- Visual intrusion of overflying aircraft or of contrails
- Loss of tranquility in remote areas caused by overflights
- Road-traffic congestion and car parking on roads around airports
- Fear of future growth and the resulting noise impacts
- Fear of air accidents
- Fear of loss of house value or the inability to sell house
- The impact of noise on the education and learning of children

---

Source: Hume et al. (2003).

TABLE 18.1 Impacts of Air Traffic Operations on Communities Surrounding Airports, as Reported by Local Residents on Noise Complaints Lines

demonstrate their commitment to minimizing them. Engaging with external stakeholders and enabling them to contribute to airport development ensures that the airport can be made more acceptable to as many people as possible. This also can be important in building the trust needed for productive dialogue. This can require the establishment of formal community consultative committees (website of the Liaison Group of UK Airport Consultative Committees: www.ukaccs.info), the development of noise performance targets in consultation with external stakeholders (Thomas et al. 2010), and the conduct of independent benchmarking to demonstrate the adoption of appropriate best practices (Francis et al. 2002).

Since it will never be possible to completely eliminate aircraft noise, airport operators need to take further action to reduce community opposition. By directing the benefits of airport operations and growth (e.g., employment or community investment) to high-noise areas and by raising awareness of the benefits of airport development among residents, some have been able to engender greater tolerance (see MAG 2007a). In this manner, airport operators can achieve a more sustainable balance between maximizing the "benefits" of airport growth that may be enjoyed by millions and minimizing the "costs" borne primarily by tens of thousands of local people.

Local Air Quality

Gaseous emissions from the operation of airports (i.e., aircraft movements, passenger and staff vehicle trips, apron activities such as refueling, and other onsite sources such as power generation), when combined with those of other nearby polluting sources, such as industrial sites or major roads, can have a significant impact on local air quality. Evidence suggests that the operation of an airport can be the most significant source of some pollutants in a particular locality and that while on the airport itself, aircraft emissions dominate, and in the area surrounding airports, road traffic can be the main source.

Regulatory controls in Europe, North America, and elsewhere that establish emissions standards or air-quality limits for a particular locality (Culberson in Ashford et al. 2011; Bennett and Raper 2010) have the potential to constrain airport growth through restrictions on road traffic or aircraft movements. A survey of European airports in 2002 confirmed that in Sweden and Switzerland (at Arlanda, Gothenburg, Zurich, and Geneva airports), local air quality (or local emissions) presents a genuine threat to future operational capacity (ACI 2010). Meanwhile, the 2010 decision by the U.K. government not to support a third runway at Heathrow airport was due in part to concerns that local air quality would fall out of compliance with European Union (EU) legislative requirements.

A number of airports have invested in public transport (i.e., bus, rail, and tram) services to reduce car use by staff and passengers

(TRB 2008) and thereby cut local emissions. To encourage airlines to operate cleaner aircraft, in 1997,[2] Zurich Airport introduced a system of emissions-related landing charges that have subsequently spread to other countries and become the subject of International Civil Aviation Organization (ICAO) guidance documentation (ICAO 2007). Infrastructure (apron and taxiway) improvements that enable more efficient handling of aircraft, the introduction of new technologies (such as fixed electrical power and preconditioned air systems on stand), and the operation of gas-powered or electric vehicles all contribute to the effective management of local air quality and therefore reduce the risk of capacity constraints developing.

The U.K. Department for Transport carried out a major study into local air quality around Heathrow Airport in response to the operator's proposal to develop an additional runway (DfT 2006). This generated a series of reports that are available online (http://webarchive.nationalarchives.gov.uk/+, /http:/www.dft.gov.uk/pgr/aviation/environmental-issues/heathrowsustain) that examine local air quality at the airport Meanwhile, ICAO (2011a) and the U.S. Federal Aviation Administration (FAA 2007) have published guidance documentation to support implementation of good-practice air quality management at airports.

Airport Carbon Management

The threat of global warming has prompted urgent action by governments around the world, many of which have set stringent carbon reduction targets to avoid "dangerous climate change" (UNFCCC 2011). While this will create significant challenges for every part of the economy, air transport faces a particular threat because it is likely to be a legacy user of fossil fuels. Forecasts suggest that carbon dioxide emissions from this sector will not decline but will increase in the short to medium term at least (CCC 2009; Lee et al. 2009b). This suggests that aviation will remain in the political spotlight for the foreseeable future, and against this background, every sector of the industry will have to demonstrate action to minimize carbon emissions.

The UN ICAO has as yet no regulatory or target-setting role on greenhouse gases (GHGs) beyond a commitment to "limit or reduce" emissions from the sector (ICAO 2010) (although new standards were being developed in 2012). The UN Framework Convention on Climate Change (UNFCCC) through the Kyoto Protocol sets carbon dioxide reduction targets, but these are not sector-specific and at present apply only to emissions from domestic flights. In part as a consequence of this, in 2012, the EU extended its emissions trading scheme (ETS) to include the air transport sector. The EU ETS sets a cap on emissions from an increasing proportion of flights in European

---

[2]"Airport Emissions Charges at Zurich Airport"; available at www.zurich-airport.com.

airspace from 2012, with further growth requiring the purchase of carbon dioxide credits from other sectors, fleet modernization, or carbon-efficient operational improvements. This action is being opposed by some ICAO Member States, making a global emissions trading scheme unlikely in the short to medium term. Meanwhile, ICAO provides support and guidance to its Member States pending the emergence of a global regulatory response (ICAO 2011b).

These drivers for reducing airline fuel consumption and carbon dioxide emissions could have important implications for airports:

- Operational improvements on the ground will potentially require new ground handling procedures, technologies, and even infrastructure (e.g., improved taxiway layout).
- Airspace or flight-path changes may be introduced to reduce sector lengths, with potential impacts on noise exposure in communities surrounding those airports.

This is important because while air transport remains in the political spotlight, airports are increasingly likely to become the focus of efforts to reduce carbon dioxide emissions. For example, in Sweden, where Stockholm-Arlanda has a cap on carbon dioxide emissions included in its environmental permit, this cap includes emissions from starting and landing of aircraft; ground transport to, from, and at the airport; and heating and cooling of airport buildings (Wigstrand 2010). Arlanda Airport has had to support the development of ground transport services to reduce car use and therefore carbon dioxide emissions from passenger-access journeys by way of compensating for increased emissions from air traffic growth. Further, legislation is anticipated in the United Kingdom that will require airports to account for their carbon emissions on top of the market measures currently established [the Carbon Reduction Commitment Energy Efficiency Scheme (DECC 2008)]. It can only be a matter of time, therefore, before legislation and/or market instruments in other countries force airports to reduce the amount of carbon dioxide arising from their operations. It is for this reason that airport operators have to understand the principles of strategic and operational carbon management, and given that such activities can be very costly, it raises the question of what sources of carbon dioxide emissions an airport could be held responsible for.

A study at Manchester Airport (Sutcliffe et al. 2005) enumerated all carbon dioxide emissions arising from its operations and those of the companies that operate there or provide services and products to the airport. The principle sources are

- Air traffic movements
- Passenger access journeys
- Staff access journeys and business travel

- Ground transport vehicles
- Cargo center movements
- Direct energy consumption (gas and electricity)
- Waste production and processing
- Food and water consumption
- Capital and revenue material consumption
- Land use

This list clearly includes sources that the airport has direct control over and those generated by third parties, over which the airport may be able to exert only some influence. The World Business Council for Sustainable Development (WBCSD) and the World Resources Institute (WRI) provide guidance entitled, *Corporate Accounting and Reporting Standards* (WBCSD and WRI 2004), for carbon dioxide. This defines *direct emissions* as sources that are owned or controlled by the reporting entity (the airport operator). *Indirect emissions* are a consequence of the activities of the reporting entity (the airport operator), but occur at sources owned or controlled by another entity (airport service partners).

This implies that an airport can reduce the climate impact of its operations by

- Reducing GHG emissions from activities over which it has direct control (e.g., its infrastructure energy consumption and vehicle fuel use—known as *scope 1 emissions*).

- Reducing indirect emissions resulting from the energy used by others on site (e.g., in buildings owned and operated by the airport company—*scope 2 emissions*).

- Working with and influencing its service partners (e.g., to reduce their infrastructure energy demand or improve aircraft operational efficiency—*scope 3 emissions*).

- Working with and influencing its customers and the traveling public (e.g., by promoting public-transport use to access the airport—*scope 3 emissions*).

Since scope 3 emissions represent a very high proportion of all carbon dioxide arising from airport operations, ACI advises airports to include them in their carbon emissions inventories despite the fact that they are not under their direct control. Indeed, ACI (2009, p. 6) defines *airport emissions* as including

All emissions from activities associated with the operation and use of an airport, including ground support equipment, power generation and ground transport. Such activities can occur inside and outside the airport perimeter fence and may be the responsibility of the airport operator or other stakeholders. Emissions from aircraft should be included in an airport inventory, although depending on the reason for the inventory, an airport operator may choose to include either the landing and takeoff (LTO) cycle or the whole of departing flight emissions.

In 2011, Manchester Airport published a carbon inventory that identified the following key contributors to its carbon footprint and, therefore, focus of its management strategy:

- Passengers' possessive journeys to and from the airport (responsible for approximately 60 percent of carbon dioxide)
- Energy used for terminal lighting and heating (approximately 20 percent)
- The movement of aircraft on the ground (approximately 20 percent) (MAG 2011)

San Francisco Airport also has published an extensive climate-change action plan detailing efforts to reduce carbon dioxide emissions from its own operations and those of its service partners (SFO 2011).

The need for greater engagement in carbon management and reporting has been further recognized with the launch of the Airport Carbon Accreditation (ACA) scheme (www.aci.org/aca) by ACI European and Asia-Pacific regions in 2009 and 2011, respectively. The European initiative has seen a rapid increase in engagement in the four years since its introduction, with 59 airports currently accredited, representing 54 percent of European air traffic.[3] A similar picture is emerging in the Asia-Pacific Region, where four large airports (Abu Dhabi, Mumbai, Singapore, and Bangalore) were accredited in the first six months of operation of the scheme.

The primary focus for many of these airports is to reduce carbon dioxide emissions from

- Airport vehicles by changing the fleet from conventional fuels to biofuel, gas and electric vehicles, and through operational efficiencies
- Airport infrastructure (primarily terminals) through
  - *Carbon avoidance*—reducing energy use (e.g., through infrastructure design)
  - *Carbon reduction* (e.g., through energy-efficiency programs)
  - Carbon substitution (e.g., by generating or purchasing energy from renewable sources)

Such actions, however, are not restricted to these regions, and examples of current good practice in carbon management can be

---

[3]A list of airports accredited by the ACI Europe ACA scheme can be found at www .airportcarbonaccreditation.org/about.html.

found in a variety of reports and on a number of websites (e.g., ACI NA 2010). Stockholm Arlanda has taken a particularly innovative approach by using the aquifer on which the airport is built to store warm water during summer for heating terminals in winter and cold water during winter for cooling buildings in summer. This reduces the airport's heating needs by 20 percent and its cooling energy requirements by at least 60 percent (www.aviationbenefitsbeyond-borders.org). Meanwhile, the "Canopy" parking facility at Denver International Airport not only provides free charging stations for plug-in hybrid and electric vehicles, but it also generates its own electricity with solar panels and wind turbines, as well as making use of geothermal heat sources (www.solaripedia.com).

In addition to immediate commercial benefits, such action helps to prepare those airports for an increasingly carbon-constrained world. In consequence, it can be confidently predicted that airports that demonstrate effective carbon management will be more favorably viewed by banks and lenders and, thus, will find it easier to secure planning approval for growth.

Energy

Airports require a guaranteed and secure supply of energy (that is appropriately priced) if they are to meet peak demand from their service partners and passengers and therefore maximize their operational capacity. The maintenance of ambient temperature and air quality within terminal buildings to ensure passenger comfort accounts for the single most significant contribution to energy at the majority of airports. This is under the direct control of airport operators, but poses a particular management challenge given the increasing reliance of airports on retail and commercial activities.

Despite increasing energy efficiency, passenger growth is resulting in growing energy consumption at many airports, and this is occurring against a background of increasing demand for power across all sectors of the economy driven by economic development. In some countries, airports put considerable pressure on grid supplies, making them vulnerable to power outages and grid failures (which have happened in even the most advanced economies, such as the United States, the United Kingdom, and Italy).

To reduce long-term operating costs and to ensure that energy demand can be met when it arises, airports are placing greater emphasis on energy-conservation measures in the design of new terminal buildings.[4] For example, the new Midfield Terminal complex under construction at Abu Dhabi Airport will be one of the first in the region to achieve the highest level of green building design and operation

---

[4]"Sustainable Airports Benefiting from UK World Class Expertise," Department for Trade and Industry, London (DTi 2009); available at www.ukti.gov.uk.

with the award of Leadership in Energy and Environmental Design (LEED) platinum status (LEED was developed by the U.S. Green Building Council and provides a rating systems for the design, construction, and operation of high-performance buildings; www.usgbc.org).

Some airports also have invested in their own power-generation systems. For example, Athens International Airport in Greece has constructed a photovoltaic installation comprising almost 29,000 photovoltaic panels with a power output of 8.05 MW, accounting for some 20 percent of its energy demand (www.aia.gr).

Airport operators work with their tenants and service partners to reduce energy use through the introduction of low-energy equipment and operating systems. Some of the key elements of a successful energy management program include promoting increased awareness and buy-in from staff, the introduction of extensive metering across the site, and the development of energy contracts that promote reduced use rather than increasing income (through recharge) for the airport operator.

Water Use

Airports consume large quantities of water to maintain essential services. For example, in 2012, water consumption at Delhi–Indira Gandhi International Airport was estimated as 20 million liters per day. Meanwhile, Paris Charles de Gaulle consumes about 2,500 million liters of drinking water per annum. The environmental and financial consequences of this are significant.

Airport operators, service partners, and passengers require water for drinking, catering, retail, cleaning, flushing toilets, system maintenance and engineering, as well as for ground maintenance and landscaping. The operational capacity of an airport and the quality of service it provides to its customers and service partners can be severely constrained if it is unable to guarantee a secure, adequate, and low-cost supply of water to meet peak demand.

Economic development is associated with increasing water consumption across all sectors of the economy (www.unwater.org). For airports, meeting the growing demand for water therefore is becoming more challenging owing to increasing competition from other sectors, especially in parts of the world where water supplies are under stress or are declining either as a result of overabstraction, excessive runoff, or a decline in rainfall resulting from the effects of a changing climate (Eurocontrol 2011).

The principle of sustainable water management involves a hierarchical approach, the most environmentally and economically effective being to minimize water use at the source by

- Raising awareness and promoting "turn off" programs
- Fitting automatic switch-off and collection systems

- The introduction of simple low-water operating practices such as the use of sand rather than water and detergents to deal with fuel spills

- The use of low-water-consumption equipment such as waterless apron sweepers

Through such approaches, airports have been able to cut their water use significantly. Atlanta Airport, for example, used 86.7 million fewer gallons of water in 2009–2010 than in 2007–2008 (www .atlanta-airport.com).

Historically, airports have been designed to make use of groundwater or municipal supplies that meet appropriate quality standards. Where this water has been used only for nonindustrial purposes (e.g., washing, cleaning, and laundry), the wastewater can be collected, treated, and reused for activities such as flushing toilets, washing, and in some cases irrigation of plants. This may require the introduction of a dual drainage system and water-purification facilities. These options can require significant investment, but are usually cost-effective in the longer term.

In 2010–2011, the wastewater treatment plant at Hong Kong International Airport processed 1.37 million cubic meters of gray water from restaurants, aircraft catering and cleaning, and bathroom sinks, a proportion of which was used for landscape irrigation (http://www.hkairport2030.com). Meanwhile, Beijing Capital International Airport produces 10 million liters of reusable water each day through the treatment of sewage waste, and this is used for toilet flushing, landscaping, and cooling at the airport's power station.

Another source of water comes from collecting (harvesting) and storing rainfall, an approach that can significantly reduce the amount of water drawn from conventional supplies and act as a reservoir to guard against water shortages. At Singapore Changi Airport, rainwater harvesting provides almost a third of the airport's water needs, saving the company approximately US$390,000 per annum in operating costs (www.changiairportgroup.com).

As with any management approach, a cost-benefit analysis can be used to reveal the most appropriate way to handle water demand at a particular site, depending upon local conditions. A comparatively low-cost but highly effective example is provided by Portland International Airport, where 400 toilets average 80,000 flushes that consume 280,000 gallons of water every day. The simple act of fitting dual-flush systems to toilets has reduced water use by 177,000 gallons of water daily (www.portofportland.com).

Historically, being national assets and of significant value to economic development, it has been assumed that where competition for resources (such as water) existed, airport demand would take priority. However, in the future, as pressures build and competition increases between the airport and other critical sectors (especially domestic and agriculture), this may no longer continue to be the case. The most sustainable approach to water management therefore is for airports to seek to become self-sufficient in water supply by maximizing opportunities for water harvesting, recycling, and minimizing consumption.

The Management of Solid Wastes

Sustainable development acknowledges the fact that the earth's resources are limited and that their extraction, consumption, and disposal has significant environmental impacts. History demonstrates that economic growth leads to greater consumption of resources and the production of wastes, a trend that is unsustainable. This is dramatically illustrated by the estimate that "99 percent of the original materials used in the production of, or contained in, goods made in the United States become waste within six weeks of sale" (Von Weizsäcker et al. 1998).

Airport operations, aircraft, and passenger handling all have the potential to generate significant quantities of waste. The magnitude of the challenge for the largest airports is considerable, as illustrated by Hartsfield-Jackson Atlanta International Airport, which handled 89 million passengers in 2010 and generated over 60 tons of solid waste every day (Hartsfield-Jackson Atlanta International Airport 2010 Environment Report; available at www.atlanta-airport.com/; accessed June 22, 2012).

Drivers for waste management include international and national regulatory requirements, the increasing cost of waste treatment and disposal, the practicalities of handling large quantities of waste, and an acknowledgment of corporate responsibility. Although waste management regulations differ between countries and even between states, they do have common features relating to

- The handling of "special" wastes that are harmful to humans or the environment or have been transported across national borders, including chemicals, clinical and radioactive wastes, and foodstuffs
- Prioritizing waste minimization in order to reduce the generation of waste in the first place, before other management options are considered
- The need to promote the reuse and recycling of materials and dissuade the dumping of waste in landfills

Sustainable waste management seeks to minimize the amount of waste generated in the first instance, but then acknowledges that waste materials, if properly segregated, are valuable resources that can deliver significant financial returns as well as environmental benefits.

The process of setting up a management strategy begins with a waste audit. The airport waste stream comprises a wide variety of

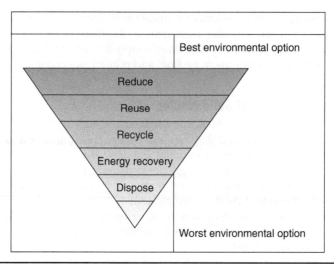

Figure 18.1   The waste management hierarchy.

materials, including glass, paper, wood, metals, chemicals, and clinical and food wastes. Different types of wastes arise from different activities and will influence the type of equipment and operating practices that need to be introduced in different parts of the airport. Figure 18.1 illustrates the principles of what is known as the *waste management hierarchy*, which seeks to minimize the production of wastes in the first instance, maximize opportunities for the reuse and recycling of materials, and minimize the amount of waste that is subsequently disposed of in landfills.

A key to *reducing waste at the source* is minimization at the point of purchase through the supply chain. This requires a review of existing and new contracts to minimize product packaging and maximize opportunities for the return of packaging. The simple process of bulk purchasing can bring important financial and environmental savings.

The *separation of wastes* on site allows the airport operator to reuse materials, sell materials to external contractors, or use them for composting or energy generation in contrast to having to pay external companies to collect and then process the untreated waste.

Airport waste management requires engagement with a large number of companies, including the airport operator, airlines, handling agents, maintenance companies, retail outlets, and other service partners. It also needs to address the specific challenges offered by airside safety (e.g., foreign-object damage) and security issues arising from the transfer of materials across the airside-landside boundary. For these reasons, a site-wide waste management system has been found to be most economically and operationally effective, with the airport operator as landlord and as the only site-wide corporate entity leading on the development of strategy, management, monitoring, and reporting systems. The airport operator also can take the lead in securing an appropriate waste-handling/ disposal contract for the airport site as a whole, thereby securing direct operational and environmental efficiencies and economic savings. The waste management programs at Frankfurt Airport (www .fraport.com) and London Heathrow Airport (www.heathrowairport .com) provide further examples of comprehensive approaches taken by airport operators.

Surface and Groundwater Pollution

The primary drivers for action to prevent surface and groundwater pollution include regulatory requirements, the need to minimize operating costs, and in the event of spills, associated environmental cleanup charges. Although the source of pollution may be an airport service partner (e.g., engineering company), the airport operator as landlord can have legal responsibility in some countries for preventing water pollution. Public health concerns, particularly about the contamination of water systems that are used for drinking or irrigation of crops, are politically very sensitive, and the airport's approach to this issue can be an important indicator of corporate responsibility. Banks and insurance companies are increasingly reluctant to take on clients with poor pollution records or will charge excessive premiums to do business with them, and the market value of an airport, when offered for sale, can be adversely affected by its legacy of groundwater pollution (Thomas 2005).

Surface water contamination can result in the death of plants and animals in rivers and can threaten human health if water is later abstracted for farming or domestic use. Damage to surface water or river systems can take many years and be very costly to repair. The quality of water (or levels of pollutants) that can be released into surface water systems is regulated in most countries, with fines being applied where pollution levels are exceeded. Avoiding groundwater contamination is particularly important; indeed, it is critical in areas of water shortage, where it is necessary to protect aquifers. For example, Tucson Airport in the United States is located on the edge of the Sonora Desert, where water is a scarce resource. The City of Tucson takes its drinking water from an aquifer that stretches under the airport. The airport operator has invested in systems designed to ensure that the water within the aquifer is protected from airport activities. This includes periodic flushing out of parts of the aquifer to ensure clean water supplies (www.azdeq.gov/environ/waste/sps/download/ tucson/tucsona.pdf).

Most activities associated with the operation of airports have potentially significant implications for surface and groundwater pollution, particularly those which take place outdoors (Table 18.2). The resulting pollutants are equally diverse and require a variety of

Activity	Potential Sources of Water Pollution
Aircraft ground handling	Fuels, oils, sewage
Aircraft, vehicle, stand washing	Detergents, oils, solids, carbon residues, heavy metals
Airfield maintenance	Runway rubber residues
	Oils, paint-stripping chemicals, kerosene, solvents
	Agricultural activities, including fertilizers, pesticides, herbicides
Winter operations	Runway and aircraft deicing and anti-icing chemicals, urea
Fire service training	Oil and firefighting foams

TABLE 18.2   The Principal Outdoor Activities Associated with Airport Operations That Have Implications for Surface and Groundwater Contamination and the Essential Measures for Minimizing Pollution

treatment systems. As with other environmental impacts, different service partners have different impacts. The airport operator has a central role in coordinating the environmental management system with which all service partners have to engage.

The effective control of surface and groundwater pollution involves the development of specific infrastructure and operational practices, the selection of appropriate materials, and the implementation of strict handling and spill-response procedures. Attempts should be made to minimize the use of more toxic chemicals through effecting planning. For example:

- Quantities of deicing materials used in winter operations can be minimized through better weather forecasting and by applying chemicals only when and where they are needed.

- Changes to operating practices can be introduced to reduce chemical use, for example, by scraping up lubricants rather than using degreasing agents, by sucking up oil spills, or by using sand rather than solvents.

- The purchase and use of less environmentally damaging (eco-friendly) materials and products that can be identified from manufacturers' technical information or by specifying particular environmental criteria when sourcing ("Schiphol Airport Purchase and Supply Chain Management"; available at www .schiphol.nl/SchipholGroup/CorporateResponsibility/ CRAtSchiphol/PurchaseAndSupplyChainManagement1 .htm; accessed June 20, 2012). It is important, however, that these new products have the same or appropriate performance characteristics, are suitable for use on airfields, and are used according to manufacturers' instructions.

Storage and handling procedures need to be clearly and strictly documented, with training and awareness-building delivered to key staff. This can reduce the risk of pollution from accidental spills chemical storage.

The airport estate is divided up into different catchment areas that feed underground drainage systems. The geographic distribution of aprons, taxiways, runways, terminal buildings, and maintenance and other infrastructure across the site can determine which catchments areas are likely to be subject to different risks in terms of surface and groundwater pollution. Underground settlement tanks or settlement ponds can be designed to retain water and allow chemical breakdown before it is released from the site. This same facility can collect highly polluted water arising from first flush (light intermittent rainfall) and store it until heavy rainfall dilutes the pollutants.

Automated monitoring and flow-diversion systems can allow water to be released into rivers and streams when levels of pollutants are within acceptable standards. The alternative can be to discharge into the foul sewer, but this can incur significant costs; as a result, on-site treatment can be most effective. In search of more sustainable solutions, airports are increasingly turning to natural water treatment systems such as planting reed beds, through which surface water is channeled (Revitt et al. 1997), thereby reducing both chemical use and industrial processes.

Adapting to a Changing Climate

The *Stern Review* (Stern 2007) concluded that the evidence of climate change is overwhelming, that climate change poses a serious global threat to human kind and is happening now, and that human activities are contributing to the phenomenon. It is now widely accepted by scientists and policy makers that if irreversible and dangerous changes in the earth's climate are to be avoided, the global mean temperature rise needs to be limited to a maximum 2°C above preindustrial levels (UNFCCC 2011). Accordingly, in 2007, the UN conference in Bali saw consensus for the first time from the whole international community on the need to cut global carbon dioxide emissions from human activities by 50 percent by 2050. In addition, some governments have started to advise all sectors of the economy to plan for a changing climate, and climate change adaptation strategies have started to emerge. For example, in 2010, the U.K. government published an order requiring the seven largest U.K. airports and all Scottish airports to assess the risk of climate change for their statutory responsibilities and bring forward plans to adapt to those threats (DEFRA 2009). A number of these have been published on the Internet and provide a climate-change risk-assessment methodology (http://archive.defra.gov.uk/environment/climate/documents/adapt-reports/08aviation/stansted-airport.pdf).

Key factors when determining the appropriate responses to this challenge are

- The consequences of climate change
- Certainty of changes and time scales over which they are likely to occur
- Implications for airport infrastructure, operations, and capacity
- The magnitude of the impact on the business
- The time taken and cost of mitigation

A considerable body of research exists into the impact of aircraft engine emissions on the climate, and attention is now turning to the likely implications of climate change itself for the future growth and development of the industry and therefore airports (Eurocontrol 2010, 2011; NATS 2011).

Although it is difficult to generalize likely changes in climate that will occur across the world, the principal consequences are summarized in Table 18.3, along with current levels of scientific certainty.

It is noteworthy that while some of these impacts (such as sea-level rise) may not become extensive for 50 to 100 years, some coastal airports are already seeing evidence of this phenomenon. Other changes listed in the table are also already manifesting at airports in different parts of the world; all are likely to have an impact on airport operations within a 30-year master planning horizon and need to be considered now.

Implications of Climate Change	Certainty
Sea-level rise	Virtually certain
Temperature rise (particularly in Arctic regions)	Virtually certain
Late onset of freeze, early thaw	Virtually certain
Increase in very hot days	Virtually certain
Increase in heat waves	Very likely
Increased extreme rainfall events	Very likely
Increases in droughts in some areas	Likely
Changes in rainfall patterns, seasons, flooding	Likely
Increases in hurricane intensity	Likely
Increased intensity of cold-season storms	Likely
Increases in winds, waves, and storm surges	Likely

Source: Adapted from IPCC (2007).

TABLE 18.3 Likely Consequences of Climate Change and Their Certainty

Global sea levels are predicted to rise 0.2 to 0.5 meter by 2100 (under the medium-emissions scenario) (IPCC 2007). This, when combined with increased storminess, would result in more frequent flooding and storm surges causing coastal erosion and land subsidence. This threat can apply not only to coastal airport infrastructure but also along ground-transport access routes. A study by Eurocontrol (2009), working with the U.K. Met Office, revealed 34 airports across Europe that would be affected; meanwhile, ICAO has identified approximately 150 airports across the world that potentially would be affected (ICAO 2010), including some large capital-city airports. Reports have emerged of significant damage already being caused to low-lying coastal airports in Norway and of proposals to move San Francisco Airport to a new location because of the long-term threat from increasing sea levels.

Rising temperatures, prolonged heat waves, and shorter winters will have a wide variety of implications for airports:

- New design specifications will be required for future airport terminals and retrofitting of existing terminals to improve thermal efficiency and reduce energy requirements for passenger comfort

- Existing infrastructure will need to be refurbished with new construction materials, for example, the risk of melting of taxiways and other infrastructure in periods of prolonged heat (already reported by one U.K. airport).

- Melting of the permafrost base will result in subsidence of runways and other airport infrastructure—particularly in Arctic and Antarctic regions. For example, in Svalbard, Norway, the depth to the permafrost has increased from 2.5 meters in 1973 to 4.5 meters in 2009, with the result that the ground is softer and the runway is starting to subside. Higher temperatures will mean a decrease in aircraft lift, requiring longer runways at some airports, changes in aircraft type or maximum payload, and potentially changes in airspace design with consequences in terms of noise impacts on surrounding communities.

Precipitation changes will similarly affect airport operations and design requirements. For example:

- Increased rainfall and extreme downpours will threaten the integrity of some airport infrastructure, causing erosion and subsidence and requiring more investment in storm water runoff and ground and surface watercourse protection.

- Changes in patterns or levels of rainfall and prolonged drought threaten water shortages for airports that, as indicated earlier, could affect capacity and levels of service quality.

- A number of airports are already reporting extreme down-pours causing flooding of aprons and terminal building, leading to delays, diversions, and disruption of operations.

- In dry areas, airports will face increasing incidences of dust storms, which again risk operational disruption.

Finally, increased storminess both at airports and in the en-route phase of flight also will generate more weather delays, diversions, and congestion, requiring better weather forecasting and more flexibility in scheduling systems.

Biodiversity

Historically, airports have tended to be built in open countryside close to major urban conurbations, and as such, many are surrounded by habitats that can be of particular value in terms of their biodiversity. Within the airport boundary, operational and safety considerations create an environment that is either hostile to wildlife or a monoculture (e.g., grassland). In the context of sustainable development, airports not only have to manage their existing estates to promote biodiversity and protect habitats, but their ability to develop new infrastructure also can be severely restricted by sensitive sites or indeed species in the surrounding countryside that are protected by international and national conventions and regulations.

Sustainable development would require that airports compensate for their adverse ecological impacts through landscape and habitat management, mitigation, and compensation programs (e.g., MAG 2007b). The challenge of protecting biodiversity while facilitating airport growth is exacerbated, however, by the issue of bird hazards to aircraft (see International Bird Strike Committee at www.int-birdstrike .org), although there are some notable examples where a successful balance has been achieved (Anderson 2004).

18.3 Environmental Management Systems

Airports have to address their environmental impacts strategically if they are to maximize their potential for long-term growth and thus their value as commercial assets. Effective and appropriate approaches can require significant financial investment in new infrastructure, technologies, and operating practices. Therefore, a thorough and systematic approach [an environmental management system (EMS); see Sheldon and Yoxon (2006)] is necessary to ensure the most cost-effective responses. An EMS allows an airport to anticipate and respond to its environmental impacts, enabling it to compensate for growth, avoid environmental capacity constraints, limit environmental liabilities, exploit commercial advantages, and maximize potential financial benefits. The International Standards Organization (ISO) defines an EMS as a tool that enables an organization to

- Identify and control the environmental impact of its activities, products or services, and to

- Improve its environmental performance continually, and to

- Implement a systematic approach to setting environmental objectives and targets, to achieving these and to demonstrating that they have been achieved[5]

A proactive and preventative approach to environmental management has been found to be more effective and less costly:

- If an airport contravenes water emission standards, there is a risk of fines, the cost of immediate remedial action, the liability for cleanup costs, and the negative commercial impact resulting from damage to corporate reputation, all of which can be avoided.

- If an airport fails to plan for road-traffic growth by reserving land for or developing mass-transit facilities for passengers, it may in the future find its operations constrained by local air-quality legislation.

Key steps in the development of an EMS are generic and described in detail in Sheldon and Yoxon (2006). The overall approach, however, is summarized by the Deming cycle (Figure 18.2) that begins with a systematic review of all environmental impacts and then establishes a cyclic system of planning, implementation, checking performance, taking corrective action, and review designed to evolve and compensate for growth.

As demonstrated earlier, airport environmental impacts are varied and often arise as a result of the activities of service partners (e.g., airline operations give rise to noise impacts, aircraft maintenance companies can pose the greatest risk of water pollution). The airport operator, however, plays a critical role in developing the most appropriate responses delivered through a site-wide integrated EMS for all key environmental impacts. The rationale for this is that the airport operator

- In some countries has legal responsibility as landlord for its service-partner activities on site, for example, a fuel spill from an aircraft

- Is best positioned to provide an overarching coordinator role

---

[5]For a description of the nature of the ISO approach to EMS, see www.iso.org/iso/iso_14000_essentials.

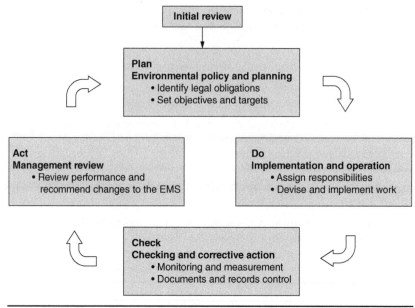

Figure 18.2   The principles of an EMS—the Deming cycle. (*Adapted from Straker, 1995.*)

- Is responsible for design of the infrastructure that determines operational practices
- Can best deliver financial and environmental benefits, for example, through reduced waste management and energy costs
- Has an interest in ensuring the sustainable development of the site and may need to make strategic decisions that require tradeoffs, for example, between environmental impacts or between the priorities of different service partners
- Is regarded by the general public as being "responsible" for all impacts associated with its operation

A successful management strategy, therefore, requires that the activities of all internal airport stakeholders are coordinated. Tens or even hundreds of different organizations can, however, be involved in the operation of a single airport, ranging from small local operators to major international companies with only a handling agent on site, and this creates additional management challenges.

The integrated nature of airport operating conditions makes it vital that all service partners are involved in the implementation of the EMS if the procedures are to be comprehensive, safe, and secure. Delivering such a complex process across so many organizations requires

- That stakeholders buy into the process
- The establishment of an airport-wide environmental forum

- The introduction of collaborative environmental management (CEM) and collaborative decision-making (CDM) systems (described in detail by the European Organization for the Safety of Air Navigation at www.eurocontrol.int)

There are currently two internationally recognized EMS certification standards, ISO 14001 (www.iso.org), which is applicable to any business sector or activity and recognized across the world and can apply to a whole organization or an individual site, and the EU's Eco-Management and Audit Scheme (EMAS; www.iema.net/ems/emas). Airports not wishing to apply for formal accreditation can refer to the wide array of published literature on how to set up an EMS (Sheldon and Yoxon 2006).

18.4   Conclusion

The link between air transport growth and economic and social development is becoming increasingly important as a result of the emerging global economy/society and the growth of international tourism. Forecasts indicate a strong and sustained increase in demand over the next two to three decades, suggesting that the development of many city regions or even countries will be linked to the development of their airports.

Sustainable development requires that this growth be compensated for through improving environmental performance, and this in itself will put new pressures on airport operators. But the challenge will be even more significant because it is clear that in the future, environmental pressures will build on the industry. These will arise from the changing climate, emerging science, new regulations, increasing costs, and changing public and political attitudes. Delivering ongoing environmental performance improvement and responding to emerging political pressures and regulations will have implications for all airport service partners, requiring a collaborative approach across the airport site.

Environmental pressures are currently most evident at larger airports, in regions of the world where aviation is most mature, where environmental pressures and regulations are most onerous, and where quality-of-life expectations are highest. But future growth is forecast to be the strongest in the emerging economies and in particular the BRIC (Brazil, Russia, India, and China) countries (Airbus 2010; Boeing 2010), leading to massive infrastructure development. Between 2011 and 2015, China plans to expand 91 airports, construct 56 new airports, and move 16 to new sites (*The Economist* 2011). In this context, it is critical that airport operators across all parts of the world (but particularly in regions of highest growth) have a good understanding of the principles of sustainable development of airports with environmental management at a strategic and operational level. Furthermore, airports need to develop cost-effective responses with their service partners to underpin ongoing growth.

References

Airbus. 2010. *Global Market Forecast 2011–2030.* Blagnac Cedex: Airbus (www.airbus .com).

Airports Council International (ACI) Europe. 2004. *York Aviation Study—The Social and Economic Impact of Airports in Europe.* Brussels: ACI (www.aci-europe.aero).

Airports Council International (ACI). 2009. *Guidance Manual: Airport Greenhouse Gas Emissions Management.* Montreal: ACI (www.aci.aero).

Airports Council International (ACI) Europe. 2010. *Effects of Aircraft Emissions on Air Quality in the Vicinity of European Airports.* Brussels: ACI.

Airports Council International (ACI) North America. 2010. *Going Greener: Minimizing Airport Environmental Impacts.* Washington, DC: ACI (www.aci-na.org).

Air Transport Action Group (ATAG). 2008. *The Economic and Social Benefits of Air Transport.* Geneva: ATAG (www.atag.aero).

Anderson, P. 2004. "Conservation Takes Off at Airports." *Land Conservation Management* 2(4): 8–11.

Ashford, N. J., S. Mumayiz, and P. H. Wright. 2011. *Airport Engineering: Planning, Design, and Engineering of 21st Century Airports,* 4th ed. Hoboken, NJ: Wiley.

Bennett, M., and D. Raper. 2010. "Impact of Airports on Local Air Quality." In R. Blockley and W. Shyy (eds.), *Encyclopedia of Aerospace Engineering.* Hoboken, NJ: Wiley, pp. 3661–3670.

Boeing. 2010. *Long Term Market, Current Market Outlook 2011–2030.* Seattle: The Boeing Company. Available at: www.boeing.com.

Committee on Climate Change (CCC). 2009. *Aviation Report.* London: CCC (www .theccc.org.uk).

Department for Environment, Food and Rural Affairs (DEFRA). 2009. *Adapting to Climate Change: Helping Key Sectors to Adapt to Climate Change: Statutory Guidance to Reporting Authorities 2009.* London: DEFRA (www.defra.gov.uk).

Department for Transport (DfT). 2006. *Project for the Sustainable Development of Heathrow: Report of the Air Quality Technical Panels.* London: DfT, HMSO.

Department for Transport (DfT). 2007. *Attitudes to Noise from Aviation Sources in England (ANASE).* London: DfT (www.dft.gov.uk).

Department of Energy and Climate Change (DECC). 2008. *Climate Change Act 2008.* London: HMSO (www.legislation.gov.uk).

Ekins, P., and S. Simon. 1998. "Determining the Sustainability Gap: National Accounting for Environmental Sustainability." In P. Vaze (ed.), *UK Environmental Accounts 1998.* London: HMSO, pp. 147–167.

Eurocontrol. 2003. *The Concept of Airport Environmental Capacity: A Study for Eurocontrol.* Brussels: Eurocontrol (www.eurocontrol.int).

Eurocontrol. 2009. *Attitudes to Aircraft Noise Around Airports (5A).* Paris: Eurocontrol (www.eurocontrol.int).

Eurocontrol. 2010. *"Challenges of Growth" Environmental Update Study: Climate Adaptation Case Studies.* Brussels: Eurocontrol (www.eurocontrol.int).

Eurocontrol. 2011. *"Challenges of Growth" Environmental Update Study: Climate Adaptation Case Studies.* Eurocontrol Commentary. Brussels: Eurocontrol (www .eurocontrol.int).

Federal Aviation Administration (FAA). 2007. *Air Quality Procedures for Civilian Airports and Air Force Bases.* Washington, DC: FAA (www.faa.gov).

Francis, G., I. Humphreys, and J. Fry. 2002. "The Benchmarking of Airport Performance." *Journal of Air Transport Management* 8: 239–247.

Hume, K. I., D. Terranova, and C. N. Thomas. 2001. "Complaints and Annoyance Caused by Airport Operations: Temporal Patterns and Individual Bias." *Noise and Health* 4(15): 45–55.

Hume, K. I., H. Morley, and C. S. Thomas. 2003. *Review of Complaints and Social Surveys at Manchester Airport: Attitudes to Aircraft Annoyance Around Airports* (EEC/SEE/2003/004). Brussels: Eurocontrol (www.eurocontrol.int).

Intergovernmental Panel on Climate Change (IPCC). 2007. "Working Group I: The Physical Science Basis." In *Fourth Assessment Report: Climate Change 2007.* Geneva: IPCC.

International Civil Aviation Organization (ICAO). 2002. *The Economic Contribution of Civil Aviation*. Montreal: ICAO (www.icao.org).

International Civil Aviation Organization (ICAO). 2008. *Recommended Method for Computing Noise Contours Around Airports* (Document No. 9911). Montreal: ICAO (www.icao.org).

International Civil Aviation Organization (ICAO). 2007. *Guidance on Aircraft Emission Charges Related to Local Air Quality* (Document No. 9884). Montreal: IACO (www.icao.org).

International Civil Aviation Organization (ICAO). 2010. Resolution A37-19: *Consolidated Statement of Continuing of ICAO Policies and Practices Related to Environmental Protection: Climate Change*. Montreal: IACO.

International Civil Aviation Organization (ICAO). 2011a. *Airport Air Quality Manual* (Document No. 9889). Montreal: IACO.

International Civil Aviation Organization (ICAO). 2011b. *Guidance Material for the Development of States' Action Plans: Towards the Achievement of ICAO's Global Climate Change Goals*. Montreal: IACO.

Lee, D. S., G. Pitari, V. Grewe, K. Gierens, J. E. Penner, A. Petzold, M. J. Prather, U. Schumann, A. Bais, T. Berntsen, D. Iachetti, L. L. Lim, and R. Sausen. 2009a. "Transport Impacts on Atmosphere and Climate: Aviation." *Atmospheric Environment* 44(37): 4678–4734.

Lee, D. S., D. W. Fahey, P. M. Foster, P. J. Newton, R. C. M. Witt, L.L. Lim, B. Owen, and R. Sausen. 2009b. "Aviation and Global Climate Change in the 21st Century." *Atmospheric Environment* 43: 3520–3537.

Manchester Airports Group (MAG). 2007a. *Community Plan: Part of the Manchester Airport Master Plan to 2030*. Manchester: MAG (www.manairport.co.uk).

Manchester Airports Group (MAG). 2007b. *Manchester Airport Environment Plan: Part of the Manchester Airport Master Plan to 2030*. Manchester: MAG (www.manairport.co.uk)

Manchester Airports Group (MAG). 2011. *Sustainability Report 2010/11*. Manchester: MAG (www.manairport.co.uk).

Natrass, B., and M. Altomare. 1999. *The Natural Step for Business: Wealth, Ecology and the Evolutionary Corporaton*. Gabriola Island, Canada: New Society Publishers.

National Air Traffic Services (NATS). 2011. Climate Change Adaptation Report, July 2011. London: NATS.

Oxford Economic Forecasting (OEF). 2006. *The Economic Contribution of the Aviation Industry in the UK*. Oxford, UK: OEF.

Revitt, D. M., R. B. E. Shutes, N. R. Llewellyn, and P. Worrall. 1997. "Experimental Reedbed Systems for the Treatment of Airport Runoff." *Water Science and Technology* 36(8): 385-390.

San Francisco International Airport (SFO). 2010. *Climate Change Action Plan 2010*. San Francisco: SFO.

Sheldon, C., and M. Yoxon. 2006. *Environmental Management Systems: A Step-by-Step Guide to Implementation*, 6th ed. London: Earthscan.

Stern, N. 2007. *The Economics of Climate Change: The Stern Review*. Cambridge, UK: Cambridge University Press.

Straker, D. 1995. *A Toolbook for Quality Improvement and Problem Solving*. London: Prentice-Hall.

Sutcliffe, M., P. D. Hooper, and C. S. Thomas. 2005. "Exploring the Potential for the Commercial Application of Ecological Footprinting Analysis: An Airport Case Study." *Proceedings of the Business Strategy and the Environment Conference*, Leeds University, September 5 and 6 (online); available at: www.crrconference.org/downloads/sutcliffe.pdf.

Thomas, C. 2005. "Environmental Issues and Their Impact upon the Market Value of Airports." *Airport Investor Monthly* 11 (www.centreforaviation.com).

Thomas, C. S., K. I. Hume, and P. D. Hooper. 2004. "Aircraft Noise, Airport Growth and Regional Development." In *Proceedings of the Royal Aeronautical Society/American Institute of Aviation Acoustics Conference*, Manchester, May 10–12.

Thomas, C. S., J. A. Maughan, P. D. Hooper, and K. I. Hume. 2010. "Aircraft Noise and Community Impacts." In R. Blockley and W. Shyy (eds.), *Encyclopedia of Aerospace Engineering*. Chichester, UK: Wiley, pp. 3599–3606.

Transportation Research Board (TRB). 2008. *Ground Access to Major Airports by Public Transportation* (ACRP Report 4). Washington, DC: TRB (www.trb.org).

United Nations Framework Convention on Climate Change (UNFCCC). 2011. *Report of the Conference of the Parties on Its Sixteenth Session, Held in Cancun from 29 November to 10 December 2010*, Part Two: "Action Taken by the Conference of the Parties at Its Sixteenth Session: Decisions Adopted by the Conference of the Parties." Bonn, Germany: UNFCCC, p. 3.

Upham, P. 2001. "A Comparison of Sustainability Theory with UK and European Airports Policy and Practice." *Journal of Environmental Management* 63(3): 237–248.

Upham, P., C. Thomas, D. Gillingwater, and D. Raper. 2003. "Environmental Capacity and Airport Operations: Current Issues and Future Prospects." *Journal of Air Transport Management* 9: 145–151.

Von Weizsäcker, E. U., A. B. Lovins, and L. H. Lovins. 1998. *Factor Four: Doubling Wealth, Halving Resource Use*. London: Earthscan Publications.

Wigstrand, I. 2010. *The ATES Project: A Sustainable Solution for Stockholm-Arlanda Airport*. Available at: http://intraweb.stockton.edu/eyos/energy_studies/content/docs/effstock09/Session_6_3_ATES_Applications/55.pdf.

World Business Council for Sustainable Development (WBCSD) and World Resources Institute (WRI). 2004. *The Greenhouse Gas Protocol: A Corporate Accounting and Reporting Standard*, Revised Edition. Geneva, WBCSD and Washington D.C., WRI.

World Commission on Environment and Development WCED. 1987. *Our Common Future*. Oxford, UK: Oxford University Press.

World Health Organization (WHO). 2011. *Burden of Disease from Environmental Noise: Quantification of Healthy Life-Years Lost in Europe*. Geneva: WHO.

Index

Note: Page numbers referencing figures or tables are *italicized*.

A

Abu Dhabi Airport, 17, 562–563
ACAP. *See Aircraft Characteristics for Airport Planning*
Accelerate-stop distance available (ASDA), 99, 101, 104
Accelerate-stop distance required (ASDR), 98–99
Access, 542. *See also* Airport access
  congestion, 411
  control within and throughout airport buildings, 277–278
  emergency road, 383
  gates, 280–281
  security with gate for staff, *279*
  vehicle, 279, *280*
Access interaction:
  check-in procedures, 419–422
  length of access time, 419
  missing flights and consequences, 422–424
  with passenger terminal operation, 418–424
  reliability of access trip, 419
Access modes:
  automobile, 424–427
  bus, *159*, 434–435
  dedicated rail systems, 435–437
  factors influencing choice of, *438*, 439
  limousine, 429–430
  rail, 430–434
  taxi, 427–429

Accidents. *See also* Airport aircraft emergencies
  aircraft, 128, 131, 347, 354–355, 379, 382, 399–400
  aircraft probability for, 377–379
  causation model, *481*
  fatality rates for, *378*
ACI. *See* Airports Council International
Adjacencies, space components and, 246–249
Administration. *See* Federal Aviation Administration; Operational administration
Administrative structure:
  of Amsterdam Schiphol Airport, *21*
  of Frankfurt Airport, *20*
Aer Rianta, 225. *See also* Irish Aviation Authority
Aerodromes:
  acceptable level of safety, 473–475
  certification and operational readiness, 119–122
  control, 345–347
  ICAO's annex 14, 467–468
  safety levels and assessments, 472–473
  SMSs, 469–475
Aerodromes, technical services:
  aeronautical information, 373–375
  air traffic control, 329–350

Aerodromes, technical services
    (*Cont.*):
  airspace wind chart, *369*
  control light signals, *346*
  DHs, *343*
  EMAS installation, *329*
  ICAO classification of airspace,
    341
  London TMA plan, *342*
  meteorology, 361–373
  planned FAB configuration for
    Europe, *335*
  probability of likelihood
    classification, *327*
  scope of, 325–326
  SID, *350*
  SMSs, 326–329
  SRA chart, *349*
  synoptic weather chart, *368*
  telecommunications, 350–361
  tolerance classification matrix,
    327
Aeronautical fee structures, *52*
Aeronautical information, 16–17
  with availability of information,
    375
  NOTAM code, 374
  scope, 373–374
  urgent operational information,
    374
Aeronautical Information Publication
    (AIP), 99
Aeroports de Paris, 38
African departure airport, 47
Agricultural produce, 239
AIP. *See Aeronautical Information
    Publication*
AIPs. *See Air Information
    Publications*
Air bridge, *157–158, 165*
Air Canada, 239
Air cargo:
  actors and relationships with, *295*
  growth rates of, 32
  interior, *305–306*
Air France, 354, 400
Air India, 257, 275
Air Information Publications (AIPs),
    90

*Air Navigation Order and
    Regulations* (CAA), 121, 337
Air quality, local, 557–578
Air traffic control (ATC), 42, 542
  classes of airspace, 338–339
  community impacted by
    operations of, *556*
  flight rules, 336
  function of, 331–333
  fundamental changes, 329–331
  general flight rules, 336–337
  IFRs, 122, 337–338
  information for, 372
  international collaboration of,
    333–336
  operational characteristics and
    procedures, 345–350
  operational structure, 343–345
  separation minima, 339–343
  VFRs, 122, 337
Air transport services, 59. *See also*
    International Air Transport
    Association
Airbags, to lift aircraft, *405*
Airbus:
  A380, 6, 114, 116–117, *158*
  Air France, 354
Aircraft. *See also* Airport aircraft
    emergencies
  accident probability for, 377–379
  accidents, 128, 131, 347,
    354–355, 379, 382, 399–400
  airbags to lift damaged, *405*
  Airbus A380, 6, 114, 116–117,
    158
  airport influences on, 90–100
  availability with airline
    scheduling, 48–49
  B747-8F, 307–308
  balance, 232
  Boeing, 6, 48–49, 72, 74, 92, 93,
    95–96, 107
  container arrangements in
    wide- and narrow-bodied,
    302
  criteria recommended for noise
    control of, *87*
  data, 97
  Douglas DC-9, 74

Aircraft (*Cont.*):
  ER, 6
  ground signalman marshalling,
    160
  isolated parking position,
    281
  loading ULDs onto, *192*
  loading with flight build,
    189–191
  measured noise levels of,
    77–78
  noise certification limits, *75*
  observations and reports,
    365–366
  quieter, 70
  recovery methods for heavy,
    404
  removal of disabled, 402–404
  rescue, 146–147
  SBR, annual peak-hourly and
    peak-day, *34–35*
  technology with two-engine ER
    aircraft, 6
  tow tractor, 160, *161*
  typical payload-range diagram,
    94
  weight, 51, 232
*Aircraft Characteristics for Airport
  Planning* (ACAP), 94
Aircraft Meteorologic Data Relay
  (AMDAR), 366
Aircraft noise, 64. *See also* Noise
  control; Noise-control
  strategies
  A-weighted day-night average
    sound level, 67–68
  community response to,
    68–69
  cumulative-event measures,
    65–68
  day-evening-night average
    sound level, 67–68
  day/night average sound
    levels, U.S., 65–66
  equivalent continuous sound
    level L_{EQ}, 66–67
  NEF in U.S., 67
  NNI, 66
  single-event measures, 65

Aircraft performance
  characteristics:
  Airbus A380, 116–117
  aircraft and, 90–100
  aircraft data, *97*
  airport influences on, 89–117
  approach and landing, 105–107
  automatic landing, 110–114
  Boeing 777-300ER payload-
    range diagram, *93*
  Boeing 777-300ER takeoff
    runway length chart,
    95–96
  Boeing 777-800 landing-distance
    performance, *107*
  climb-path segments, 102, *103*
  departure, 100–105
  DHs and RVRs for precision-
    approach runways, *111*
  gross climb-gradient
    requirement, *102*
  mass data for Boeing's highest
    gross weight, *92*
  obstacle-assessment surfaces,
    113
  operational and performance
    regulation tradeoff, *110*
  operations in inclement
    weather, 114–116
  phase of flight data, *108*
  runway declared distances,
    100
  safety considerations, 107–110
  typical payload-range diagram,
    94
Aircraft ramp servicing:
  catering, 166
  cooling/heating, 164–166
  deicing and washing, 164, *165*
  fault servicing, 162
  fueling, 162–163
  ground handling and,
    162–166
  ground power supply, 163
  onboard servicing, 166
  other servicing with, 166
  wheels and tires, 163
Aircraft Rescue Firefighting
  Facility (ARFF), 547, *548*

Airfield construction, 148–150
Airfield inspections, 138–140
Airline-related operations
    functions:
  aircraft weight and balance,
    232
  balance/trim, 235
  in flight, 232–233
  flight-crew briefing, 235–238
  flight dispatch, 230–231
  flight planning, 231–232
  flight watch, 239
  landing, 234–235
  loading, 235
  takeoff, 232
Airline-related passenger services,
    228–230
Airlines. *See also* Scheduling,
    airline; *specific airlines*
  alliances, 14–15
  designated check-in area,
    155
  passenger steps, *157*
  scheduling within, 52–56
  as system actor, 1–2, *3*
Airmen's meteorological
    information (AIRMETs),
    369–370
Airport access:
  bus bays at LHR, *435*
  capacity constraints on
    outbound airport
    throughput, *412*
  by car or taxi by selected
    airports, *417*
  check-in times for passengers
    prior to departure, *423*
  conclusions, 439–440
  curbside access activity times,
    428
  curbside access length per
    million passengers per
    year, *426*
  factors influencing access-mode
    choice, *438*, *439*
  flight length on passenger
    terminal dwell times, *421*
  flight type on passenger
    terminal dwell times, *421*

Airport access (*Cont.*):
  highway and air passenger
    traffic patterns, *418*
  in-town and other off-airport
    terminals, 437–438
  interaction with passenger
    terminal operation,
    418–424
  missed flights and influence on
    passenger terminal dwell
    times, *423*
  modal split by bus, *440*
  modes, 424–437
  parking demand with annual
    and peak-hour originating
    passengers, *425*
  parking requirement
    recommendations, *426*
  as part of airport system,
    411–413
  passenger terminal dwell times
    and influence on, *420*
  passenger terminal dwell times
    for long- and short-access
    journeys, *420*
  passengers, workers, visitors,
    senders/greeters in
    proportion, *415*
  passengers with origin or
    destination in CBD, *416*
  ranked importance of selected
    attributes in passengers'
    choice of access mode, *439*
  rapid transit in relation to
    terminals at LHR, *432*
  rapid transit on access modal
    split to LHR, *433*
  remote bus park for Caribbean
    resort airport, *436*
  users and modal choice,
    414–418
Airport aircraft emergencies:
  accident fatality rates, *378*
  accident site layout, *397*
  aircraft lifted with airbags,
    405
  airport categorization, *380*
  airport emergency plan,
    389–398

Airport aircraft emergencies (*Cont.*):
  all-terrain RFFS vehicle, *388*
  communication and alarm
    requirements, 384–385
  emergency plan checklist,
    406–409
  emergency plan document,
    391–394
  failure probability by type, *378*
  firefighting and rescue
    procedures, 398–400,
    540–541
  flow-control chart for accident
    off airport, *396*
  flow-control chart for accident
    on airport, *395*
  foaming of runways, 381–382,
    400–402
  frequency of, *380*
  general, 377
  level of protection required,
    380–383
  minimum characteristics for
    rescue and firefighting, *386*
  minimum number of vehicles,
    381
  minimum usable amounts of
    extinguishing agents, *381*
  personnel requirements,
    388–389
  probability of accidents,
    377–379
  recovery methods of heavy
    aircraft, *404*
  removal of disabled aircraft,
    402–404
  rescue and firefighting vehicles,
    381, 385–388
  rescue equipment on rescue
    vehicles, *387*
  summary, 404–409
  types of, 379–380
  water supply and emergency
    access roads, 383
Airport and Airways Trust Fund, 6
Airport authorities:
  BAA, 6–8, 14, 33–34, 462–463
  Brazil, 17
  Dublin, 17

Airport authorities (*Cont.*):
  INFRAERO, 17, *306*
  Ireland, 17
  Metropolitan Washington, 5
  PANYNJ, 17, 23, *25*
Airport buildings, access control
  of, 277–278
Airport carbon management,
  558–562
Airport duty manager, AOCCs,
  523
Airport Emergency Plan:
  command, 394–397
  communications, 398
  coordination, 398
  document, 391–394
  explanation, 389–390
Airport Engineering (Ashford,
  Mumayiz and Wright), 90, 443
Airport industry:
  benchmarking, 463–464
  commercialization and impact
    on, 459–461
  with economic regulatory
    oversight, 462–463
Airport operations, 464
Airport Operations Control
  Centers (AOCCs):
  AOCS dynamics, 505–507, *506*
  AOCS users, 508
  AT&T Global Network
    Operations Center, 499,
    500, 501
  Auckland Airport, *526*
  Auckland Airport with
    customer service, 525
  Beijing Capital City Airport, *527*
  Beijing Capital International
    Airport and best practices,
    525–528
  best practices for
    implementation, 534–536
  concept of, 497–505
  coordination function, *509*
  design and equipment
    considerations, 516–520
  Dublin Airport, queue
    management dashboard,
    529

Airport Operations Control
    Centers (AOCCs) (*Cont.*):
Dublin Airport, with automated
    LOS measurement, 528
ergonomics, *519–520*
Fort Lauderdale-Hollywood
    International Airport,
    528–530, *530*
incident management system,
    515
Kuala Lumpur International
    Airport, 530, *531*
LAX, 530–532, *531*
leading, 524–534
management philosophy,
    502
Munich Airport, 533
organizational and human
    resources, 520–524
organizational position/
    structure, *522*
origins to present, 499–502
performance-monitoring
    function, 513–516
physical layout, 516–517
regulatory requirement for,
    504–505
SITA Command Center,
    500, 501
snow removal communication
    chart, *513*
strategic significance, 502–504
systems and equipment,
    518–519
typical layout, *517*
Zagreb Airport, 533–534, *534*
Airport Operations Control
    Systems (AOCS):
dynamics, 505–507
users, 508
Airport Operations Coordination
    Function:
anticipating passenger-
    processing problems,
    511–512
applications, 511–513
purpose, 508–510
snow-removal operations
    control, 512–513

Airport Operations Manual:
ARFF index classification, *548*
classification of airports by air
    carrier operations, *545*
distribution, 544–545
format, 538–544
function, 537
U.S. example, 545–551
Airport operators:
challenges facing, 31–32
services for flight crew
    members and, 371–372
as system actor, 1–2, *3*
Airport peaks. *See* Peaks, airport
Airport performance-monitoring
    function:
application, 515–516
purpose, 513–515
Airport security:
access control within and
    throughout airport
    buildings, 277–278
access gate for staff, *279*
advantages and disadvantages
    of centralized and
    decentralized search areas,
    267–268
aircraft isolated parking
    position and parking area,
    281
annex 17 standards, 259–260, 282
baggage search and screening,
    275–277
carry-on baggage x-ray
    machine, *274*
explanation of, 257–258
explosive trace detector, *272*
freight and cargo search and
    screening, 277
freight x-ray machine, *278*
general depictions of security
    areas, *263*
ICAO framework of
    international regulations,
    258–259
inline hold baggage x-ray
    scanning, *276–277*
with passenger and carry-on
    baggage, 265–275

Airport security (*Cont.*):
  perimeter control for
    operational areas, 279–281
  planning cycle, *261*
  planning in U.K., *265*
  planning outside U.S., 264–265
  program, 261–262, 541
  security program for typical
    airport, 281–287
  SSCP layout, *270*
  sterile and public areas, *268–269*
  structure of planning for,
    260–261
  TSA checkpoint elements, *270*
  U.S. federal involvement in
    aviation and, 262
  U.S. structure for, 262–264
  vehicle access and vehicular
    identification, 279
  vehicle barriers for controlled
    access, *280*
  whole-body scanner, *273*
  WTMD, *271*
Airport systems. *See also* British
    National Airport System; U.S.
    National Airport System
  airport access as part of,
    411–413
  NPIAS, 5
  in U.K., 6–8
Airports. *See also specific airports*
  air carrier operations and
    classification of, *545*
  aircraft performance
    characteristics influenced
    by, 89–117
  function of, *8*, 9–10
  nonaeronautical activities at,
    16–17
  organizational structure of
    traditional, *449*
  peak differences, *38–39*
  personnel requirements in
    electrical shop of category
    II, *147*
  revenue capability of, 15–17
  scheduling viewpoint of,
    59–60
  schematic layout of, *442*

Airports (*Cont.*):
  security program for typical,
    281–287
  slip, 48
  strategic business plan,
    444, *445*
  as system, 1–8
Airports Council International
    (ACI), 29, 326
Airspace:
  classes of, 338–339
  FAB, *335*
  ICAO classification of, *341*
  restricted, 333
  SESAR, 334
  wind chart, *369*
Air waybill, 297, *298*
AIS. *See* Automatic identification
    system
AKE, ULDs, *190*
AKH, ULDs, *190*
Alarms, with aircraft emergencies,
    384–385
All-freight operations, 301
Alliances, airline, 14–15
AMDAR. *See* Aircraft
    Meteorologic Data Relay
American Airlines, 15, 184
Amsterdam Schiphol Airport, 15,
    21, *52*, *70*, *74*, 194, 215, 223
ANA. *See* Civil Aviation Authority
    of Portugal
Animals, 128–132, 292
Annex 3, 325. *See also* Meteorology
Annex 10, 325. *See also*
    Telecommunications
Annex 11, 325. *See also* Air traffic
    control
Annex 14, 112, 121, 326, 467–468.
    See also Aerodromes;
    Aerodromes, technical
    services
Annex 15, 325. *See also*
    Aeronautical information
Annex 16, 63, 65, 74
Annex 17, 259–260, 282. *See also*
    Standards
AOCCs. *See* Airport Operations
    Control Centers

AOCS. *See* Airport Operations
 Control Systems
Appearance profiles, baggage
 handling, 204, *205*
Approach:
 aircraft performance with
  landing and, 105–107
 airport operations and tactical,
  447–449
 control, 347–350
 DHs and RVRs for precision-
  approach runways, *111*
 lighting and maintenance
  schedule, *143*
 with noise-control strategy, 72
 SRA guidance, *349*
Aprons:
 cable electrical supply, *164*
 cargo, 307–314
 dispatch breakdown and delay,
  170
 hydrant system, *163*
 inspections, 139
 mobile apron engine air-start
  vehicle, *161*
 mobile apron fuel tanker, *162*
 mobile lounge for passenger
  transport across, *159*
 parking requirements, 51
 passenger transport via bus,
  159
Aquaplaning, 115
ARFF. *See* Aircraft Rescue
 Firefighting Facility
Arrivals:
 board, *247*
 delivery performance, 209–211
 immigration desk area, *240*
 reclaim, 191–193
ASDA. *See* Accelerate-stop
 distance available
ASDR. *See* Accelerate-stop
 distance required
Ashford, N.J., 90, 443
Asia. *See also specific countries in
  Asia*
 with cargo market and
  globalization, 291
 runways in, 47

ATC. *See* Air traffic control
Athens Spata Airport, 18, *52*, 336,
 563
Atlanta Hartsfield-Jackson
 International Airport, 194,
 216, *251*, 339
AT&T Global Network Operations
 Center, 499, *500*, 501
Auckland Airport, 525, *526*
Automatic identification system
 (AIS), 372
Automatic landing, 110–114
Automobiles, *417*, 424–427. *See
 also* Bus; Service vehicles;
 Trucks; Vehicles
Average daily peak. *See* Peak-
 profile-hour
Aviation and Transportation
 Security Act of 2001, 262
Aviation security, 262. *See also*
 Airport security
Aviation Security Improvement
 Act of 1990, 261

B

B747-8F maximum payload,
 307–308
BAA, 14, 332
 timeline and summary of
  regulatory issues, 462–463
Bag drop:
 with baggage-handling
  processes, 184–188
 check-in and, 194–195
 self-service, *187*
 staffed, *186*
Baggage, search and screening,
 275–277
Baggage handling:
 AKE and AKH ULDs, *190*
 appearance profiles, 204,
  205
 bag storage, 189, 198–200
 batch and compressed build
  process, *202*
 bingo-card method, 191
 build lateral, *200*
 check-in and bag drop,
  194–195

Baggage handling (*Cont.*):
  check-in-desk configurations,
    195
  context, history and trends,
    181–183
  crane-served bag store, *199*
  with customer complaints,
    182
  designated system for, *230*
  equipment, systems and
    technologies, 194–204
  flatbed, *203*
  flexible check-in options, *185*
  flight build, 200–202
  flow system schematic, *214*
  hold-baggage screening,
    188–189, 197, *198*
  hold-baggage screening
    equipment, *198*
  inclined, *204*
  management and performance
    metrics, 207–211
  multilevel screening protocol,
    188
  organization, 206–207
  process and system design
    drivers, 204–206
  process diagram, *183*
  processing times, *206*
  reclaim, 203–204
  self-service bag drop, *187*
  self-service kiosks, *185*
  sorting, 196–197
  staffed bag drop, *186*
  systems configurations, 194
  tilt-tray sorter, *196*
  tote-based system, *197*
  transfer ratios, 205
  tug and dolly train, *202*
  ULDs loaded onto aircraft, *192*
  x-ray machines, *274*, *276–278*
Baggage-handling processes:
  arrivals reclaim, 191–193
  bag drop, 184–188
  bag storage, 189, 198–200
  flight build and aircraft loading,
    189–191
  hold-baggage screening,
    188–189, 197, *198*

Baggage-handling processes (*Cont.*):
  interterminal transfers,
    193
  overview, 183–184
  transfer input, 193
Bahamas, 50
Balance:
  aircraft, 232
  trim, 235
Banks, 61
Beacons, 356
Beijing Capital International
    Airport, 525–526, *527*, 528
Belgium, 333
Benchmarking, 463–464
BHR. *See* Busy-hour rate
Bingo-card method, with baggage
    handling, 191
Biodiversity, 572
Biometrics, 279
Bird-strike control, as operating
    constraint, 128–132
Birmingham, 7, 215
Bisignani, Giovanni, 182
Blind landings, 124
Boeing, 6, 289
  DC-8, 74
  mass data for highest gross
    weight variant of, *92*
  MSG-3, 48–49
  noise levels for 747, *72*
  707, 74
  777-300ER and takeoff runway
    length chart, *95–96*
  777-300ER with payload-range
    diagram, *93*
  777-800 with landing-distance
    performance, *107*
Bombs, 257, 272, 281
Boston Logan, 339, 415
Braking, efficiency with coefficient
    of friction, *136*
Brazil, 304
Brazil airport authority, 17
Bridge:
  elevating passenger air, *157*
  with fixed ground cooling unit
    attached, *165*
  three-loading, *158*

Bridgetown Airport, *52*
Briefing, flight-crew, 235–238
Bristol, 7
British Airports Authority, 6, 119,
   332
   laissez-faire system and, 7–8
   SBR and, 33–34
   timeline and summary of
      regulatory issues, 462
British Airways, 15, 60, 239
British National Airport System,
   6–7
Broadcast services, 360–361
Brussels Airport, *52*, 333, 415
BTH. *See* Busiest timetable hour
Build:
   aircraft loading and flight,
      189–191
   baggage handling and flight,
      200–202
   batch and compressed build
      process, *202*
   lateral, *200*
Buildings, access control of
   airport, 277–278
Bus, 250
   as access mode, 434–435
   air, 116–117, *158*
   apron passenger transport,
      159
   bays at LHR, *435*
   Caribbean resort airport and
      remote park for, *436*
   modal split by, *440*
Busiest timetable hour (BTH), 37
Business:
   CBD, *416*
   plan for airports, 444, *445*
   SBUs, *450*
Busy-hour rate (BHR), 36, 37

═══ **C** ═══

CAA. *See* U.K. Civil Aviation
   Authority
Cairo, 47
Carbon management, airport,
   558–562
Cardiff Airport, 18

Cargo, search and screening, 277
Cargo market:
   air waybill, 297, *298*
   cargo types, 292
   cost, 290
   GDP, 289–290
   globalization of trade and Asian
      development, 291
   just-in-time logistics, 291
   miniaturization, 290
   patterns of flow, 293–295
   regulations loosened, 291
   relationships among actors with
      air cargo, *295*
   rising consumer wealth, 291
   technological improvements,
      290
   variations in, *292–294*
Cargo operations:
   air cargo interior, *305–306*
   airplane servicing arrangement,
      311
   cargo apron operation with,
      307–314
   container arrangements in
      wide- and narrow-bodied
      aircraft, *302*
   examples of modern cargo
      terminal design and,
      317–320
   facilitation, 314–317
   fixed mechanized, 305–307
   flow through terminal, 297–298,
      299–300, 301
   freight container loader with,
      312–313
   Gantt chart for turnaround for
      large all freight flight, 308,
      309–310
   HACTL terminal, 299–300, *319*,
      320
   handling within terminal,
      304–307
   by integrated carriers, 320–323
   low mechanization/high
      manpower, 304
   Lufthansa terminal schematic,
      318
   movement expedited, 295–297

Cargo operations (*Cont.*):
open mechanized, 305
spoke and hub terminal
schematic, *322*
stages of exporting and
importing freight, *315*
ULD compatibility, *303*
ULDs, 301–304
Caribbean, 50, *436*
Carriers:
airport classification by air,
545
cargo operations by integrated,
320–323
LCC, 39
Carry-on baggage:
centralized and decentralized
screening, 266–269
search and screening, 265–275
security screening checkpoint,
269–275
x-ray machine, *274*
Cars. *See* Automobiles; Service
vehicles; Trucks; Vehicles
Catchment area, 40
Categories, *147*
aircraft emergencies and
airport, *380*
of British National Airport
System, 6–7
definitions of U.S. airport, *5*
Catering, 166, *167*
CBD. *See* Central Business District
CDG. *See* Paris Charles de Gaulle
Center of gravity (CG), 91
Central Business District (CBD),
416
Centralized passenger terminal
systems, 10–15, 215
Centralized screening, 266–269
Centralized search areas, *267–268*
Certification:
aerodrome, 119–122
FAA manual on airport, 545–551
limits with aircraft noise
certification, *75*
noise, 74–76
CFR. *See Code of Federal Regulations*
CG. *See* Center of gravity

Challenges:
of airport operators, 31–32
sustainable development,
555
Changi Singapore, 60
Charleston Airport, 328, *329*
Charlotte, NC, 61
Charter flights, 39
Charts:
airspace wind, *369*
Boeing 777-300ER takeoff
runway length, *95–96*
functional adjacency, *249*
Gantt, 308, *309–310*
off-airport emergency flow-
control, *396*
on-airport emergency flow-
control, *395*
snow-removal communication,
513
SRA, *349*
SWCs, 366–368
synoptic weather, *368*
Check-in:
airline-designated area for,
155
bag drop and, 194–195
curbside, 184
desk configurations, *195*
flexible options for baggage
handling, *185*
Munich Airport's CUTE
passenger, *156*
procedures, 419–422
showing area under lease to
airline, *229*
times for passengers prior to
departure, *423*
Checkpoints, TSA, *270. See also*
Airport security
Checks, MSG-3, 48–49
Chicago O'Hare Airport, 3, 13, *44*,
82, 194, 215, 413
China, 291
CIP facilities. *See* Commercially
important persons facilities
Civil Aeronautics Board, 6
Civil Aviation Authority (ANA) of
Portugal, 24–25

U.K. Civil Aviation Authority
  (CAA), 71, 109, 326, 328, 332
certification, 121
with runway slots, 47
Classifications. *See also* Aerodromes,
  technical services
of airports by air carrier
  operations, *545*
ARFF index, *548*
noise control and major land-
  use guidance zone, *86*
of U.S. airport system, *4*
Climate change:
consequences and certainty of,
  570
with sustainable development,
  569–572
Climatologic information, 371
Climb-path segments, 102, *103*
Club areas, 229
Code of Federal Regulations (CFR),
  121, 262–263
Command:
airport emergency plan, 394–397
SITA center of, *500*, 501
Commercialization, of airport
  industry, 459–461
Commercially important persons
  (CIP) facilities, 229
Common travel, 215
Common-use terminal equipment
  (CUTE), *156*, 230
Communications, 542. *See also*
  Telecommunications
with Airport Emergency Plan,
  398
charts for snow-removal, *513*
meteorology and use of, 372
radio, 142
requirements with airport
  aircraft emergencies,
  384–385
SMSs, 489, 494–495
technology, 330
Community:
air traffic operations and impact
  on, *556*
response to aircraft noise, 68–69
Complaint reports, 174

Concessions:
operational mode of, *226–227*
summary of, 17–18
Congestion, access, 411
Constraints:
airline scheduling factors and,
  44–52
airport access and capacity, *412*
long-haul, 48
operating, 122–132
terminal, 47–48
Construction, airfield, 148–150
Consumer wealth, cargo market
  and rising, 291
Contaminants, 132
Context:
baggage handling, 181–183
operational administration and
  strategic, 441–446
SMSs regulatory, *476*
Control. *See also* Air traffic control;
  Noise control; Noise-control
  strategies
aerodrome, 345–347
airport security and perimeter,
  279–281
bird-strike, 128–132
definition, 344
ground handling and
  departure, 169–170
ground handling and efficiency,
  174–175
night curfews and noise, 83
operations program, 457–458
with snow-removal operations,
  512–513
within and throughout airport
  buildings, 277–278
Controller, AOCCs, 523
Conventions, on security, 258–259
Cooling/heating, 164–166
Coordination:
airport emergency plan, 398
airport operations coordination
  function, 508–513
Copenhagen Airport, 7
*Corporate Accounting and Reporting
  Standards* (WBCSD and WRI),
  560

Cost:
  cargo market, 290
  LCC, 39
Crane-served bag store, *199*
Crewing, availability with airline
  scheduling, 48
Crosswind effects, 124–128
Curbside access, *426, 428*
Curbside check-in, 184
Curfews, noise control and night,
  83
Customers. *See also* Passengers
  complaints with baggage
    handling, *182*
  experience map, 450, *451*
CUTE. *See* Common-use terminal
  equipment

================ **D** ================

Dallas–Fort Worth (DFW), 13, 60,
  194, 215, *216*, 339
Day/night average sound levels,
  65–66
Deaths. *See* Fatalities
Debris, 133, 137–138
Decentralized passenger terminal
  systems, 10–15, 215
Decentralized screening, 266–269
Decentralized search areas,
  267–268
Decision heights (DHs), *111,
  343*
Declared capacity, 47–48
Dedicated rail systems, as access
  mode, 435–437
Deficit operations, 1
Deicing, washing and, 164, *165*
Delta, 184
Deming cycle, 573, *574*
Department of Homeland Security
  (DHS), 262
Departure, 47
  aircraft performance with,
    100–105
  check-in times for passengers
    prior to, *423*
  control, 169–170
  SID, *350*

Deregulation:
  of large airports, 4
  in U.K., 7–8
  in U.S., 6, 17
Design:
  of AOCCs, 516–520
  baggage handling and system
    design drivers, 204–206
  terminal, 317–320
Desks, full-service, 187
Deutsche Flugsicherung (DFS), 332
Development:
  globalization, cargo market and
    Asian, 291
  sustainable, 553–578
DFS. *See* Deutsche Flugsicherung
DFW. *See* Dallas–Fort Worth
DHs. *See* Decision heights
DHS. *See* Department of
    Homeland Security
Direct passenger services, 222–228
Dispatch:
  breakdown and delay with
    apron, *170*
  flight, 230–231
Domestic/international ratio, 39
Dominican Republic, 17
Douglas DC-9, 74
Dry operating weight, 232, 234
Dubai, 209, 225, 299
Dublin Airport, 528, *529*
Dublin Airport Authority, 17
Dusseldorf Airport, *52*
Duty-free shops, *223–224*, 254
Dynamics, AOCS, 505–507
Dynatest, runway friction test, *135*

================ **E** ================

EAP. *See Electronic Aeronautical
    Publication; Electronic Aviation
    Publication*
EASA. *See* European Aviation
    Safety Agency
EC. *See* European Commission
Eckerson, W. Wayne, 457
Economic regulatory oversight,
    462–463
EDS. *See* Explosive-detection
    systems

Effective period of movement, 58
Electrical maintenance, 142–146,
   164
Electrical shops, *147*
Electronic Aeronautical Publication
   (EAP), 99
Electronic Aviation Publication
   (EAP), 90
EMAS. *See* Engineered Material
   Arresting Systems
Emergencies. *See also* Airport
   aircraft emergencies; Airport
   Emergency Plan
   access road, 383
   aircraft rescue and firefighting,
      146–147
   IATA policy and, 58
   services, 42
   types of, 379–380
   water supply in, 383
Emergency plan document,
   391–394
Emissions charge, 51
Employees:
   flight crew, 48, 94, 235–238,
      371–372
   number at large airports, 3
   personnel, *147*, 388–389, 489–490
   staff departments, 18, *19*
   staff structure, *20, 21,* 27
   staffing and, *186,* 206–207, *247,*
      279, 522–524
   TSA, 189
Employment:
   European airports and types of
      on-site, *28*
   WLUs and high-density, 29
EMS. *See* Environmental
   Management System
Energy, 562–563
Engine failure, 109, 114
Engineered Material Arresting
   Systems (EMAS), *329*
Environment:
   LEED, 563
   management systems with,
      572–575
   sustainable development and,
      553–578

Environmental Management
   System (EMS), *574*
Environmental Protection Agency
   (EPA), 67
Equipment. *See also specific types of
   equipment*
   AOCC, 516–520
   CUTE, *156,* 230
   hold-baggage screening, *198*
   IATA policy and changes, 58
   performance results of
      passenger sensitive, *120*
   rescue, *387*
   types with RVR measurement,
      123
Equipment, baggage-handling:
   bag storage, 189, 198–200
   check-in and bag drop, 194–195
   flight build, 200–202
   hold-baggage screening, 198
   reclaim, 203–204
   sorting, 196–197
   system configurations, 194
Equivalent continuous sound
   level L_{EQ}, 66–67
ER aircraft. *See* Extended-range
   aircraft
Ergonomics, AOCCs, *519–520*
ETD. *See* Explosive trace detector
ETP. *See* Explosive trace portal
EU. *See* European Union
Europe:
   on-site employment types in
      airports, *28*
   runways in, 47
European Aviation Safety Agency
   (EASA), 48–49, 328, 334
European Commission (EC), 70,
   171, 174, 206, 221, 333
European Union (EU), 29, 171,
   221, 333
Exclusive rights, 228
Explosive-detection systems
   (EDS), 188
Explosive trace detector (ETD),
   272
Explosive trace portal (ETP),
   272
Extended-range (ER) aircraft, 6

F

FAA. *See* Federal Aviation Administration
FAB. *See* Functional Airspace Block
Facilitation, of cargo operations, 314–317
Facilities:
    ARFF, 547–548
    CIP, 229
    parking, 13
    VIP lounge, *242*
Failure, engine, 109, 114
FANS. *See* Future Air Navigation Systems
FAR. *See Federal Aviation Regulations*
Faro Airport, *52*
Fatalities. *See also* September 11, 2001
    aircraft accident rates, *378*
    in aircraft accidents, 128, 347, 354–355, 379, 382
    Pan Am, 257, 275
Fault servicing, 162
FCOM. *See* Flight crew operations manual
Federal Aviation Administration (FAA), 36
    with aircraft availability, 48–49
    airport certification manual, 545–551
    banks and, 61
    certification and, 121
    FAR, 67, 74, 336
    hubs and, 60
    legislation, 331
    with noise control, 65
    with runway slots, 47
    with TPHP, 37
Federal Aviation Regulations (FAR), 67, 74, 336
FedEx, 320
Fees:
    landing-fee pricing policies, 50–52
    structure of aeronautical, *52*
    tariffs on handling, 51
    variation of, *53*

Fencing, perimeter control, 279–280
Fines, noise control, 83
Firefighting. *See also* Operational readiness
    with aircraft rescue, 146–147
    ARFF, 547, *548*
    extinguishing agents, *381*
    foaming runways, 381–382, 400–402
    minimum characteristics for rescue and, *386*
    rescue procedures with aircraft, 398–400, 540–541
    vehicles for rescue and, *381*, 385–388
Fixed services, telecommunications and, 351–352
Flatbed reclaim, *203*
Fleet utilization:
    airline scheduling and, 56–58
    for short- and medium-haul operations, *56–57*
Flight build:
    aircraft loading and, 189–191
    baggage handling and, 200–202
Flight crew. *See also* Employees
    airline scheduling and availability of, 48
    briefings for, 235–238
    member services, 371–372
Flight crew operations manual (FCOM), 94
Flight dispatch, 230–231
Flight information, VDUs for, *245*
Flight plan, international, *233*
Flight planning, 231–232
Flight rules. *See also* Regulations
    general, 336–337
    instrument, 122, 337–338
    visual, 122, 337
Flight watch, 239
Flights, missed, 422–424
Florida, 50
Foaming:
    of runways, 400–402
    types, 381–382

Forecasts:
aircraft noise and noise
exposure, 67
SIGWXs and SWCs, 366–368
TAFs, 366, *367*
upper-air grid-point data, 368–
369
WAFS, 362
Foreign Airport Security Act, 261
Fort Lauderdale-Hollywood
International Airport, 528–
530, *530*
Frankfurt Airport, 13, *52*, 194, 221,
223, 415
administrative and staff
structure of, *20*
Lufthansa at, 60
Freight, *315*
all-freight operations, 301
cargo differentiated from, 289
container loader, *312–313*
search and screening, 277
x-ray machine, *278*
Friction:
coefficient of braking efficiency
and, *136*
Dynatest for runway, *135*
Fuel, 163
dispenser, mobile aircraft, *163*
loads, 91
oil-price increases and, 31–32,
290
reserves, 104
spill hazard reduction, 401
takeoff and trip, 234
tanker, mobile apron, *162*
zero-fuel weight, 232
Full emergency, 379, *380*
Full-service desks, 187
Functional adjacency chart, *249*
Functional Airspace Block (FAB),
335
Future Air Navigation Systems
(FANS), 325, 330

G

Galileo, 360
Gantt chart, 308, *309–310*
Gates, access, 280–281

Gateway international airports, in
U.K., 6, 7
Gatwick Airport, 7, 50
GDP. *See* Gross domestic product
General aviation airports, in U.K.,
6, 7
General flight rules, 336–337
Geographic location:
with aircraft availability, 49
peaks with, 40
Germany, 17, 38
Glasgow Airport, 281
Global Positioning System (GPS),
101, 344, 358–360
Globalization, 291
Glonass, 359–360
GMT. *See* Greenwich Mean Time
Government:
aviation security and U.S., 262
with deregulation of airports,
17–18
requirements with passenger
terminal operations,
239–240
GPS. *See* Global Positioning
System
Greenwich Mean Time (GMT),
239
Gross domestic product (GDP),
289–290
Ground handling:
aircraft marshalled by ground
signalman, *160*
aircraft ramp servicing and,
162–166
aircraft tow tractor, 160, *161*
airline-designated check-in
area, *155*
airline passenger steps, *157*
apron cable electrical supply,
164
apron-dispatch breakdown and
delay, *170*
apron passenger transport bus,
159
catering truck in loading
position, *167*
checklist for monitoring
efficiency of, *176–178*

Ground handling (*Cont.*):
  critical path of turnaround, 169
  CUTE passenger check-in desks, *156*
  deicing/washer vehicle, 164, *165*
  departure control, 169–170
  division of responsibilities, 171, *172–173*, 174
  efficiency control, 174–175
  elevating passenger air bridge, *157*
  fixed ground cooling unit with air bridge, *165*
  general, 175–178
  mobile aircraft fuel dispenser, *163*
  mobile apron engine air-start vehicle, *161*
  mobile apron fuel tanker, *162*
  mobile lounge for passenger transport across apron, *159*
  passenger handling and, 153–158
  ramp handling and, 158–162
  ramp layout, 166–167, *168*, 169
  scope of, *154*
  three-loading bridge serving A380, *158*
Ground power supply, 163
Ground signalman, *160*
Groundwater pollution, 567–569
Growth rate:
  air cargo, 32
  passengers and world wide, 31–32
GRU. *See* São Paulo Guarulhos International Airport
Grupo Ferrovial, 7
Guidance, 345

━━━ **H** ━━━

HACTL. *See* Hong Kong Air Cargo Terminal
Handheld metal detector (HHMD), 271

Handling:
  baggage, 181–211
  fees, 51
  ground, 153–179
  passenger, 153–158
  ramp, 42, 158–162
  runway, 42
  taxiway, 42
Hazards:
  fuel spill reduction, 401
  identification, 482–483
Heathrow. *See* London Heathrow
Heating/cooling, 164–166
Heavy maintenance visit (HMV), 49
HHMD. *See* Handheld metal detector
Hierarchical system, of airport relationships, 2
Hijacking, 257, 258, 287
Historical precedence, IATA policy and, 58
History, baggage handling, 181–183
HMV. *See* Heavy maintenance visit
Hold-baggage screening, 188–189, 197, *198*
Hong Kong, 184
Hong Kong Air Cargo Terminal (HACTL), 299–300, *319*, 320
Hubs:
  with airline scheduling, 60–61
  hubbing considerations, 254
  terminal schematics, *322*
Human resources, AOCCs:
  airport duty manager, 523
  controller, 523
  management structure and reporting relationships, 520–522
  operations analyst, 523–524
  senior duty manager, 523
  staffing and key competencies, 522–524
Hydrant system, apron, *163*

I

IATA. *See* International Air Transport Association
Iberia, 15
IBM, 498–499
ICAO. *See* International Civil Aviation Organization
ICARUS system, 316
Ice. *See* Deicing, washing and
Identification:
  AIS, 372
  hazard, 482–483
  SIDA, 277–278
  vehicular, 279
IFRs. *See* Instrument flight rules
Immigration, 239, *240*
In flight, as operational function, 232–233
In-town terminals, 437–438
Inclined reclaim, *204*
India, 291
Industry. *See* Airport industry
Information:
  aeronautical, 16–17, 90, 99, 373–375
  ATC, SAR and AIS, 372
  MANTIS, 80
  meteorology, 370–372
  passenger systems, 241–246
  SIGMETs/AIRMETs, 369–370
  signs in terminals, *244*
  SMSs, 480–481
  staffed desks for, 247
  urgent operational, 374
  VDUs, *245*
INFRAERO airport authority, 17, *306*
Inline hold baggage x-ray scanning, *276–277*
Inspections, airfield, 138–140
Instrument flight rules (IFRs), 122, 337–338
Insulation, as noise-control strategy, 73–74
Integrated carriers, 320–323

Interaction. *See* Access interaction
Interlining, 301
International Air Transport Association (IATA), 13, 47, 58–59, 182
International Civil Aviation Organization (ICAO), 244, 504
  aerodrome certificate, 119–122
  with airport noise, 63
  airspace classification, *341*
  annex 14 aerodromes, 467–468
  commercialization and, 461
  framework of international regulations, 258–259
  SARPs, 467
  stance on member state's safety programs, 468
International collaborations, with ATC, 333–336
International flight plan, *233*
International regulations, ICAO framework of, 258–259
International Security and Development Act of 1985, 261
Interterminal transfers, 193
Iran Air, 257
Ireland, 15, 17
Irish Aviation Authority, 17, 225

J

Japan, 291
JAR. *See* Joint Aviation Regulations
JFK Airport, 45, 47, 60, 64, 339
Joint Aviation Regulations (JAR), 98–99, 114
Jurisdiction areas, SMSs, *486–487*
Just-in-time logistics, with cargo market, 291

K

Kanki, B. G., 489
Kansas City, 13, 215
Key performance indicators (KPIs), 445, 453–454, 457–459
Kiosks, baggage-handling, *185*
KLM, *229*

Knock Airport, Ireland, 17
Korea, 250, 291, 418
Kowloon, 184
KPIs. *See* Key performance
    indicators
Kuala Lumpur International
    Airport, 418, 530, *531*
Kyoto Protocol, 558

━━━ **L** ━━━

LaGuardia Airport, 425
Laissez-faire system, 7–8. *See also*
    Deregulation
Land purchase, as noise-control
    strategy, 73–74
Land use, noise compatibility and,
    84–88
Landing:
    aircraft performance with
        approach and, 105–107
    aircraft performance with
        automatic, 110–114
    as airline-related operational
        function, 234–235
    blind, 124
    MLS, 357–358
Landing distance available (LDA),
    99, 109
Landing distance required (LDR),
    98–99, 109
Landing-fee pricing policies, 50–52
Large airports. *See also specific large
    airports*
    deregulation of, 4
    employee numbers at, 3
    organization of, 8, *9*, 10
    organizations influenced by, 3
    privatization of, 4–6, 25–26
    in Washington, D.C., 5
LAWA. *See* Los Angeles World
    Airports
LAX, 3, 74, 339, 427
    AOCC, 530, *531*
    United Airlines at, 60
LCC. *See* Low-cost carrier
LDA. *See* Landing distance
    available

LDR. *See* Landing distance
    required
Leadership in Energy and
    Environmental Design
    (LEED), 563
Legislation. *See also specific
    legislation*
    FAA, 331
    security, 261–262
Level of service (LOS), 2, 42, 60
LHR. *See* London Heathrow
Lighting, 142, 541–542
    aerodrome control light signals,
        346
    maintenance schedule for
        centerline and touchdown-
        zone, *144*
    maintenance schedule for
        medium-intensity
        approach, *143*
    with switchover times in power
        failure, *145*
Lima, Peru, 18
Limousine, as access mode,
    429–430
Line departments, 18, *19*
Livestock, 292
Load and trim sheet, *236*
Load factors:
    with airline scheduling, 45
    passenger, 51
Loading:
    as airline-related operational
        function, 235
    flight build and aircraft,
        189–191
    instructions, *237*
Loads, fuel, 91
Local airports, in U.K., 6, *7*
Local standby, 379, *380*
Logistics, just-in-time, 291
London Gatwick, *52*, 339, 416
London Heathrow (LHR), 3, 7,
    13–14, 18, 45, 47, 74, 83, 194, 223
    British Airways at, 60, 239
    bus bays at, *435*
    impact of reduced visibility on
        potential regularity at, *124*

London Heathrow (LHR) (*Cont.*):
  limits on air transport services at, 59
  monthly passenger traffic variations at, *40*
  rapid transit in relation to terminals at, *432*
  rapid transit on access modal split to, *433*
  scheduling windows for east- and westbound flights into, *46*
  TMA plan, *342*
  traffic with peak tariffs at, *50*
  variations in hourly traffic volumes, *41*
  variations in passenger flows in peak week, *41*
London Stock Exchange, 7, 462
Long-haul crewing constraints, 48
Long-haul flights, 39
Long-haul windows, 45–47
LOS. *See* Level of service
Los Angeles International Airport. *See* LAX
Los Angeles Van Nuys, 5
Los Angeles World Airports (LAWA), 22
Louisville, KY, 320
Lounge:
  mobile, *159*
  VIP facilities, *242*
Low-cost carrier (LCC), 39
Lufthansa, 60, 112, 317, *318*
Luton Airport, U.K., 18
Luxembourg, 333

M

Madrid Barajas Airport, *14, 15, 52*
Maintenance. *See also* Operational readiness
  with aircraft rescue and firefighting, 146–147
  airfield construction and, 148–150
  electrical, 142–146
  management, 140–142

Maintenance (*Cont.*):
  with operational readiness, 140–150
  preventative, 142
  regularly scheduled inspection checklist, *141*
  with safety aspects, 147–148
  schedule for centerline and touchdown-zone lighting systems, *144*
  schedule for medium-intensity approach lighting, *143*
Maintenance Steering Group 3 (MSG-3), 48–49
Malaysia, 291
Management. *See also* Safety Management Systems
  airport carbon, 558–562
  AOCC incident management systems, *515*
  environmental systems, 572–575
  maintenance, 140–142
  of operational performance, 452–464
  operational structures and, 17–29
  performance metrics with baggage handling and, 207–211
  philosophy of AOCCs, 502
  risk, 264, 477, 481–484
  SMSs, 326–329
  of solid wastes, 166, 565–567
  structure or staff and line functions, 18, *19*
  terminal, *220,* 221
  terminal philosophies, 221
Managers, AOCCs, 523
Manchester Airport Noise and Tracking Information System (MANTIS), 80
Manchester International Airport, 7, 17, *52,* 73, 76, 194, 559–560
  noise and monthly infringement report at, *81*
  noise-measuring points at, *79*
MANTIS. *See* Manchester Airport Noise and Tracking Information System

Manuals:
  airport operations, 537–557
  FAA airport certification,
    545–551
  FCOM, 94
  SMSs, 475–484
Maps, customer experience, 450,
  451
Marketability, with airline
  scheduling, 49–50
Marshalling, 159–160
MATRA. See Multi-Agency Threat
  and Risk Assessment
Maximum takeoff weight
  (MTOW), 91–93, 102, 232
McDonnell Douglas, 289
Medium-haul operations, 56–57
Meeters, 245
Memphis, TN, 320
Metal detectors, 271
METAR/SPECI:
  decoding tables for, 364
  explanation of, 362–363
  local routine met reports and,
    363–365
Meteorology:
  aircraft observations and
    reports, 365–366
  climatologic information, 371
  function of, 361–362
  information for ATC, SAR and
    AIS, 372
  METAR and SPECI, 362–365
  meteorologic observations and
    reports, 362
  services and trends, 372–373,
    542
  services for operators and flight
    crew members, 371–372
  SIGMETs/AIRMETs, 369–370
  SIGWXs and SWCs, 366–368
  TAFs, 366, 367
  upper-air grid-point data
    forecasts, 368–369
  use of communication, 372
  WAFS, 362
  weather information support
    for general aviation,
    370–371

Metropolitan Washington Airports
  Authority, 5
Miami Airport, 52, 251, 339
Microwave landing system (MLS),
  357–358
Middle East, 290, 358
Miniaturization, with cargo
  market, 290
"Miracle on the Hudson," 128,
  131
Missed flights, consequences of,
  422–424
MLS. See Microwave landing
  system
Mobile aircraft fuel dispenser,
  163
Mobile apron engine air-start
  vehicle, 161
Mobile apron fuel tanker, 162
Mobile lounge, 159
Mobile services,
  telecommunications and,
  352–355
Modal choice:
  access users and, 414–418
  factors influencing, 438, 439
Models, 481
Modes. See also Access modes
  change of, 7–8, 213
  concessions and operational,
    226–227
Mombasa, 281
Montego Bay Airport, 52
Moscow Domodedovo Airport,
  281
Movement:
  cargo operations and expedited,
    295–297
  IATA policy and effective
    period of, 58
  type and change, 7–8, 213
MSG-3. See Maintenance Steering
  Group 3
MTOW. See Maximum takeoff
  weight
Multi-Agency Threat and Risk
  Assessment (MATRA), 264
Mumayiz, S., 90
Munich Airport, 50, 156, 336, 533

═══ N ═══

Naples (NAP), *40*
National Aeronautics and Space
  Administration (NASA), 70,
  113
National Air Traffic Services
  (NATS), 332
National airport systems:
  U.K., 6–8
  U.S., 4–6, 17
National Plan of Integrated
  Airport Systems (NPIAS), 5
NATS. *See* National Air Traffic
  Services
Navigation:
  radio, 355–358
  satellite, 358–360
  technology, 330
NEF. *See* Noise-exposure forecast
The Netherlands, 15, 17, 333
New Jersey. *See* Newark
  International; Port Authority
  of New York and New Jersey
New Seoul International Airport
  (NSIA), 250, 418
New York. *See* JFK Airport;
  LaGuardia Airport; Port
  Authority of New York and
  New Jersey
Newark International, 339
Newcastle Airport, 7, 17
Night curfews, with noise, 83
NNI. *See* Noise and number index
Noise Advisory Council, 71
Noise and number index (NNI), 66
Noise control:
  aircraft noise and, 64–68
  aircraft noise certification limits,
    75
  average monthly noise levels,
    80
  with Boeing 747 noise levels, *72*
  with community response to
    aircraft noise, 68–69
  criteria recommended for
    aircraft, *87*
  with degree of annoyance, *69*
  distance from start of roll, *72*

Noise control (*Cont.*):
  explanation of, 63–64
  fines, 83
  historic and predicted future
    noise contours, *82*
  location of noise-measurement
    points, *75*
  major land-use guidance zone
    classifications, *86*
  Manchester Airport and noise-
    measuring points, *79*
  Manchester Airport's monthly
    infringement report, *81*
  measured aircraft noise levels,
    77–78
  night curfews and, 83
  noise certification with, 74–76
  noise compatibility and land
    use with, 84–88
  noise-measure relationships, *68*
  with noise-monitoring
    procedures, 76–82
  power cutback on ground noise
    levels, *71*
  scale of noise and sound, *64*
  typical airport noise patterns, *85*
Noise-control strategies:
  approach, 72
  explanation of, 69
  insulation and land purchase,
    73–74
  noise-preferential runways, 70–71
  operational noise-abatement
    procedures, 71–72
  quieter aircraft, 70
  runway operations, 73
  takeoff, 71–72
Noise-exposure forecast (NEF), 67
Noise impacts, 555–557
Noise levels, 51
Noise-monitoring procedures,
  76–82
Non-passenger-related airport
  authority functions, 240–241
Nonaeronautical activities, 16–17
Nonuser, *3*
Notices to Airmen (NOTAMS), 100,
  149, 235, 238, 352, 374, 479, 512

NPIAS. *See* National Plan of
Integrated Airport Systems
NSIA. *See* New Seoul International
Airport

━━ **O** ━━

Oakland International Airport,
394
Off-airport terminals, 437–438
Oil prices, increases, 31–32, 290
Olympics, 526
Onboard servicing, 166
One World airline alliance, 15
Open-skies policies, 291
Operating constraints:
bird-strike control, 128–132
crosswind effects, 124–128
with operational readiness, 122–
132
visibility, 122–124
Operation styles, 49
Operational administration:
of airport operations and
tactical approach, 447–449
airport strategic business plan,
444, *445*
customer experience map, 450,
451
key success factors for high
performance with, 464
with operational performance
managed, 452–464
organizational considerations,
449–452
SBU-based airport management
structure, *450*
schematic airport layout, *442*
service delivery plan, *453*
service effectiveness
determinants, *446*
strategic context, 441–446
traditional airport
organizational structure,
449
Operational areas, 132–138
Operational cycles, 49
Operational noise-abatement
procedures, 71–72

Operational performance:
airport economic regulatory
oversight with, 462–463
airport industry
commercialization with,
459–461
control of, 457–458
execution of, 455–457
external assessment, 459–464
industry benchmarking and,
463–464
internal assessment, 458–459
management of, 452–464
with operational
administration, 441–464
planning for, 452–455
Operational readiness:
aerodrome certification, 119–122
airfield inspections, 138–140
bird-strike control and, 128–132
coefficient of friction and
braking efficiency, *136*
crosswind effects and, 124–128
ground-based radio aids and
switchover times with
power failure, *146*
impact of reduced visibility on
potential regularity, *124*
instrumented pickup truck with
retractable fifth wheel, *135*
lighting and switchover times
with power failure, *145*
maintaining, 140–150
maintenance schedule for
centerline and touchdown-
zone lighting systems, *144*
maintenance schedule for
medium-intensity
approach lighting, *143*
operating constraints with, 122–
132
operational areas with, 132–138
passenger sensitive equipment
performance results, *120*
PC readout of Dynatest runway
friction test, *135*
personnel requirements in
electrical shop of category
II airport, *147*

Operational readiness (*Cont.*):
  regularly scheduled inspection
    checklist, *141*
  RVR measurement equipment
    types, *123*
  snow blower, *137*
  visibility and, 122–124
  wind rose, *128*
  wind table, *126–127, 129–130*
Operational structures:
  ATC, 343–345
  management and, 17–29
Operational system, airport as:
  centralized and decentralized
    passenger terminal
    systems, 10–15, 215
  complexity of, 15–17
  explanation of, 1–3
  function of airport, *8,* 9–10
  as hierarchical system, *2*
  hierarchical system of
    relationships with, *2*
  management and operational
    structures, 17–29
  national airport systems, 4–8
Operations. *See also* Airport
    Operations Control Centers;
    Airport Operations
    Coordination Function;
    Airport Operations Manual;
    Cargo operations
  AOCS, 505–508
  AT&T Global Network
    Operations Center, 499,
    500, 501
  community impacted by ATC,
    556
  deficit, 1
  fleet utilization for short- and
    medium-haul, *56–57*
  inclement weather and aircraft
    performance, 114–116
  noise-control strategies for
    runway, 73
  organizations influenced by
    large airport, *3*
  passenger terminal, 213–255,
    418–424

Operations analyst, AOCCs,
  523–524
Operators. *See* Airport operators
Organization:
  baggage-handling, 206–207
  of large airports, 8, 9, 10
  of LAWA, *22*
  with operational
    administration, 449–452
  of PANYNJ, *25*
  of Portuguese civil aviation
    system, *26*
  of private company with
    multinational airport
    interests, *27*
  of Sacramento Airport, *22*
  of San Francisco Airport, *23*
  of scheduling within typical
    airline, *55*
  SMSs Manual and, 477–479
  with structure of traditional
    airports, *449*
  for three-airport multimodal
    planning, *24*
Organizations. *See also specific
    organizations*
  large airport operations
    influencing, *3*
Orlando Airport, *40, 52,* 223

━━━ **P** ━━━

Palmer, M. T., 489
Pan Am, 257, 275
PANYNJ. *See* Port Authority of
    New York and New Jersey
Paris Charles de Gaulle (CDG), 13,
    215, 418
Parking area, 281
  access and requirement
    recommendations, *426*
  demand with annual and peak-
    hour originating
    passengers, *425*
Parking facilities, 13
Parking requirements, apron, 51
Passenger terminal operations:
  aides to circulation, 249–253
  airline-related operational
    functions, 230–239

Passenger terminal operations
(*Cont.*):
  airline-related passenger
    services, 228–230
  arrivals, immigration desk area,
    240
  arrivals board, 247
  Atlanta Hartsfield, 216
  baggage flow system schematic,
    214
  bank of VDUs for flight
    information, 245
  centralized processing, 217–218
  check-in procedures, 419–422
  check-in showing area under
    lease to airline, 229
  Dallas–Fort Worth, 216
  designated baggage-delivery
    system, 230
  direct passenger services, 222–
    228
  duty-free shops, 223–224
  dwell times for long- and short-
    access journeys, 420
  flight-crew briefing sheet, 238
  flight length and type on dwell
    times, 421
  functional adjacency chart, 249
  functions of, 213–219
  governmental requirements,
    239–240
  hubbing considerations, 254
  information signs in terminal,
    244
  international flight plan, 233
  length of access time, 419
  load and trim sheet, 236
  loading instructions, 237
  missed flights and
    consequences, 422–424
  non-passenger-related airport
    authority functions,
    240–241
  operational mode of
    concessions, 226–227
  passenger information systems,
    241–246
  people-mover control room, 253
  people mover station, 252–253

Passenger terminal operations
(*Cont.*):
  processing VIPs, 241
  reliability of access trip, 419
  road signs, 243, 246
  space components and
    adjacencies, 246–249
  staffed information desk, 247
  terminal functions, 219–221
  terminal management
    philosophies, 221
  terminal management structure,
    220
  terminal space distribution, 248
  VIP lounge facilities, 242
  walkway tunnel, 251
Passenger terminal systems:
  centralized and decentralized,
    10–15, 215
  function of, 7–8, 213–219
Passenger traffic:
  monthly variations in, 40
  SBR with peak-hour and
    annual, 43
  variations in hourly volumes,
    41
  variations in peak week, 41
  volumes, hourly, 33
Passengers:
  air bridge for, 157–158, 165
  airport staff and annual
    throughput of, 27
  bags per, 204–205
  with centralized and
    decentralized screening,
    266–269
  check-in times prior to
    departure, 423
  complaints with baggage
    handling, 182
  customer experience map for,
    450, 451
  CUTE check-in desks for, 156
  flow at Chicago O'Hare, 44
  handling with ground handling,
    153–158
  information systems, 241–246
  load, 51
  mobile lounge for, 159

Passengers (*Cont.*):
with origin or destination in CBD, *416*
parking area demand with annual and peak-hour originating, *425*
search and screening, 269–275
security screening checkpoint and, 269–275
sensitive equipment performance results, *120*
steps, *157*
TPHP, 36–37
walking distances for, 13
with workers, visitors and senders/greeters, *415*
world wide rate of growth for air, 31–32
Pavement surface conditions, 132–138
Peak-hour tariffs, *50*, 51
Peak-profile-hour (PPH), 37–38
Peaks, airport:
airline scheduling and, 31–61
airport differences with, 38–39
BHR and, 36, *37*
BTH and, *37*
methods to describe, 33–42
nature of, 39–42
other methods with, 38
PPH and, 37–38
problem with, 31–33
with SBR, annual peak-hourly and peak-day aircraft, *34–35*
SBR and, 33–36
TPHP and, 36–37
volume variations and implications for, 42–44
Pedestrian walkways, 250
People movers, 250–251, *252–253*
Performance. *See* Aircraft performance characteristics; Airport performance-monitoring function; Operational performance; Operational readiness

Performance metrics:
arrivals delivery, 209–211
baggage system, 209
management and, 207–211
Perimeter control:
access gates, 280–281
fencing, 279–280
for operational areas, 279–281
Personnel competency, SMSs, 489–490
Personnel requirements:
in electrical shop of category II airport, *147*
in emergencies, 388–389
Philadelphia, 61
Philosophies:
AOCCs and management, 502
of terminal management, 221
Phoenix Deer Valley, 5
Pictograms, *243*
Pilot Reports (PIREPs), 365
Pittsburgh, 61
Planning:
airport security and structure of, 260–261
flight, 231–232
operational performance, 452–455
organization for three-airport multimodal, *24*
outside U.S., with airport security, 264–265
with SMSs Manual, 477, 479
Policies:
IATA, 58–59
landing-fee pricing, 50–52
open-skies, 291
with scheduling, 44–52
in SMSs Manual, 477–478
Pollution, surface and groundwater, 567–569
Port Authority of New York and New Jersey (PANYNJ), 17, 23, *25*, 241
Portuguese civil aviation system, *26*
Power cutback, *71*

Power failure:
    ground-based radio aids with
        switchover times in, *146*
    lighting with switchover times
        in, *145*
Power supply, ground, 163
PPH. *See* Peak-profile-hour
Preventative maintenance, 142
Prices, oil, 31–32, 290
Pricing, landing-fee policies with,
    50–52
Private company, with
    multinational airport
    interests, *27*
Privatization, 17–18
    of large airports, 4–6, 25–26
    ten key lessons for successful
        airport, 460
    in U.K., 6
Problems:
    with airport peaks, 31–33
    passenger-processing, 511–512
Procedures:
    ATC operational characteristics
        and, 345–350
    check-in, 419–422
    firefighting and rescue, 398–400,
        540–541
    noise-monitoring, 76–82
    operational noise-abatement,
        71–72
Protection, emergencies and level
    of, 380–383
Punctuality reports, 174
Punta Cana Airport, Dominican
    Republic, 17

━━━ **Q** ━━━
Qantas, 15, 45, 186
Quality:
    local air, 557–578
    of service, 228

━━━ **R** ━━━
Radio:
    aids with switchover times in
        power failure, *146*
    communications, 142
    navigation services, 355–358

Rail:
    as access mode, 430–434
    dedicated systems, 435–437
    rapid transit at LHR, *432, 433*
Ramps, 42
    ground handling and,
        158–162
    layout, 166–167, *168*, 169
Rates:
    BHR, 36, *37*
    cargo, 296–297
    passenger growth and world
        wide, 31–32
    SBR, 33–36
    short-landed, 208
Readiness. *See also* Operational
    readiness
    aircraft rescue and firefighting,
        146–147
    with airfield construction,
        148–150
    electrical maintenance and,
        142–146
    with maintenance management,
        140–142
    preventative maintenance and,
        142
    safety aspects of, 147–148
Reason, James, 480, *481*, 488
Reclaim:
    arrivals, 191–193
    baggage handling, 203–204
    flatbed, *203*
    inclined, *204*
Regional airports, in U.K., 6, 7
Regulations. *See also* Deregulation
    *Air Navigation Order and
        Regulations*, 121, 337
    aircrafts with operational
        and performance tradeoff,
        110
    cargo market and loosening,
        291
    CFR, 121, 262–263
    FAR, 67, 74, 336
    flight rules and, 336–338
    ICAO framework of
        international, 258–259
    JAR, 98–99, 114

Regulatory:
  AOCCs requirements, 504–505
  BAA issues, 462–463
  SMSs context, *476*
Reliability:
  of access trip, 419
  with airline scheduling, 45
Removal, of disabled aircraft,
    402–404
Reports:
  aircraft observations and, 365–366
  METAR and SPECI, 362–365
  meteorologic observations and,
    362
  monthly complaint and
    punctuality, 174
  noise at Manchester Airport and
    monthly infringement, *81*
  PIREPs, 365
Requirements:
  airport aircraft emergencies and
    alarm, 384–385
  airport aircraft emergencies and
    personnel, 388–389
  airport aircraft emergencies and
    protection, 380–383
  AOCCs and regulatory, 504–505
  apron parking, 51
  passenger terminal operations
    and government, 239–240
  security, 51
Rescue. *See also* Operational
    readiness
  aircraft, 146–147
  equipment, *387*
  procedures with firefighting,
    398–400, 540–541
  SAR, 372
  vehicles, *381*, 385–388
Restricted airspace, 333
Revenue capability, of airports,
    15–17
RFFS vehicle, all-terrain, *388*
Risk management, 477
  application of, 484
  definition, 481–482
  hazard identification, 482–483
  MATRA, 264
  with risk mitigation, 483–484

Roads:
  emergency access, 383
  signs, *243*, *246*
Rogers, Cal, 128
Routing, 345
Rules:
  flight, 336
  general flight, 336–337
  instrument flight, 337–338
  visual flight, 122, 337
Runway visual ranges (RVRs):
  measurement equipment types,
    123
  precision-approach runways
    with DHS and, *111*, *343*
Runways:
  aquaplaning on, 115
  Boeing 777-300ER takeoff length
    chart for, *95–96*
  crosswind effects with, 124–128
  declared distances, *100*
  DHs and RVRs for precision-
    approach, *111*, *343*
  foaming of, 381–382, 400–402
  handling, 42
  in inclement weather, 114–115
  inspections, 139
  noise-preferential, 70–71
  with operational noise-abatement
    procedures, 71–72
  operations as noise-control
    strategy, 73
  PC readout of Dynatest friction
    test for, *135*
  slots with airline scheduling, 47
  surface conditions, 133
  takeoff and landing slots, 47
  visibility on, 122–124
RVRs. *See* Runway visual ranges

S

Sabotage, 258
Sacramento Airport, 17, *22*
Safety:
  considerations with aircraft
    performance, 107–110
  positive culture of, 488–489
  readiness and aspects of,
    147–148

Safety Management Systems
(SMSs):
aerodrome technical services
and, 326–329
aerodromes and, 469–475
communication, 489
communication technology,
494–495
complexity, 485–490
framework, 465–468
guidance and resources,
490–493
ICAO Annex 14 aerodromes,
467–468
implementation, 485–493
integration, 493–494
issues, 485
jurisdiction areas, *486–487*
key success factors, 493–495
model, *471*
Reason's model of accident
causation, *481*
regulatory context, *476*
regulatory framework, 466
safety promotion and culture,
488–489
training and personnel
competency, 489–490
typical table of contents, *490–493*
Safety Management Systems
(SMSs) Manual:
goals and indicators, 480
hazard identification, 482–483
overview, 475–476
policy, organization, strategy
and planning, 477
risk management, 477
risk management application, 484
risk management definition,
481–482
risk mitigation, 483–484
safety assurance, 477
safety information and
documentation, 480–481
safety organization, 478–479
safety planning, 479
safety policy, 478
safety promotion, 477
safety standards, 479–480

Salzburg, Austria, 109
San Francisco International
Airport, 339, 416, 571
highway and air passenger
traffic patterns near, *418*
organizational structure of, *23*
Santiago (SCL), *40*
São Paulo Guarulhos (GRU)
International Airport, *41*,
306
SAR. *See* Search and rescue
SARPs. *See* Standards and
Recommended Practices
Satellites:
GPS, 101, 344, 358–360
navigation, 358–360
SBR. *See* Standard busy rate
SBUs. *See* Strategic Business Units
Scanners, whole-body, *273*
Scheduling, airline:
aircraft availability with, 48–49
within airlines, 52–56
airport peaks and, 31–61
airport runway slots and, 47
airport viewpoint on, 59–60
fleet utilization and, 56–58
general crewing availability
with, 48
hubs, 60–61
IATA policy on, 58–59
international airport charges, *54*
landing-fee pricing policies
with, 50–52
long-haul crewing constraints
with, 48
long-haul windows with, 45–47
marketability with, 49–50
organization of, *55*
policies with factors and
constraints, 44–52
reliability with, 45
short-haul convenience with, 48
summer-winter variations with,
50
terminal constraints and, 47–48
turnaround charges by type, *55*
utilization and load factors, 45
variation of fees, *53*
SCL. *See* Santiago

Screening:
  baggage search and, 275–277
  centralized and decentralized,
    266–269
  hold-baggage, 188–189, 197,
    198
  hold-baggage equipment for,
    198
  multilevel protocol for, *188*
  passenger and carry-on
    baggage search and,
    265–275
  SSCP, *270*
Search and rescue (SAR), 372. *See
  also* Rescue
Searches:
  of areas with airport security,
    267–268
  baggage, 275–277
  of carry-on baggage, 265–275
  of freight and cargo, 277
  of passengers, 269–275
Security. *See also* Airport security;
    Department of Homeland
    Security
  conventions on, 258–259
  legislation, 261–262
  program for typical airport,
    281–287
  requirements, 51
  screening checkpoint,
    269–275
  structure of planning for,
    260–261
Security executive group (SEG),
    264
Security identification area
    (SIDA), 277–278
Security screening checkpoint
    (SSCP) layout, *270*
SEG. *See* Security executive group
Self-service:
  bag drop, *187*
  baggage-handling kiosks, *185*
Senior duty manager, AOCCs,
    523
Separation minima, 339–343
September 11, 2001, 257, 262, 287,
    465

Service vehicles:
  aircraft tow tractor, 160, *161*
  all-terrain RFFS, *388*
  mobile apron engine air-start
    vehicle, *161*
  mobile apron fuel tanker, *162*
  mobile lounge, *159*
  pickup truck with retractable
    fifth wheel, *135*
  snow blowers, *137*
Services:
  aerodrome technical, 325–376
  air transport, 59
  airline-related passenger,
    228–230
  broadcast, 360–361
  delivery plan for, *453*
  direct passenger, 222–228
  effectiveness determinants for,
    446
  emergency, 42
  LOS, 2, 42, 60
  for operators and flight crew
    members, 371–372
  quality of, 228
  radio navigation, 355–358
  telecommunications and fixed,
    351–352
  telecommunications and
    mobile, 352–355
  terminal, 42
  trends in meteorologic, 372–373,
    542
Servicing:
  aircraft ramp, 162–166
  cargo operations and airplane,
    311
  fault, 162
  onboard, 166
SESAR Program. *See* Single
    European Sky Airspace
    Research Program
Shanghai International Airport,
    12, 418
Shannon Airport, 15
Sheldon, C., 573
Short-haul convenience, 48
Short-haul flights, 39
Short-haul operations, *56–57*

Short-landed rate, 208
SID. *See* Standard instrument departure
SIDA. *See* Security identification area
SIGMETs. *See* Significant meteorological information
Signage:
  directional, 222
  information signs in terminals, *244*
  for roads, *243, 246*
Signalman, ground, *160*
Significant meteorological information (SIGMETs), 369–370
Significant Weather Charts (SWCs), 366–368
Significant Weather Forecasts (SIGWXs), 366–368
Singapore Changi, 13, 223
Single European Sky Airspace Research (SESAR) Program, 334
SITA Command Center, *500*, *501*
Ski resorts, 50
Sky Team Alliance, 15
Slip airports, 48
SMSs. *See* Safety Management Systems; Safety Management Systems Manual
Snow blower, *137*
Snow clearance, 136
Snow-removal operations control, 512–513
Solid wastes, 166, 565–567
Sorting, baggage handling and, 196–197
Sound, scale of noise and, *64*
South Korea, 291. *See also* New Seoul International Airport
Soviet Union, 359–360
Space components, adjacencies and, 246–249
Spain, 17
SPECI/METAR. *See* METAR/SPECI
Spoke terminals, 322

SRA. *See* Surveillance-radar-approach guidance
SSCP. *See* Security screening checkpoint layout
Staff departments, 18, *19*
Staff structure:
  of Amsterdam Schiphol Airport, *21*
  annual passenger throughput in relation to, *27*
  of Frankfurt Airport, *20*
Staffing. *See also* Employees
  AOCCs, 522–524
  with bag drop, *186*
  baggage handling and, 206–207
  at information desks, *247*
  security and access gate for, *279*
Standard busy rate (SBR):
  hourly passenger traffic volumes and, *33*
  location of, *36*
  peak-day aircraft, annual peak-hourly and, *34–35*
  with peak-hour passenger volume and annual passenger volume, *43*
Standard instrument arrival route (STAR), 348
Standard instrument departure (SID), 348, *350*
Standards:
  annex 17, 259–260, 282
  Corporate Accounting and Reporting, 560
  SARPs, 63, 467
  SMSs, 479–480
Standards and Recommended Practices (SARPs), 63, 467
STAR. *See* Standard instrument arrival route
Star Alliance, 15
Steps, passenger, *157*
Stern Review, 569
Storage, baggage handling and, 189, 198–200
Strategic Business Units (SBUs), *450*

Strategies:
  AOCCs and significance of,
    502–504
  business plan for airports, 444,
    445
  noise-control, 69–74
  for operational administration,
    441–446
  SBUs, 450
  SMSs Manual, 477
Suicide bombing, 281
Summer-winter variations, with
    airline scheduling, 50
Surface pollution, 567–569
Surface shipments, 300–301
Surveillance, 344
Surveillance-radar-approach
    (SRA) guidance, 330–331, 349
Sustainable development:
  air traffic operations and impact
    on communities, 556
  airport carbon management,
    558–562
  biodiversity, 572
  challenges, 555
  climate change, 569–572
  climate change consequences
    and certainty, 570
  EMS principles and Deming
    cycle, 574
  energy, 562–563
  with environmental capacity of
    airports, 553–578
  environmental management
    systems and, 572–575
  issues, 555–572
  local air quality, 557–578
  noise impacts, 555–557
  solid waste management, 166,
    565–567
  surface and groundwater
    pollution, 567–569
  waste management hierarchy,
    566
  water-polluting sources, 568
  water use, 563–565
SWCs. See Significant Weather
    Charts

System actors, 1–2, 3
System design drivers, baggage
    handling and, 204–206

T

TAFs. See Terminal Airport
    Forecasts
Tail-to-tail transfer, 193
Taiwan, 291
Takeoff:
  as airline-related operational
    function, 232
  Boeing 777-300ER length chart
    for, 95–96
  fuel, 234
  landing slots and, 47
  MTOW, 91–93, 102, 232
  with noise-control strategy,
    71–72
Takeoff distance available
    (TODA), 99–101, 104
Takeoff distance required (TODR),
    98–99, 101
Takeoff run available (TORA),
    99–101, 104
Takeoff run required (TORR), 98,
    101
Tampa International Airport, 11,
    251, 422
Tankering, 91
Tariffs:
  duty-free shops, 223–224,
    254
  on handling fees, 51
  peak-hour, 50, 51
Taxes. See Tariffs
Taxi, 417, 427–429
Taxiways, 42, 139
Technologies:
  baggage-handling, 194–204
  cargo market and improvement
    with, 290
  communications, 330
  navigation, 330
  SMSs and communication,
    494–495
  two-engine ER aircraft, 6

Telecommunications:
  broadcast services,
    360–361
  explanation of, 350–351
  fixed services, 351–352
  mobile services,
    352–355
  radio navigation services,
    355–358
  satellite navigation,
    358–360
Terminal Airport Forecasts (TAFs),
    366, *367*
Terminal maneuvering area
    (TMA), *342*
Terminals:
  breakdown of traffic in,
    14, 15
  cargo operations and flow
    through, 297–298, *299–300*,
    301
  cargo operations and handling
    within, 304–307
  constraints with airline
    scheduling, 47–48
  design, 317–320
  equipment, *156*
  fixed mechanized, 305–307
  HACTL, 299–300, *319*, 320
  hubs, 60–61, 254, *322*
  in-town and other off-airport,
    437–438
  information signs in, *244*
  LHR rapid transit in relation to,
    432
  low mechanization/high
    manpower, 304
  management, *220*, 221
  open mechanized, 305
  operations for passenger,
    213–255
  services, 42
  space distribution, *248*
  spoke, *322*
  transfers, inter-, 193
Terrorists, 257, 262, 264,
    465
Tilt-tray sorter, *196*

Time. *See also* Airport access
  baggage-handling processing,
    206
  GMT, 239
  passenger terminal operation
    and length of access, 419
  variations and demand levels
    with, 31
Timetable hour, 37
Tires, 163
TMA. *See* Terminal maneuvering
    area
TODA. *See* Takeoff distance
    available
TODR. *See* Takeoff distance
    required
Toilet holding tanks, 166
TORA. *See* Takeoff run available
Toronto International Airport,
    239
TORR. *See* Takeoff run required
Tote-based system, *197*
TPHP. *See* Typical peak-hour
    passengers
Tractor, aircraft tow, 160, *161*
Traffic patterns, *418*. *See also*
    Passenger traffic
Training, SMSs, 489–490
Trains, tug and dolly, *202*. *See also*
    Rail
Transfers:
  input and interterminal, 193
  ratios with baggage handling
    and, 205
Transport Canada, 48–49
Transportation Security
    Administration (TSA), 262,
    269
  checkpoint, *270*
  employees, 189
Trends:
  baggage handling, 181–183
  in meteorologic services,
    372–373, 542
Trim, balance, 235
Trips:
  fuel, 234
  reliability of access, 419

Trucks:
  catering, *167*
  pickup, *135*
TSA. *See* Transportation Security
    Administration
Tug and dolly train, *202*
Tunnel, walkway, *251*
Typical peak-hour passengers
    (TPHP), 36–37

U

U.K. *See* United Kingdom
Unit load devices (ULDs), 182,
    189, 290
  AKE and AKH, *190*
  with cargo operations,
    301–304
  compatibility of, *303*
  loading onto aircraft, *192*
United Airlines, 239, 499
  flight-crew briefing sheet, *238*
  at LAX, 60
United Kingdom (U.K.):
  airport security planning in, *265*
  CAA, 47, 71, 109
  certification in, 121
  laissez-faire system in, 7–8
  national airport system in, 6–8
  on noise control, 71
  privatization in, 6
United States (U.S.), 6
  airport operations manual,
    545–551
  airport security planning
    outside, 264–265
  certification in, 121
  deregulation in, 6, 17
  federal involvement in aviation
    security, 262
  hub terminals in, 61
  NEF in, 67
  number of airports in, 4
  structure of airport security
    program in, 262–264
UPS, 320
Urgent operational information,
    374
U.S. *See* United States

U.S. National Airport System:
  explanation of, 4–6
  with revenue capability, 17
Users:
  AOCS, 508
  modal choice with access, 414–
    418
  as system actor, 1–2, *3*
Utilization factors:
  with airline scheduling, 45
  fleet, 56–58

V

Variations:
  airport peaks with volume,
    42–44
  demand levels with time, 31
  scheduling and summer-winter,
    50
VDUs. *See* Visual Display Units
Vehicles. *See also* Service vehicles
  access, 279, *280*
  all-terrain RFFS, *388*
  rescue and firefighting, *381*,
    385–388
Vehicular identification, 279
Very Important Persons (VIPs),
    241, *242*
VFRs. *See* Visual flight rules
VIPs. *See* Very Important Persons
Virgin Atlantic, *155*
Visibility:
  London Heathrow and impact
    of reduced, *124*
  as operating constraint,
    122–124
Visitors, 414, *415*
Visual Display Units (VDUs), 244,
    245
Visual flight rules (VFRs), 122, 337
Volumes:
  airport peaks and variations in,
    42–44
  SBR and hourly passenger
    traffic, *33*
  SBR with peak-hour passenger
    and annual passenger, *43*
  variations in hourly traffic, *41*

W

WAFS. *See* World Area Forecast System
Walk-through metal detector (WTMD), *271*
Walking distances, for passengers, 13
Walkway tunnel, *251*
Walkways, pedestrian, 250
Washing, deicing and, 164, *165*
Washington, D.C., National Airport, 5, 59, 339
Water:
    pollution of ground, 567–569
    pollution sources, *568*
    supply in emergencies, 383
    use, 563–565
Waybill, air, 297, *298*
WBCSD. *See* World Business Council for Sustainable Development
Wealth, rising consumer, 291
Weather. *See also* Meteorology
    aircraft performance in inclement, 114–116
    climate change, 569–572
    crosswind effects, 124–128
    information support for general aviation, 370–371
    SIGWXs and SWCs, 366–368
    snow blowers, *137*
    snow clearance, 136
    snow-removal operations control, 512–513
    synoptic weather chart, *368*
    wind rose, 125, *128*
    wind table, *126–127, 129–130*
Weight:
    aircraft, 51, 232
    dry operating, 232, 234
    maximum takeoff, 91–93, 102, 232
    zero-fuel, 232

Wheels, *135*, 163
Whole-body scanner, *273*
Wiley, John R., 24
Wind:
    airspace wind chart, *369*
    rose, 125, *128*
    shear, 115–116
    table, *126–127, 129–130*
The Wings Club, 182
Workload units (WLUs), 29
World Area Forecast System (WAFS), 362
World Business Council for Sustainable Development (WBCSD), 560
World Resources Institute (WRI), 560
World Trade Center, 262
Worldwide Scheduling Guidelines (IATA), 58
WRI. *See* World Resources Institute
Wright, P. H., 90
WTMD. *See* Walk-through metal detector

X

X-ray machines, with baggage, 188, *274, 276–278*

Y

Yoxon, M., 573

Z

Zagreb Airport, 533–534
Zero-fuel weight, 232
Zurich Airport, 47, 336, 415, 558

CPSIA information can be obtained
at www.ICGtesting.com
Printed in the USA
BVHW01*1301121217
502140BV00009B/2/P